Lecture Notes in Mathematics 2178

More information about this series at http://www.springer.com/series/304

Steven R. Costenoble • Stefan Waner

Equivariant Ordinary Homology and Cohomology

Springer

Steven R. Costenoble
Department of Mathematics
Hofstra University
Hempstead
New York, USA

Stefan Waner
Department of Mathematics
Hofstra University
Hempstead
New York, USA

ISSN 0075-8434 ISSN 1617-9692 (electronic)
Lecture Notes in Mathematics
ISBN 978-3-319-50447-6 ISBN 978-3-319-50448-3 (eBook)
DOI 10.1007/978-3-319-50448-3

Library of Congress Control Number: 2016963346

Mathematics Subject Classification (2010): Primary 55N91, Secondary 55P91, 57R91, 55N25, 55P42, 55P20, 55R70, 55R91

Printed on acid-free paper

This Springer imprint is published by Springer Nature
The registered company is Springer International Publishing AG
The registered company address is: Gewerbestrasse 11, 6330 Cham, Switzerland

Introduction

Poincaré duality in ordinary homology is a powerful tool in the study of manifolds. In the presence of a smooth action of a compact Lie group, it has not been clear what the appropriate analogue is or whether there even is one. The fundamental question that must be addressed is what one means by equivariant ordinary homology and cohomology.

Historically, one of the earliest candidates was Borel homology, which, for a G-space X, is the nonequivariant homology of $EG\times_G X$, where EG is a nonequivariantly contractible free G-space [3]. This is still what many people mean when they use the term equivariant homology. May comments, in [46]: "This theory has the claim of priority and the merit of ready computability, and many very beautiful results have been proven with it. However, it suffers from the defects of its virtues. Precisely, it is 'invariant', in the sense that a G-map $f: X \to Y$ which is a nonequivariant homotopy equivalence induces an isomorphism on Borel cohomology." (G-maps of this sort are sometimes called *pseudoequivalences*.) He concludes that no theory invariant in this sense "is powerful enough to support a very useful theory of characteristic classes." More pertinent to the goals of this book, Borel homology satisfies Poincaré duality only for free G-manifolds. For example, the Borel homology of an orbit G/H, for H nontrivial, is the group homology of H, while the cohomology is the group cohomology of H, and these are generally not dual because each is nontrivial and concentrated in nonnegative degrees. On the other hand, as we will discuss in Sect. 1.18, Borel homology is a special case of the ordinary theories we will define, just not the most useful one.

Integer-graded Bredon-Illman homology, introduced in [4] and [31], has several advantages over Borel homology. It obeys a dimension axiom for all orbits, meaning that the homology of any orbit G/H is concentrated in degree 0, hence deserves the name *ordinary* homology; Borel homology obeys a dimension axiom only for the free orbit G/e and, in particular, fails to obey one for a single, fixed point. Also, Bredon-Illman homology detects equivariant homotopy type, in the sense that it induces isomorphisms only for G-maps $f: X \to Y$ such that each fixed set map $f^H: X^H \to Y^H$ is a nonequivariant homology equivalence; it does not take all pseudoequivalences to isomorphisms. Bredon-Illman homology also exhibits Poincaré duality for a larger class than just free manifolds: It exhibits duality for all manifolds in which the local representations are trivial. (By the local representations, we mean, for each point x in the manifold, the linear action of the stabilizer G_x on the tangent plane at x.) Examples of such manifolds include those on which G acts trivially or, for finite groups, manifolds of the form $G/H \times M$, where G acts trivially on M.

However, Poincaré duality generally fails in Bredon-Illman homology for other types of manifolds. For example, consider the group $G = \mathbb{Z}/p$ acting on $V = \mathbb{R}^2$ by rotation, and let $M = S^{2V}$, the one-point compactification of $V \oplus V$. We can calculate the Bredon homology and cohomology of M explicitly with, say, constant $\mathbb{Z}$ coefficients and see that they simply are not dual to one another.

It is not hard to see why we should not expect Poincaré duality to hold in Bredon-Illman homology in general: Conceptually, Poincaré duality is based on a one-to-one correspondence between the simplices of a triangulation and its dual cells. Bredon-Illman homology is based on cells of the form $G/H \times D^n$, where G acts trivially on D^n, as appear in a triangulation, but, as we shall see, the dual cells have the more general form $G \times_H D(W)$, where $D(W)$ denotes the unit disc in what is usually a nontrivial H-representation W. (See Example 1.1.5(6).)

Related to this problem is the fact that Bredon-Illman homology is not fully stable, in the sense that, while it has suspension isomorphisms for suspensions by spheres with trivial G-action, in general it lacks suspension isomorphisms for suspensions by spheres of arbitrary representations. Thus, we cannot represent it by a proper equivariant spectrum (only a so-called naive spectrum), and we cannot take advantage of equivariant Spanier-Whitehead duality, which says that the V-dual of a G-manifold M embedded in a G-representation V (necessarily nontrivial if the action of G on M is nontrivial), with normal bundle ν, is $T\nu$; put another way, the stable dual of M is $\Sigma^{-V}T\nu$. Related to this, we cannot expect the existence of transfer maps for G-fibrations.

Theories that have suspension isomorphisms for all representations can be considered to be graded on the representation ring $RO(G)$, the free abelian group generated by the irreducible orthogonal representations of G. Bredon-Illman homology has previously been extended to an $RO(G)$-graded theory (see [27, 38, 39, 41], and [47, Chaps. X, XI, and XIII]) with the desired suspension isomorphisms. Moreover, when G is finite, the $RO(G)$-graded theory exhibits Poincaré duality for a larger class of G-manifolds, including S^V: ones that are locally modeled on a fixed representation of G and have a suitable notion of orientation (see [34, 56], and [67]). However, the documentation of $RO(G)$-graded ordinary homology is sparse and incomplete. In particular, its behavior with regard to products, restriction to fixed sets, and restriction to subgroups is nowhere discussed in detail. (An outline of the product structure is given in [47, XIII.5].) One of our purposes in writing this book is to fill this gap in the literature.

When G is infinite, Poincaré duality will generally fail in the $RO(G)$-graded theory, even for manifolds modeled on a single representation (see Sect. 1.16 for an example). The underlying reason one should not expect it to hold is, as noted by May in [47, XIII], that ordinary homology and cohomology need not be dual when G is an infinite compact Lie group—the spectra representing the two theories have different characterizations and are generally not equivalent. Although ordinary homology and cohomology are not dual to one another, each possesses a dual theory, defined by its representing spectrum. These theories, which we refer to as *dual cohomology and homology,* are mentioned in [47] but not described in any detail. Here we shall give these theories a geometric description and discuss them in full. We can then say that Poincaré duality holds for manifolds modeled on a single representation, in the sense that the ordinary cohomology of such a manifold is isomorphic to the dual homology with the appropriate change of grading, and similarly the ordinary homology is isomorphic to the dual cohomology.

An underlying problem when trying to extend Poincaré duality further, to all manifolds, is the question of where the fundamental class should live. Nonequivariantly, it lives in the nth homology group, where n is the dimension of the manifold. In the $RO(G)$-graded case, for a manifold modeled on a single representation V, it lives in the Vth homology group, and we think of such a manifold as V-dimensional. Following where the geometry lead us, we need a grading that includes the "dimensions" of arbitrary G-manifolds, using a notion of dimension that captures the possibly varying local representations that appear on the manifold. Such a notion of dimension was defined in [20] as a "representation of the fundamental groupoid" of the manifold; we review this notion in Sect. 2.1. We write $\Pi_G X$ for the fundamental groupoid of a G-space X and $RO(\Pi_G X)$ for its ring of (orthogonal) representations. In [14], we showed that, when G is finite, there is an extension of the $RO(G)$-graded theory to an $RO(\Pi_G X)$-graded theory in which Poincaré duality holds for all smooth compact G-manifolds. The $RO(\Pi_G X)$-graded theory is inherently twisted, so no orientability assumptions are required. In [15] and [17], we applied the theory to obtain π-π theorems for equivariant Poincaré duality spaces and equivariant simple Poincaré duality spaces. In other words, our equivariant ordinary theory is sensitive enough to yield a surgery theory that parallels the nonequivariant theory [63] modulo the usual equivariant transversality issues [55, 70].

Our main goals in this book are to give a more complete and coherent account of what is already known, to extend the $RO(\Pi_G X)$-graded theories to actions by arbitrary compact Lie groups, and to show that Poincaré duality holds in these theories for arbitrary smooth G-manifolds.

In Chap. 1, we give a reasonably complete account of $RO(G)$-graded ordinary homology and cohomology as well as their associated dual theories, both geometrically (from the cellular viewpoint) and homotopically (from the represented viewpoint). The cell complexes we use to define the ordinary groups in gradings $V+*$ have cells of the form $G\times_H D(V\pm\mathbb{R}^n)$; we consider such a cell to be $(V\pm n)$-dimensional. When $V=\mathbb{R}^k$, these are exactly the G-CW complexes used to define Bredon-Illman homology. The cell complexes used to define the dual groups have cells of the form $G\times_H D(V\pm\mathbb{R}^n-\mathscr{L}(G/H))$, where $\mathscr{L}(G/H)$ is the tangent representation to G/H at the identity coset; we consider such a "dual" cell to again be $(V\pm n)$-dimensional—notice that $V\pm n$ can be thought of as the *geometric* dimension of the whole cell, including the orbit G/H. These are the cell complexes that arise naturally as the duals of G-triangulations of manifolds. In order to handle both kinds of complex at the same time, as well as intermediate kinds that arise when discussing products, we describe a more general type of complex, of which the ordinary and dual complexes appear as extreme cases.

After defining and discussing the basic properties of the resulting homology and cohomology theories, we say a word about their representing spectra, including the Eilenberg-MacLane spectra representing cohomology and the "MacLane-Eilenberg" spectra representing homology. For infinite groups, these are different spectra: The spectrum representing ordinary homology does not represent ordinary cohomology. However, the spectrum representing ordinary homology does represent dual cohomology, and the spectrum representing ordinary cohomology also represents dual homology. From the point of view of stable homotopy, this bifurcation results from the fact that, when G is finite, its orbits are self-dual, while for G infinite that is no longer the case, but each orbit G/H is dual to itself "desuspended" by the representation $\mathscr{L}(G/H)$ [42, II.6.3]. This is the homotopical analogue of the observation above that the cells dual to an ordinary G-CW(V) structure on a manifold are dual cells, having again a shift in dimension of $\mathscr{L}(G/H)$.

We then discuss various change-of-groups results as well as products. We go on to show that, for general compact Lie group actions, Poincaré duality holds for manifolds modeled on a single representation, in that if M is closed, equivariantly orientable (in a sense to be made precise), and modeled locally on the G-representation V, one has duality isomorphisms like $H_G^W(M) \cong \mathscr{H}_{V-W}^G(M)$, where $\mathscr{H}_*^G$ is the dual theory. We end the chapter with some examples, a survey of calculations, and remarks.

Chapter 2 is largely about parametrized spaces and spectra, which raises the obvious question: Why should we be interested in parametrized objects in the first place? Indeed, in [14] and other related papers, we developed $RO(\Pi_G X)$-graded theories (for finite group actions) without mentioning parametrized objects at all. However, to handle various technical aspects of the theory, including naturality and representability, it seems convenient to fix a space B and a representation of its fundamental groupoid and consider spaces parametrized by B. Our earlier approach (which did not address representability) then amounts to parametrizing each space over itself. The connection between these approaches is discussed in more detail in Remarks 3.8.5.

Our discussion of parametrized homotopy theory relies heavily on work by May and Sigurdsson [51]. However, we emphasize a point of view not taken by them: that, rather than concentrating on maps that strictly commute with projection to the base space, we should take seriously maps that commute only up to specified homotopies in the base space. We call these *lax maps*. That homology should extend to the category of lax maps is implicit in Dold's axiom **CYL** in [21] and is consistent with the discrete nature of representations of the fundamental groupoid.

It is also in Chap. 2 that we review basic facts about the equivariant fundamental groupoid $\Pi_G X$ and its representations, as developed in detail in [20]. In the case of $RO(G)$-graded ordinary theories, it is well-known that the coefficient systems we use must be Mackey functors, i.e., functors on the *stable* orbit category. In the

$RO(\Pi_G X)$-graded theory, the fundamental groupoid replaces the orbit category, so we must also have a stable version of $\Pi_G X$, which is developed in Sect. 2.6. We also discuss parametrized homology and cohomology theories in general and how they are represented by parametrized spectra. We then discuss a notion of duality most appropriate for parametrized homology and cohomology.

It is in Chap. 3 that we construct the $RO(\Pi_G X)$-graded theories, culminating in the Thom isomorphism and Poincaré duality theorems. This chapter parallels Chap. 1 as much as possible and in many cases relies on results from that chapter. A word about orientability: The $RO(\Pi_G X)$-graded theories are, by their nature, twisted, so that the Thom isomorphism and Poincaré duality theorems require no orientability assumptions. Indeed, restricting to orientable bundles or manifolds in this context gives us no advantage—the fundamental groupoid is woven so deeply into the theory of equivariant orientations [20] that there is no significant simplification in considering the orientable case.

Whereas calculations in $RO(G)$-graded ordinary homology are sparse, they are as yet almost nonexistent for the $RO(\Pi_G X)$-graded theory. The first author has preliminary results that we summarize in Sect. 3.12. We also have work in progress to generalize [15] to the compact Lie case, based on the results of Chap. 3.

How to Read This Book

We have made an effort to keep the discussion as geometric and elementary as we can. Among other things, that means that we have avoided using model categories where we might have; when we need to discuss spectra, we have kept the discussion as concrete as possible; and we have given a number of simple examples to try to motivate our definitions.

For a first reading that focuses on the geometry of the constructions and avoids technicalities about defining the theories on general spaces, representability, change of ambient group, and products, we recommend reading, in Chap. 1, Sects. 1.1–1.8, 1.11, and, if you will accept that the usual sorts of products work, Sects. 1.15–1.17. In Chap. 2, read Sects. 2.1, 2.2, 2.6 (read the statement of Theorem 2.6.4 but skip its proof), and Sect. 2.7. In Chap. 3, read Sects. 3.1–3.3, 3.6, and, subject to faith in products, Sects. 3.11 and 3.12.

For a more in-depth reading that concentrates on spaces rather than spectra, the reader can skip material related to spectra without losing the thread of the discussion.

Of course, the ambitious reader or researcher is encouraged to read everything!

History

This project has been in development for long enough that it is worth saying a few words about its history and relation to other works. We have long believed that equivariant Thom isomorphism and Poincaré duality theorems require homology and cohomology theories that take into account the varying local representations given by the fibers of a general equivariant vector bundle. This belief led to the work with Peter May on equivariant orientations that became [20]; early versions of the orientation theory, as well as versions of $RO(\Pi_G X)$-graded ordinary homology and cohomology, appeared in [14] and [16]. (See [49] for a different approach to Thom isomorphisms suggested by May.)

While we were working on [20], we had this work in mind as well. Early on, we did not give parametrized homotopy theory a prominent role, but as we worked out the details, it became clearer that we should, particularly when discussing how the theories are represented. In an earlier version of this manuscript [18], we used parametrized spectra as the representing objects, but our naive belief was that the theory of parametrized spectra was an easy generalization of the nonparametrized theory. At about the same time we posted that preprint, May had started looking at parametrized homotopy theory for other reasons, and he warned us that he had already found serious pitfalls and that the parametrized theory was definitely *not* an easy generalization of the nonparametrized case. May was joined in his efforts by Johann Sigurdsson, and their work became [51]. We are pleased that they incorporated some of the ideas from our preprint, particularly our notion of homological duality, and we are indebted to them for providing a firm foundation on which we could build this work. Once [51] appeared, it became clear that we needed to substantially rewrite our earlier manuscript. This version uses parametrized homotopy theory much more extensively, and we have attempted to do so with the care that May has enjoined us to take. In several places, [51] gives better proofs of results from our earlier preprint, so we have replaced our proofs with references to theirs.

One thing that took us a long time to fully appreciate is the difficulty of trying to construct a good theory of equivariant CW parametrized spectra, which May and Sigurdsson warned about in the final chapter of [51]. After the failures of several attempts to construct such a theory, we have retreated in this work to discussing only CW parametrized spaces, not spectra. Thus, the homology and cohomology theories we discuss throughout this work are restricted to be defined only on spaces, not spectra.

Acknowledgments

We are, of course, highly indebted to Peter May for our joint work on equivariant orientation theory, which serves as a conceptual basis for the current work; to May and Sigurdsson for their monumental work on equivariant parametrized homotopy theory; and to Gaunce Lewis for discussions over many years, before his untimely death, about compact Lie group actions as they relate to ordinary homology and Mackey functors, among many other things. We are also indebted to the reviewers of the penultimate draft of this book for their many helpful suggestions toward making it more readable.

February 14, 2016

Contents

Chapter 1
RO(G)-Graded Ordinary Homology and Cohomology

A construction of $RO(G)$-graded ordinary homology and cohomology was announced in [41] and ultimately given in [27]; the construction appears also in [47, XIII.4]. That construction proceeds by defining integer-graded analogues for G-spectra of Bredon-Illman homology and cohomology, using G-CW spectra, then uses the fact that such theories are representable by G-spectra, therefore extend to $RO(G)$-graded theories on G-spectra and hence also on G-spaces. In this chapter we give another construction, versions of which first appeared in [64] for finite groups, in [38] for compact Lie groups, and in outline form in [47, X]. This construction is more geometric and applies directly to G-spaces, so gives a better intuitive understanding of what the groups indexed by representations mean. It will also generalize to the context of parametrized spaces in Chap. 3, whereas the theory of G-CW spectra does not appear to do so. In this chapter we will also give more details about the behavior of the $RO(G)$-graded theories. Some of this material is well-known, but a good deal is new, particularly the material on dimension functions, on change of groups, and on products.

As mentioned, in Chap. 3 we shall generalize this construction to give ordinary homology and cohomology theories defined on spaces parametrized by a fixed base space, and graded on a larger group. In a very precise sense, the theory discussed in this chapter is the local case of the more general one. We spend significant time on details in this chapter, where the setting is somewhat simpler, to allow us to concentrate later on just those things that need to be changed in the parametrized context.

In Sect. 1.1 below we will give definitions and examples of various kinds of cell complexes that have arisen in the literature and that we will use in constructing ordinary homology. For Bredon-Illman integer-graded equivariant cellular homology, the appropriate notion of a cell complex is a G-CW complex in the sense of [4] and [31]. This is a G-space built from cells of the form $G/H \times D^n$ where $H \leq G$ and G acts trivially on the unit n-disc D^n. The cell complexes we use to construct the $RO(G)$-graded extension have cells of the form $G \times_H D(W)$ where $D(W)$ is the unit

S.R. Costenoble, S. Waner, *Equivariant Ordinary Homology and Cohomology*, Lecture Notes in Mathematics 2178, DOI 10.1007/978-3-319-50448-3_1

disc in a (possibly nontrivial) representation W of H. There are several motivations for doing this that also provide suggestions for what representations we should use.

One simple thought is that generalizing from integer grading to $RO(G)$ grading should involve replacing integer discs $D(\mathbb{R}^n)$ with discs of representations. Evidence that this should work comes from thinking about what happens when we suspend a G-CW complex by a non-trivial G-representation V—we turn cells of the form $G/H \times D^n$ into cells of the form $G/H \times D(V + n)$.

Particularly compelling to us, though, is to consider equivariant Poincaré duality from a geometric point of view when G is infinite. Consider a smooth compact G-manifold M and suppose, for the purposes of this chapter, that it is V-dimensional. That is, there is a representation V of G such that, at each point $x \in M$, the tangent plane at x is isomorphic to V as a representation of the stabilizer G_x of x. Illman showed that we can triangulate M, making it a G-CW complex with cells of the form $G/H \times D^n$ as above. We can form a "dual cell structure" by taking as the top-dimensional cells the orbits of the closed stars of the original vertices in the first barycentric subdivision of the triangulation, while the lower dimensional cells are intersections of these. These cells have interesting actions. For example, if x is a vertex in the original triangulation, then the dual cell centered at x will have the form $G \times_{G_x} D(W)$, where $W \oplus \mathscr{L}(G/G_x) \cong V$ as representations of G_x, with $\mathscr{L}(G/G_x)$ denoting the tangent representation of the orbit G/G_x at the identity coset. We can write this as $W \cong V - \mathscr{L}(G/G_x)$. In general, the dual cells will have the form $G \times_{G_x} D(V - \mathscr{L}(G/G_x) - k)$ where x is the center of a k-cell in the original triangulation. Notice that the *total* dimension of such a cell, including the dimension of the orbit, is $V - k$, which is appropriate for a cell dual to a k-dimensional simplex in a V-dimensional manifold. This differs from the convention in Bredon-Illman homology, where a cell of the form $G/H \times D^n$ is considered to be n-dimensional. The difference is in what we consider the orbit to be contributing to the dimension of the cell. In the case of Bredon-Illman homology, the orbit contributes nothing to the dimension, but in the case of our dual cells, it contributes its whole geometric dimension. In order to consider both possibilities and treat them in a uniform way, we introduce the notion of a *dimension function* for G, which assigns to each orbit G/H an H-representation $\delta(G/H)$ that tells us how much the orbit is to contribute to the dimension of a cell. For each such δ there will be corresponding $RO(G)$-graded cellular homology and cohomology theories. In the case of Bredon-Illman homology we let $\delta(G/H) = 0$, but in the case of the "dual homology" suggested above we let $\delta(G/H) = \mathscr{L}(G/H)$. It will also be useful, particularly when discussing products, to allows cases between these two extremes, but the reader is invited to think about only these two cases on a first reading of most of this chapter.

We now fix some assumptions and terminology: Throughout we understand all spaces to be k-spaces, i.e., spaces X such that a subspace $A \subset X$ is closed if and only if $A \cap K$ is closed in K for every compact subspace $K \subset X$. It is more common to restrict to weak Hausdorff k-spaces, also known as compactly generated spaces (though that term is used for various notions) but the extra generality is needed in subsequent chapters when we discuss parametrized spaces (see also [51]). Our ambient group G will always be a compact Lie group and subgroups are understood to be closed. We write $\mathscr{K}$ for the category of k-spaces, $G\mathscr{K}$ for the category of G-spaces and G-maps, $\mathscr{K}_*$ for the category of based k-spaces, and $G\mathscr{K}_*$ for the category of based G-spaces (with G-fixed basepoints) and basepoint-preserving G-maps. If X is an unbased G-space we write X_+ for X with a disjoint basepoint adjoined. If Y is a based G-space and V is a representation of G, we write $\Sigma^V_G Y = \Sigma^V Y$ for $Y \wedge S^V$, the smash product with the one-point compactification of V. We write $\Omega^V_G Y = \Omega^V Y = F(S^V, Y)$ for the space of based, nonequivariant maps from S^V to Y, with G acting by conjugation.

Two based G-maps $f, g: X \to Y$ are said to be *G-homotopic* (or simply homotopic) if there is *G-homotopy* between them, i.e., a based G-map $h: X \wedge I_+ \to Y$ such that $h(-,0) = f$ and $h(-,1) = g$. We write $\pi G\mathscr{K}_*$ for the category obtained from $G\mathscr{K}_*$ by identifying G-homotopic maps. We say that a based G-map $f: X \to Y$ is a *weak G-equivalence* if all of its restrictions to fixed sets $f^H: X^H \to Y^H$ are nonequivariant weak equivalences. It is possible to invert all weak G-equivalences to obtain the *homotopy category* $\mathrm{Ho}\, G\mathscr{K}_*$. The most common way of doing this is to show that weak G-equivalences between G-CW complexes are G-homotopy equivalences, and that any G-space can be approximated up to weak G-equivalence by a G-CW complex, so we can model $\mathrm{Ho}\, G\mathscr{K}_*$ by the full subcategory of $\pi G\mathscr{K}_*$ consisting of the G-CW complexes. We will reprove these results for the more general complexes we consider in Sect. 1.4.

1.1 Examples of Equivariant Cell Complexes

We give a brief survey of the kinds of equivariant cell complexes that have appeared in the literature. These will all be special cases of the class of complexes we will define in Sect. 1.4.

1.1.1 G-CW Complexes

These are the classical equivariant cell complexes defined by Bredon [4] and Illman [32]. The cells all have the form $G/H \times D^n$:

Definition 1.1.1

(1) A *G-CW complex* is a G-space X together with a decomposition $X = \mathrm{colim}_n X^n$ such that

  (a) X^0 is a disjoint union of G-orbits G/H.
  (b) X^n is obtained from X^{n-1} by attaching a disjoint union of cells $G/H \times D^n$ along their boundary spheres $G/H \times S^{n-1}$.

(2) A *relative* G-CW complex is a pair (X, A) where $X = \mathrm{colim}_n X^n$, X^0 is the disjoint union of A with orbits as in (a) above, and cells are attached as in (b).

(3) A *based* G-CW complex is a relative G-CW complex $(X, *)$ where $*$ denotes a G-fixed basepoint.

Examples of G-CW complexes include Illman's equivariant simplicial complexes, which we can define as follows [32]. Write Δ_n for the standard nonequivariant n-simplex and let $\Delta_{n-1} \to \Delta_n$ denote the inclusion of the front face.

Definition 1.1.2

(1) Let $H_0 \geq H_1 \geq \cdots \geq H_n$ be a sequence of subgroups of G. The *standard equivariant n-simplex of type* $(H_0, \ldots, H_n)$, denoted $\Delta_n(G; H_0, \ldots, H_n)$, is the colimit of the following diagram:

$$\begin{array}{ccccccccc} & G/H_1 \times \Delta_0 & & G/H_2 \times \Delta_1 & & & G/H_n \times \Delta_{n-1} & \\ \swarrow & & \searrow \quad \swarrow & & \searrow & \swarrow & & \searrow \\ G/H_0 \times \Delta_0 & & G/H_1 \times \Delta_1 & & \cdots & & & G/H_n \times \Delta_n, \end{array}$$

where the maps going to the left are induced by the quotient maps of orbits $G/H_k \to G/H_{k-1}$ and the maps going to the right are induced by the inclusions $\Delta_{k-1} \to \Delta_k$. (The effect is that $\Delta_n(G; H_0, \ldots, H_n)$ is the quotient of $G/H_n \times \Delta_n$ in which the front face is identified to $\Delta_{n-1}(G; H_0, \ldots, H_{n-1})$. This could be used as a recursive definition.)

(2) A *G-simplicial complex* is a G-space X with a G-CW structure such that the induced cell structure on the orbit space X/G makes it a nonequivariant simplicial complex, and such that, if $p: X \to X/G$ is the projection and σ is any simplex of X/G, then $p^{-1}(\sigma)$ is G-homeomorphic to a standard equivariant n-simplex $\Delta_n(G; H_0, \ldots, H_n)$ in such a way that the induced homeomorphism $\Delta_n(G; H_0, \ldots, H_n)/G = \Delta_n \to \sigma$ is linear.

A G-space with the structure of a G-simplicial complex is also said to have a *G-triangulation.* We introduce the definition here so that we can use it in examples, but we shall not use it in the development of the theory that follows.

Illman, in [32], showed that all smooth G-manifolds can be given G-triangulations. G-CW complexes model many other types of G-space as well. In particular, The second author, in [65], showed that a large class of reasonable G-spaces have the G-homotopy types of G-CW complexes, directly generalizing the corresponding nonequivariant result. Most generally, every G-space has the weak G-homotopy type of a G-CW complex—see [65, 3.7] and the references given there. This will also follow from our general results in Sect. 1.4.

Here are some specific examples of G-CW complexes that the reader should keep in mind.

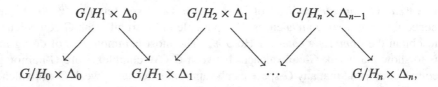

Examples 1.1.3

(1) Let $X = G/H$, an orbit of G. This is a G-CW complex with a single, 0-dimensional cell, regardless of the geometric dimension of G/H. Particularly interesting examples are given by the action of $SO(n)$ on S^{n-1}.
(2) If B is a nonequivariant CW complex and $X \to B$ is a principal G-bundle, then the CW structure on B pulls back to a G-CW structure on X, in which every cell is free.
(3) Consider the action of $G = S^1$ on the sphere S^2 by rotation, with fixed points the north and south poles. It has a G-CW structure with two G-fixed 0-cells, being the poles, and one free 1-cell connecting them, the orbit of a line of longitude.
(4) Let $G = \mathbb{Z}/n$ be any finite subgroup of S^1 and consider the restriction of the action of S^1 on S^2 from the preceding example. For a G-CW structure, we can take two fixed 0-cells, the north and south poles, one free 1-cell, consisting of n evenly spaces lines of longitude, and one free 2-cell, filling in between those lines.
(5) The preceding G-CW structure is not a triangulation of the sphere, but the following is: For the 0-cells, take the north and south poles, plus one free 0-cell consisting of n evenly distributed points along the equator. There are three free 1-cells, consisting of the lines connecting the north pole to the chosen points on the equator, the similar lines from the south pole, and the line segments along the equator connecting the points. There are two free 2-cells, one consisting of the evident n triangles in the northern hemisphere, the other consisting of those in the southern hemisphere.
(6) There is another interesting S^1-CW structure on S^2: As 0-cells, take the two poles and the equator. There are then two free 1-cells, corresponding to the north and south hemispheres. This structure is a G-triangulation, whereas the structure in (3) is not.
(7) The cell structure in the preceding example gives a cell structure on the projective plane with the induced action of S^1: It has two 0-cells, one being fixed by S^1 and the other being a copy of $S^1/(\mathbb{Z}/2)$. It has a single, free 1-cell, attached to the fixed point in the obvious way and attached to the other 0-cell by the quotient map (double covering).
(8) Let X be the Möbius strip, thought of as the disc bundle of the nontrivial line bundle over a circle $S^1/(\mathbb{Z}/2)$. With $G = S^1$, we can think of X as a G-manifold with boundary. It has a G-CW structure with two 0-cells, one being the copy of $S^1/(\mathbb{Z}/2)$ in the middle and the other being the boundary, a copy of S^1. There is one 1-cell connecting them, attached to the middle 0-cell by the quotient map. This structure is not only a G-triangulation, but gives the Möbius strip the structure of a standard equivariant 1-simplex!

1.1.2 G-CW(V) Complexes

The second author, in the unpublished [64] and later in [69], and Lewis in [38], generalized G-CW complexes by introducing cell complexes with cells of the form $G \times_H D(V \pm n)$ for a fixed representation V of G. These G-CW(V) complexes were used in the cellular definition of $RO(G)$-graded ordinary homology and cohomology and our approach will generalize that one. Here's the precise definition.

Definition 1.1.4 Let V be a representation of G. A G-CW(V) *complex* is a G-space X together with a decomposition $X = \mathrm{colim}_n X^n$ such that

(1) X^0 is a disjoint union of G-orbits G/H for which H acts trivially on V.
(2) X^n is obtained from X^{n-1} by attaching a disjoint union of cells of the form $G \times_H D(V + (n - |V|))$ along their boundary spheres where, if $n - |V| < 0$, we require that $|V^H| \geq |V| - n$, so that the apparently virtual representation $V + (n - |V|)$ is, in fact, an actual representation of H. (Here, we write $|V|$ for the dimension of V.)

Relative and based complexes are defined similarly.

For notational convenience, and to remind ourselves of the role of V, we shall also write X^{V+n} for $X^{|V|+n}$. Thus, a $(V+n)$-cell, that is, one added in forming X^{V+n}, has the form $G \times_H D(V + n)$, where $|V^H| + n \geq 0$.

Examples 1.1.5

(1) Let V be any nontrivial representation of G. Then the point G/G is *not* a G-CW(V) complex. It is this fact that makes the $RO(G)$-graded homology of a point nontrivial in dimensions $V + n$. (Stong gave the first calculations of the homology of a point, for $G = \mathbb{Z}/p$, in circulated but unpublished notes; his calculation was published by Lewis in [37].) In general, an orbit G/H is a G-CW(V) complex if and only if H acts trivially on V.
(2) Consider the action of $G = S^1$ on the sphere S^2 by rotation, with fixed points the north and south poles. Let V be the two-dimensional representation on which S^1 acts by rotation. Then S^2 has the following G-CW(V) structure: There is one free $(V-2)$-cell $S^1 = S^1 \times D(V-2)$, the equator. There are two V-cells of the form $D(V) = S^1 \times_{S^1} D(V)$, the north and south hemispheres, attached to the equator via the identification of $S(V)$ with S^1.
(3) Consider the action of a finite subgroup $G = \mathbb{Z}/n$ of S^1 on S^2, with V the restriction of the rotation representation used in the preceding example. Then S^2 has the following G-CW(V) structure: There is one free $(V-2)$-cell, consisting of n equally spaced points on the equator, one free $(V-1)$-cell, consisting of line segments on the equator connecting those points, and two V-cells of the form $D(V) = G \times_G D(V)$, the north and south hemispheres.
(4) Consider the Möbius strip with its action of $G = S^1$. We would like to give it a cell structure in which one of the cells is $G \times_{\mathbb{Z}/2} D(L)$, where L is the nontrivial representation of $\mathbb{Z}/2$ on a line. However, L does not extend to a representation of S^1, so this would not be an example of a G-CW(V) complex. It will, however,

be an example of the general kind of complex that we will consider in Chap. 3. A similar problem arises with the projective space we get by identifying the boundary of the Möbius strip to a point.

(5) Suppose that B is a G-CW complex. Let $p: E \to B$ be a V-*bundle*, that is, a G-vector bundle such that, for each $b \in B$, the fiber E_b over b is isomorphic to V as a representation of the stabilizer G_b. Then the Thom space $T(p)$ is a based G-CW(V) complex with the compactification point as basepoint and with a cell of dimension $V + n$ corresponding to each cell of dimension n in B. This correspondence underlies the Thom isomorphism we shall discuss in Sect. 1.15.

(6) Assume that G is finite. As in [34, 56], and [67], we say that a smooth G-manifold M is a V-*manifold* if its tangent bundle is a V-bundle. (This will be true for some V if, for example, the fixed sets M^H of M are all nonempty and connected.) Fix a triangulation of M. We can then form the "dual cell complex" in the usual way, taking as the top-dimensional cells the orbits of the closed stars of the original vertices in the first barycentric subdivision of the triangulation, while the lower dimensional cells are intersections of these. Corresponding to each simplex in the original triangulation will be one cell in the dual complex, the intersection of the stars of its vertices, which intersects the simplex normally at its center. If x is the center point of an n-simplex, then the corresponding dual cell has the form $G \times_{G_x} D(V - n)$. Thus, the dual cell complex gives a G-CW(V) structure on M. The correspondence between the simplices (cells) of the triangulation and the cells in the dual structure is the geometry underlying Poincaré duality, which we shall discuss in Sect. 1.15.

However, when G is an infinite compact Lie group, the dual cell structure is *not* a G-CW(V) structure, but is an example of a "dual" cell structure, which we discuss in the following subsection.

(7) As an example of a dual cell complex obtained as in the preceding item, consider $G = \mathbb{Z}/n$ acting on S^2 and the triangulation given in Example 1.1.3(5). Let V be the rotation representation of G, so that S^2 is a V-manifold. (It helps to draw a picture while reading the following description.) The dual cell complex then has two free $(V - 2)$-cells, one consisting of n evenly distributed points along a line of latitude in the northern hemisphere, the other of n such points in the southern hemisphere, with the two cells aligned along lines of longitude. There are three free $(V - 1)$-cells: One consists of the line segments connecting adjacent points in the $(V - 2)$-cell in the northern hemisphere; one consists of the similar line segments in the southern hemisphere; and the third consists of the line segments connecting each point in the $(V - 2)$-cell in the northern hemisphere to the point in the $(V - 2)$-cell in the southern hemisphere on the same line of longitude. Finally, there are three V-cells: One is the free cell filling in the squares just drawn that lie across the equator; one is the cell $D(V) = G \times_G D(V)$ covering the north pole and descending to the latitude given by the $(V - 1)$ cell in the northern hemisphere; and the last is the similar cell at the south pole.

(8) The G-CW(V) structure given in item (3) above is dual to the G-CW structure in Example 1.1.3(4) in much the same way, even though that G-CW structure is not a G-triangulation.

The notion of a G-CW(V) complex extends to that of a G-CW($V - W$) complex, useful for discussing the $V - W$ part of $RO(G)$-graded homology. We will take this approach in Sect. 1.4.

1.1.3 Dual G-CW(V) Complexes

In Example 1.1.5(6), we saw that G-CW(V) complexes are inadequate to describe geometric duality when G is infinite. Let's look at the "dual" structure we get more closely. We will use the following notation here and through the rest of this book.

Definition 1.1.6 If G is a compact Lie group, let $\mathscr{L}(G)$ denote its tangent space at the identity (i.e., its Lie algebra, but we won't use the Lie algebra structure here). If H is a subgroup of G, let $\mathscr{L}(G/H)$ denote the tangent space to G/H at the identity coset eH, considered as a representation of H.

Let M be a closed V-manifold, i.e., a manifold in which, for each $x \in M$, the tangent representation T_xM at x is isomorphic to V as a representation of G_x. Consider an equivariant triangulation of M. Form the dual cells as mentioned in Example 1.1.5(6): The top-dimensional cells are the orbits of the closed stars of the original vertices in the first barycentric subdivision of the triangulation, while the lower dimensional cells are intersections of these. If x is the center point of an n-simplex, then its tangent plane decomposes as

$$V \cong TM_x \cong \mathbb{R}^n \oplus \mathscr{L}(G/G_x) \oplus W$$

for some G_x-representation W, where $\mathbb{R}^n$ consists of the tangents along the simplex while $\mathscr{L}(G/G_x)$ consists of the tangents along the orbit of x. The corresponding dual cell then has the form $G \times_{G_x} D(W)$. This leads us to the following definition.

Definition 1.1.7 Let V be a representation of G. A *dual G-CW(V) complex* is a G-space X together with a decomposition $X = \operatorname{colim}_n X^n$ such that

(1) X^0 is a disjoint union of G-orbits G/H for which G/H is a finite set and H acts trivially on V.
(2) X^n is obtained from X^{n-1} by attaching a disjoint union of cells of the form $G \times_H D(W)$ along their boundary spheres, where W is an actual representation of H such that $\mathscr{L}(G/H) + W$ is stably isomorphic to the possibly virtual representation $V + (n - |V|)$. (Here, we again write $|V|$ for the dimension of V.)

Relative and based complexes are defined similarly.

For notational convenience, we shall also write X^{V+n} for $X^{|V|+n}$. Thus, a dual $(V + n)$-cell, that is, one added in forming X^{V+n}, has the form $G \times_H D(W)$, where

$\mathscr{L}(G/H) + W \cong V + n$ as H-representations. One way of looking at this is that we take as the dimension of a cell $G \times_H D(W)$ its *geometric* dimension, which is $\mathscr{L}(G/H) + W$. (Notice that its tangent bundle is $G \times_H (\mathscr{L}(G/H) + W)$.) In contrast, in a G-CW(V) complex we take the dimension of a cell $G \times_H D(V+n)$ to be $V + n$.

Examples 1.1.8

(1) When G is finite, dual G-CW(V) complexes are, of course, the same thing as G-CW(V) complexes. Dual complexes are only needed when G is infinite.
(2) An orbit G/H is a dual complex with a single cell, if $\mathscr{L}(G/H)$ is the restriction of a representation of G. For example, if H is a normal subgroup of G, then $\mathscr{L}(G/H)$ is a trivial representation of H, hence the restriction of the trivial representation of G. Thus, if G is abelian, every orbit is a dual complex with a single cell.
(3) As in Example 1.1.5(2), consider the action of $G = S^1$ on the sphere S^2 by rotation, with fixed points the north and south poles, and let V be the two-dimensional representation on which S^1 acts by rotation. It has a dual G-CW(V) structure with one dual $(V-1)$-cell $S^1 \times D(V-2)$, the equator, and two dual V-cells $S^1 \times_{S^1} D(V)$, the northern and southern hemispheres. This structure is intuitively dual to the G-CW structure given in Example 1.1.3(3). Note that this dual structure uses the same topological cells as the G-CW(V) structure in Example 1.1.5(2), but the equator is considered to have dimension one higher. Comparing dimensions with those of the cells in Example 1.1.3(3), the cells in the dual G-CW(V) structure are in the right dimensions to be dual to that G-CW structure, whereas the cells in Example 1.1.5(2) are not.
(4) Example 1.1.5(6) gives a triangulation of the action of S^1 on S^2, hence we can form the dual cell structure as described at the beginning of this subsection. We leave it as an exercise for the reader to check that the resulting structure has two dual $(V-1)$-cells, being lines of latitude in the northern and southern hemispheres, and three dual V-cells, being a thickened equator and caps covering each pole.
(5) The ancient geographers new another dual G-CW(V) structure on the sphere: The dual V-cells are the tropical zone, the two temperate zones, and the arctic and antarctic zones. The dual $(V-1)$-cells are the Tropics of Cancer and Capricorn, and the arctic and antarctic circles.
(6) We gave a G-triangulation of the Möbius strip in Example 1.1.3(8). Its dual structure is *not* a dual G-CW(V) structure for any representation V of S^1, for the reasons pointed out in Example 1.1.5(4). It will, again, be an example of a more general kind of complex.

Here is one more thing we will want to do with CW complexes, when we discuss products in cohomology. Suppose that X is a G-CW(V) complex and Y is a dual H-CW(W) complex. Their product $X \times Y$, considered as a $G \times H$-space, has a natural cell structure: The product of a $(V+m)$-cell $G \times_J D(V+m)$ and a dual $(W+n)$-cell

$H \times_K D(W - \mathscr{L}(H/K) + n)$ has the form

$$\begin{aligned} G \times_J D(V + m) \times H \times_K D(W - \mathscr{L}(H/K) + n) \\ = (G \times H) \times_{J \times K} D(V + W - \mathscr{L}(H/K) + m + n), \end{aligned}$$

and should be considered to be $(V + W + m + n)$-dimensional. However, this is neither an ordinary cell nor a dual cell, as we appear to be considering the orbit $(G \times H)/(J \times K)$ to be $\mathscr{L}(H/K)$-dimensional, which may be strictly between 0 and $\mathscr{L}((G \times H)/(J \times K))$, the geometric dimension of the orbit $(G \times H)/(J \times K)$. This suggests that we need to consider a general way of assigning a "dimension" to each orbit of the ambient group, to be used when defining cell complexes. This we take up in the following section.

1.2 Dimension Functions

In the preceding section we saw different possible contributions of an orbit to the dimension of a cell $G \times_H D(V)$. The least an orbit can contribute is nothing and the most is its geometric dimension. The definition of a "dimension function" below will allow us to handle simultaneously G-CW complexes, dual complexes, and products of different kinds of complexes.

Recall the representations $\mathscr{L}(H/K)$ defined in Definition 1.1.6. We organize all the spaces $\mathscr{L}(H/K)$, $K \le H \le G$, as follows. We assume chosen an invariant metric on G so that $\mathscr{L}(G)$ is an inner product space. The (left) conjugation action of G on itself induces an orthogonal action of G on $\mathscr{L}(G)$ (well-known as the *adjoint representation* of G). We consider $\mathscr{L}(H) \subset \mathscr{L}(G)$ for $H \le G$, with $\mathscr{L}(H)$ identified with the subspace of vectors in $\mathscr{L}(G)$ tangent to H. $\mathscr{L}(H)$ is an $N_G H$-subspace of $\mathscr{L}(G)$, where $N_G H$ denotes the normalizer of H in G. If $K \le H \le G$, we then identify

$$\mathscr{L}(H/K) \cong \mathscr{L}(H) - \mathscr{L}(K),$$

the orthogonal complement of $\mathscr{L}(K)$ in $\mathscr{L}(H)$. Writing $N_G(H/K)$ for $N_G H \cap N_G K$, $\mathscr{L}(H/K)$ is an $N_G(H/K)$-subspace of $\mathscr{L}(G)$. For any $g \in G$, we have $g \cdot \mathscr{L}(H/K) = \mathscr{L}(H^g/K^g)$ as subsets of $\mathscr{L}(G)$, where $H^g = gHg^{-1}$.

Note that $N_G(H/K)$ acts on H/K by conjugation. The induced action on $\mathscr{L}(H/K)$ is the same as the action specified above. Restricting to K, the conjugation action of K on H/K is the same as its action by left multiplication, and the induced action on $\mathscr{L}(H/K)$ is the usual action of K on the fiber of the tangent bundle $H \times_K \mathscr{L}(H/K)$ of H/K over the coset eK.

Note also that, if $K \le H \le G$, then we have $\mathscr{L}(G/H) \subset \mathscr{L}(G/K)$ and

$$\mathscr{L}(G/K) - \mathscr{L}(G/H) = \mathscr{L}(H/K).$$

More generally, if $L \le K \le H \le G$, we have $\mathscr{L}(H/K) \subset \mathscr{L}(H/L)$ and

$$\mathscr{L}(H/L) - \mathscr{L}(H/K) = \mathscr{L}(K/L).$$

In the definition of dimension function below, we use a general collection of subgroups of G. The most interesting case is the collection of all subgroups, but the more general case will be needed at times.

Definition 1.2.1 A *dimension function* δ for a group G consists of a subset $\mathscr{F}(\delta)$ of subgroups of G (the *domain* of δ), closed under conjugation (but not necessarily under taking subgroups), and an assignment to each $H \in \mathscr{F}$ of a sub-H-representation $\delta(G/H) \subset \mathscr{L}(G/H)$ such that

(1) for each $K \le H$ such that $K, H \in \mathscr{F}$,

$$\delta(G/H) \subset \delta(G/K)$$

and

$$\mathscr{L}(G/H) - \delta(G/H) \subset \mathscr{L}(G/K) - \delta(G/K)$$

and

(2) for each $H \in \mathscr{F}$ and $g \in G$, $\delta(G/H^g) = g \cdot \delta(G/H)$ (as subspaces of $\mathscr{L}(G)$).

Examples 1.2.2

(1) One obvious example of a dimension function is $\delta = 0$, the function assigning 0 to every orbit of G; its domain consists of all subgroups of G. This is the dimension function we will associate with ordinary G-CW(V) complexes.
(2) The other obvious example is $\delta = \mathscr{L}$, which assigns the representation $\mathscr{L}(G/H)$ to every orbit G/H; again, its domain contains all subgroups of G. This is the dimension function we will associate with dual G-CW(V) complexes.
(3) If δ is a dimension function for G and ϵ is a dimension function for another group H, we let $\delta \times \epsilon$ be the dimension function for $G \times H$ defined as follows.

$$\mathscr{F}(\delta \times \epsilon) = \{J \times K \mid J \in \mathscr{F}(\delta) \text{ and } K \in \mathscr{F}(\epsilon)\}$$

and

$$(\delta \times \epsilon)((G \times H)/(J \times K)) = \delta(G/J) \oplus \delta(H/K).$$

We leave to the reader to show that this is, indeed, a dimension function. In particular, the dimension function $0 \times \mathscr{L}$ will be associated with the induced structure on the product of an ordinary G-CW(V) complex and a dual H-CW(W) complex.

(4) If δ is a dimension function for G, we can define another dimension function ϵ with the same domain by letting

$$\epsilon(G/H) = \mathscr{L}(G/H) - \delta(G/H).$$

We call this the *dual* function to δ, and usually write $\mathscr{L} - \delta$ for ϵ. Note that 0 and $\mathscr{L}$ are dual to one another.

Remarks 1.2.3 In the following remarks, let $K \leq H$ with $K, H \in \mathscr{F}(\delta)$.

(1) We could replace the requirement that $(\mathscr{L} - \delta)(G/H) \subset (\mathscr{L} - \delta)(G/K)$ with the equivalent requirement that

$$\delta(G/K) - \delta(G/H) \subset \mathscr{L}(G/K) - \mathscr{L}(G/H).$$

We leave the equivalence of the two conditions as an exercise for the reader.

(2) Rather than think of $\mathscr{L}(G/H) \subset \mathscr{L}(G/K)$, we could consider the projection $\mathscr{L}(G/K) \twoheadrightarrow \mathscr{L}(G/H)$ induced by the projection $G/K \to G/H$. The first conditions in the definition are then equivalent to the conditions that $\delta(G/K)$ maps onto $\delta(G/H)$ under the projection and that $(\mathscr{L} - \delta)(G/K)$ maps onto $(\mathscr{L} - \delta)(G/H)$.

(3) Carrying the preceding idea further, we can consider a dimension function as an assignment to each orbit G/H (with $H \in \mathscr{F}(\delta)$) of a subbundle $G \times_H \delta(G/H)$ of the tangent bundle $G \times_H \mathscr{L}(G/H)$. To each map of orbits $G/K \to G/H$ we then require that the projection

$$G \times_K \mathscr{L}(G/K) \twoheadrightarrow G \times_H \mathscr{L}(G/H)$$

splits as the sum of two projections,

$$G \times_K \delta(G/K) \twoheadrightarrow G \times_H \delta(G/H)$$

and

$$G \times_K (\mathscr{L} - \delta)(G/K) \twoheadrightarrow G \times_H (\mathscr{L} - \delta)(G/H).$$

We find it more convenient, however, to think in terms of inclusions as in the definition above.

Definition 1.2.4 Let $\mathscr{F}$ be a collection of subgroups of G closed under conjugation.

(1) Let $\bar{\mathscr{F}}$ denote the closure of $\mathscr{F}$ under taking subgroups, i.e.,

$$\bar{\mathscr{F}} = \{H \leq G \mid H \leq K \text{ for some } K \in \mathscr{F}\}.$$

(2) Let $\mathcal{O}_{\mathcal{F}}$ denote the category of orbits G/H, $H \in \mathcal{F}$, and all G-maps between them.
(3) Let $\mathcal{O}_G$ denote the category of all orbits of G.

Definition 1.2.5 Let δ be a dimension function for G.

(1) If $\mathcal{F}(\delta)$ is closed under taking subgroups (i.e., $\mathcal{F}(\delta) = \overline{\mathcal{F}(\delta)}$), so is a *family* of subgroups, we say that δ is *familial.*
(2) If $\mathcal{F}(\delta)$ is the collection of all subgroups of G, we say that δ is *complete*.

Product dimension functions $\delta \times \epsilon$ are generally not familial, even if δ and ϵ are. 0 and $\mathcal{L}$ are complete.

Conjecture 1.2.6 0 and $\mathcal{L}$ are the only complete dimension functions.

It's not hard to show that this conjecture is true for tori, hence for all groups for which the component of the identity is a torus.

If $K \leq H \leq G$ with K and H in $\mathcal{F}(\delta)$, we let

$$\delta(H/K) = \delta(G/K) - \delta(G/H).$$

Then $\delta(H/K)$ is a representation of K and

$$\delta(G/K) = \delta(G/H) \oplus \delta(H/K).$$

Further, as noted in Remarks 1.2.3, we have

$$\delta(H/K) \subset \mathcal{L}(H/K).$$

Remarks 1.2.7

(1) If $K \leq J \leq H \leq G$ with $K, J, H \in \mathcal{F}(\delta)$, then

$$\delta(H/K) = \delta(H/J) \oplus \delta(J/K).$$

We sometimes refer to this as the *additivity* of δ.
(2) Let $L \leq K \leq J \leq H$ with $K, J, H \in \mathcal{F}(\delta)$. From the fact that $\delta(H/J)$ is a subrepresentation of $\delta(H/K)$ and also a subrepresentation of $\mathcal{L}(H/J)$, we get the following inequalities that will be used later:

$$|\delta(H/J)^L| \leq |\delta(H/K)^L|$$

$$|\delta(H/J)^L| \leq |\mathcal{L}(H/J)^L|$$

Finally, the following relation will be useful.

Definition 1.2.8 Let δ and ϵ be dimension functions for G. We say that δ *dominates* ϵ, and write $\delta \succcurlyeq \epsilon$, if $\mathscr{F}(\delta) \subset \overline{\mathscr{F}(\epsilon)}$ and, for each $K \in \mathscr{F}(\delta)$ and $H \in \mathscr{F}(\epsilon)$ such that $K \leq H$, we have that $\epsilon(G/H)$ is K-isomorphic to a subrepresentation of $\delta(G/K)$. (Note that we do not require it to be a subspace.)

This relation is a partial order on familial dimension functions, but it is not transitive on the set of all dimension functions for G. Also note that, if ϵ is familial, the last condition of the definition can be replaced with the requirement that, for each $K \in \mathscr{F}(\delta)$, we have that $\epsilon(G/K)$ is isomorphic to a subrepresentation of $\delta(G/K)$.

1.3 Virtual Representations

Following common usage, we shall say that our homology theories are graded on $RO(G)$, however, that is really a misnomer and, if taken too seriously, leads to sign ambiguities (meaning, equivariantly, ambiguities up to units in the Burnside ring). Rather, these theories are functors on a category of virtual representations, with automorphisms of representations inducing possibly nonidentity isomorphisms of homology groups. There is no canonical way of "decategorifying" homology theories to functions on isomorphism classes of representations. See, for example the discussion in [47, §XIII.1], where homology theories are defined as functors on a category of actual representations, and then extended to virtual representations via the suspension isomorphism. In the context of Chap. 3, with varying local representations, that approach won't be adequate, so we prefer to use here another approach, to define a category of virtual representations on which homology groups are functorial.

So, we want to view virtual representations $V \ominus W$ as first-class objects in their own right. The following definition is a restriction of one given in [20, §19]. Let $\mathscr{U}$ denote a complete G-universe, i.e., the sum of countably many copies of each irreducible representation of G.

Definition 1.3.1 The category of *virtual representations of G* has as its objects the pairs (V, W) of finite-dimensional representations of G; we think of and often write a pair (V, W) as a formal difference $V \ominus W$. A *(virtual) map* $V_1 \ominus W_1 \to V_2 \ominus W_2$ is the equivalence class of a pair of orthogonal G-isomorphisms

$$f: V_1 \oplus Z_1 \to V_2 \oplus Z_2$$
$$g: W_1 \oplus Z_1 \to W_2 \oplus Z_2$$

where Z_1 and Z_2 are finite-dimensional G-subspaces of $\mathscr{U}$. The equivalence relation on such pairs (f, g) is generated by two basic relations, the first being G-homotopy through orthogonal maps. The second relation is as follows. Let T_1 be a finite-dimensional G-subspace of $\mathscr{U}$ orthogonal to Z_1 and T_2 a finite-dimensional G-subspace of $\mathscr{U}$ orthogonal to Z_2, and let $k: T_1 \to T_2$ be an orthogonal G-

isomorphism. Then (f, g) is equivalent to the "suspension" $(f \oplus k, g \oplus k)$ where

$$f \oplus k \colon V_1 \oplus (Z_1 + T_1) \to V_2 \oplus (Z_2 + T_2)$$
$$g \oplus k \colon W_1 \oplus (Z_1 + T_1) \to W_2 \oplus (Z_2 + T_2).$$

Composition of virtual maps is defined by suspending until the pairs can be composed as pairs of orthogonal G-isomorphisms.

That this gives a well-defined category is not difficult to show; see [20, 19.2]. We say that two virtual representations are *stably equivalent* if they are equivalent in this category. If $\alpha = V \ominus W$, we write $|\alpha| = |V| - |W|$ for the integer dimension of α.

Remark 1.3.2 We would get an isomorphic category if, instead of using a pair of maps (f, g) as above, we used a single orthogonal isomorphism $F \colon V_1 \oplus W_2 \oplus Z \to V_2 \oplus W_1 \oplus Z$ and the equivalence relation generated by G-homotopy and suspension by identity maps.

If V is an (actual) representation of G, we consider V as the virtual representation $V \ominus 0$ and we write $-V$ for $0 \ominus V$. If n is an integer, we write $V \ominus W + n$ for $(V \oplus \mathbb{R}^n, W)$ if $n \geq 0$ and $(V, W \oplus \mathbb{R}^{|n|})$ if $n < 0$. Note that there will be times where we will have to be careful to specify exactly what copy of $\mathbb{R}^n$ we are using. In general, we write $(V_1 \ominus W_1) + (V_2 \ominus W_2) = (V_1 \oplus V_2) \ominus (W_1 \oplus W_2)$ and $-(V \ominus W) = W \ominus V$.

1.4 Cell Complexes

Much of this section is based on May's elegant approach to CW complexes, as in [45] and [48]. However, the presence of a group action and, particularly, dimension functions and virtual representations, result in some interesting complications.

If V is an orthogonal representation of G, continue to write $D(V)$ for its unit disc and $S(V)$ for its unit sphere.

Definition 1.4.1 Let $\alpha = V \ominus W$ be a virtual representation of G and let δ be a dimension function for G with domain $\mathscr{F}(\delta)$.

(1) A subgroup H of G is *δ-α-admissible*, or simply *admissible*, if $H \in \mathscr{F}(\delta)$ and $\alpha - \delta(G/H) + n$ is stably equivalent to an actual H-representation for some integer n.
(2) A *δ-α-cell* is a pair of G-spaces of the form $(G \times_H D(Z), G \times_H S(Z))$ where $H \in \mathscr{F}(\delta)$ and Z is an actual representation of H such that Z is stably equivalent to $\alpha - \delta(G/H) + n$ for some integer n. Note that H must, by definition, be admissible for such a cell to exist. We say that the *dimension* of such a cell is $\alpha + n$.

As a more concise notation, we shall write

$$G \times_H \bar{D}(Z) = (G \times_H D(Z), G \times_H S(Z)).$$

Note that the dimension of $G \times_H \bar{D}(Z)$ is $Z + \delta(G/H)$. So, we think of $\delta(G/H)$ as the contribution of G/H to the total dimension of $G \times_H \bar{D}(Z)$.

Definition 1.4.2 Let α be a virtual representation of G and let δ be a dimension function for G.

(1) A *δ-G-CW(α) complex* is a G-space X together with a decomposition $X = \operatorname{colim}_n X^n$ such that

  (a) X^0 is a disjoint union of G-orbits G/H such that H acts trivially on α, and $H \in \mathscr{F}(\delta)$, $\delta(G/H) = 0$.
  (b) X^n is obtained from X^{n-1} by attaching δ-α-cells of dimension $\alpha - |\alpha| + n$ along their boundary spheres. Notice that the boundary sphere may be empty and the disc simply an orbit, in which case attaching the cell means taking the disjoint union with the orbit.

  For notational convenience, and to remind ourselves of the role of α, we shall also write $X^{\alpha+n}$ for $X^{|\alpha|+n}$.

(2) A *relative* δ-G-CW(α) complex is a pair (X, A) where $X = \operatorname{colim}_n X^n$, X^0 is the disjoint union of A with orbits as in (a) above, and cells are attached as in (b).
(3) A *based* δ-G-CW(α) complex is a relative δ-G-CW(α) complex $(X, *)$ where $*$ denotes a G-fixed basepoint.
(4) If we allow cells of any dimension to be attached at each stage, we get the weaker notions of absolute, relative, or based *δ-α-cell complex* (or δ-G-α-cell complex if we need to specify the group).
(5) If X is a δ-G-CW(α) or δ-α-cell complex with cells only of dimension less than or equal to $\alpha + n$, we say that X is *$(\alpha + n)$-dimensional.*

A simple observation, but useful: If (X, A) is a relative cell complex, then $(X/A, *)$ is a based cell complex.

Examples 1.4.3

(1) A 0-G-CW(0) complex is simply a G-CW complex, as defined in Definition 1.1.1.
(2) If V is an actual representation of G, then a 0-G-CW(V) complex is a G-CW(V) complex, as defined in Definition 1.1.4.
(3) If V is an actual representation of G, then an $\mathscr{L}$-G-CW(V) complex is a dual G-CW(V) complex, as defined in Definition 1.1.7.
(4) If δ is a dimension function for G, ϵ is a dimension function for H, α is a virtual representation of G, and β is a virtual representation of H, then a product of a δ-G-CW(α) complex with an ϵ-H-CW(β) complex is a $(\delta\times\epsilon)$-$(G\times H)$-CW($\alpha+\beta$) complex.

Definition 1.4.4 Let δ and ϵ be dimension functions for G. If (X, A) is a relative δ-G-CW(α) complex and (Y, B) is a relative ϵ-G-CW(α) complex, then we say that a map $f: (X, A) \to (Y, B)$ is *cellular* if $f(X^{\alpha+n}) \subset Y^{\alpha+n}$ for each n. As special cases, we have the notions of cellular maps of absolute or based CW complexes. We write $G\mathscr{W}^{\delta,\alpha}$ for the category of δ-G-CW(α) complexes and cellular maps; we write $G\mathscr{W}_*^{\delta,\alpha}$ for the category of based δ-G-CW(α) complexes and based cellular maps.

Definition 1.4.5 Let δ be a dimension function for G and let $H \in \mathscr{F}(\delta)$. Define $\delta|H$, the *restriction of δ to H*, to be the dimension function with

$$\mathscr{F}(\delta|H) = \{K \in \mathscr{F}(\delta) \mid K \leq H\}$$

and

$$(\delta|H)(H/K) = \delta(H/K) = \delta(G/K) - \delta(G/H)$$

for $K \in \mathscr{F}(\delta|H)$.

It is easy to check that $\delta|H$ is a dimension function for H. Where the meaning is clear, we will usually write δ again for $\delta|H$.

The following observation is key to change-of-group homomorphisms and also simplifies some arguments.

Proposition 1.4.6 *Let δ be a dimension function for G and let $H \in \mathscr{F}(\delta)$. If X is a δ-H-$(\alpha - \delta(G/H))$-cell complex, then $G \times_H X$ is a δ-G-α-cell complex with corresponding cells. Applying this construction to CW complexes and cellular maps, we get functors*

$$G \times_H -: H\mathscr{W}^{\delta,\alpha-\delta(G/H)} \to G\mathscr{W}^{\delta,\alpha}$$

and

$$G_+ \wedge_H -: H\mathscr{W}_*^{\delta,\alpha-\delta(G/H)} \to G\mathscr{W}_*^{\delta,\alpha}.$$

Proof If X is a δ-$(\alpha - \delta(G/H))$-cell complex, a typical cell has the form

$$H \times_K \bar{D}(\alpha - \delta(G/H) - \delta(H/K) + n)$$

for $K \in \mathscr{F}(\delta|H)$. Applying $G \times_H -$ gives a corresponding cell of the form

$$G \times_H H \times_K \bar{D}(\alpha - \delta(G/H) - \delta(H/K) + n) \cong G \times_K \bar{D}(\alpha - \delta(G/K) + n),$$

using that $\delta(G/K) \cong \delta(G/H) \oplus \delta(H/K)$. Hence, $G \times_H X$ is a δ-α-cell complex. If X is CW, then so is $G \times_H X$ and, if $f: X \to Y$ is cellular, then so is $G \times_H f$. □

Definition 1.4.7 If n is an integer, a G-map $f: X \to Y$ is a δ-*$(\alpha + n)$-equivalence* if, for each admissible H and each actual representation V of H stably equivalent to

$\alpha - \delta(G/H) + i$, with $i \leq n$, every diagram of the following form is homotopic to one in which there exists a lift $G \times_H D(V) \to X$:

$$\begin{array}{ccc} G \times_H S(V) & \longrightarrow & X \\ \big\downarrow & & \big\downarrow f \\ G \times_H D(V) & \longrightarrow & Y \end{array}$$

We say that f is a *δ-weak$_\alpha$ equivalence* if it a δ-$(\alpha + n)$-equivalence for all n.

Shortly we shall characterize δ-weak$_\alpha$ equivalences for a familial dimension function δ in terms of their behavior on fixed points, but if not all H are δ-α-admissible, δ-weak$_\alpha$ equivalences are not in general weak G-equivalences.

We have the following variant of the "homotopy extension and lifting property" of [45].

Lemma 1.4.8 (H.E.L.P.) *Let $r\colon Y \to Z$ be a δ-$(\alpha + n)$-equivalence. Let (X, A) be a relative δ-α-cell complex of dimension $\alpha + n$. If the following diagram commutes without the dashed arrows, then there exist maps $\tilde{g}$ and $\tilde{h}$ making the diagram commute.*

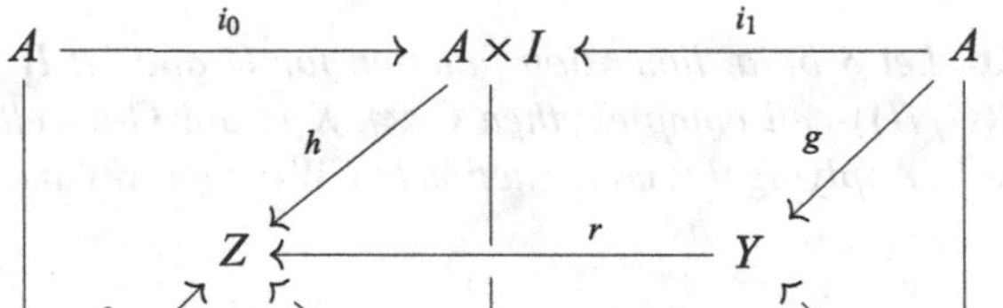

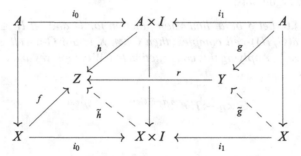

The result remains true when $n = \infty$.

Proof The proof is by induction over the cells of X not in A, so it suffices to consider the case $(X, A) = G \times_H \bar{D}(V)$ where V is stably equivalent to $\alpha - \delta(G/H) + i$ with $i \leq n$. As in the picture on the left of Fig. 1.1 (where $D = D(V)$ and $S = S(V)$), identify $D(V) \cup (S(V) \times I)$ with $D(V)$ and use f and h to define a map $G \times_H D(V) \to Z$

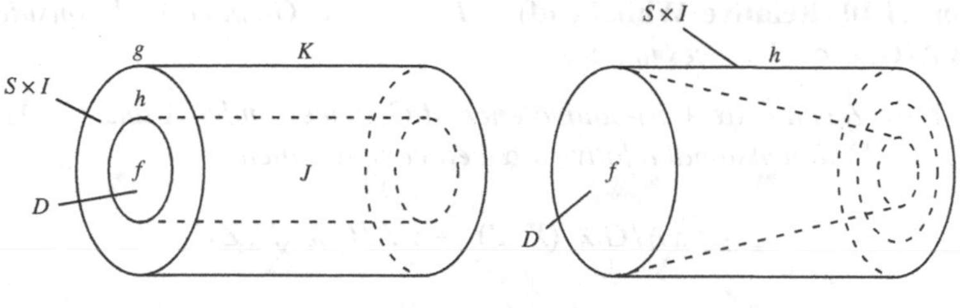

Fig. 1.1 Proof of H.E.L.P

that lifts to g on its boundary:

$$\begin{array}{ccc} G \times_H S(V) & \xrightarrow{g} & Y \\ \downarrow & & \downarrow r \\ G \times_H D(V) & \xrightarrow[f \cup h]{} & Z. \end{array}$$

By assumption, this diagram is homotopic to one in which we can find a lift $k: G \times_H D(V) \to Y$. Let K be the homotopy of g and J be the homotopy of $f \cup h$, as in Fig. 1.1. Now distort the cylinder in that figure to become the picture on the right. On the far right end of this new cylinder, k and K combine to give a map $\tilde{g}: G \times_H D(V) \to Y$, while the whole cylinder describes a map $\tilde{h}: G \times_H D(V) \times I \to Z$. It is easy to see that these maps make the diagram in the statement of the lemma commute. □

Theorem 1.4.9 (Whitehead)

(1) *If $f: Y \to Z$ is a δ-$(\alpha + n)$-equivalence and X is an $(\alpha + n - 1)$-dimensional δ-α-cell complex, then*

$$f_*: \pi G\mathscr{K}(X, Y) \to \pi G\mathscr{K}(X, Z)$$

is an isomorphism, where $\pi G\mathscr{K}(-,-)$ denotes the set of homotopy classes of G-maps. It is an epimorphism if X is $(\alpha + n)$-dimensional.

(2) *If $f: Y \to Z$ is a δ-weak$_\alpha$ equivalence and X is a δ-α-cell complex, then*

$$f_*: \pi G\mathscr{K}(X, Y) \to \pi G\mathscr{K}(X, Z)$$

is an isomorphism. Therefore, any δ-weak$_\alpha$ equivalence of δ-α-cell complexes is a G-homotopy equivalence.

Proof For the first part, apply the H.E.L.P. lemma to the pair $(X, \emptyset)$ to get surjectivity and to the pair $(X \times I, X \times \partial I)$ to get injectivity. The second part follows in the same way from the last statement of the H.E.L.P. lemma. □

Theorem 1.4.10 (Relative Whitehead) *Let A be a G-space and consider the category $A/G\mathscr{K}$ of G-spaces under A.*

(1) *If $f: Y \to Z$ is a δ-$(\alpha + n)$-equivalence of G-spaces under A and (X, A) is an $(\alpha + n - 1)$-dimensional relative δ-α-cell complex, then*

$$f_*: \pi A/G\mathscr{K}(X, Y) \to \pi A/G\mathscr{K}(X, Z)$$

is an isomorphism, where $\pi A/G\mathscr{K}(-,-)$ denotes the set of homotopy classes of G-maps under A. It is an epimorphism if (X, A) is $(\alpha + n)$-dimensional.

(2) *If $f: Y \to Z$ is a δ-weak$_\alpha$ equivalence of G-spaces under A and (X, A) is a relative δ-α-cell complex, then*

$$f_*: \pi A/G\mathscr{K}(X, Y) \to \pi A/G\mathscr{K}(X, Z)$$

is an isomorphism. Therefore, any δ-weak$_\alpha$ equivalence of relative δ-α-cell complexes $(X, A) \to (Y, A)$ is a G-homotopy equivalence rel A.

Proof For the first part, apply the H.E.L.P. lemma to the pair (X, A) to get surjectivity and to the pair $(X \times I, X \times \partial I \cup A \times I)$ to get injectivity. The second part follows in the same way from the last statement of the H.E.L.P. lemma. □

Note that, when we take $A = *$ in the relative Whitehead theorem, we get the special case of based G-spaces and based cell complexes. One reason for stating many of the results of this section in relative form is to include both the based and unbased cases.

Thus far, the results are essentially formal. To get more substantive results we need to know more about what maps are δ-$(\alpha + n)$-equivalences. From this point on we need to assume that δ is familial when discussing δ-equivalences. We start with the following important example, which uses the relation defined in Definition 1.2.8. Note that it applies to the most important case $\epsilon = \delta$ because $\delta \succcurlyeq \delta$; we will need the more general statements in later sections.

Proposition 1.4.11 *Let δ be a familial dimension function and let ϵ be another dimension function (not necessarily familial) with $\delta \succcurlyeq \epsilon$. If (X, A) is a relative ϵ-G-$CW(\alpha)$ complex then the inclusion $X^{\alpha+n} \to X$ is a δ-$(\alpha + n)$-equivalence.*

Proof By the usual induction, this reduces to showing that, if Y is obtained from B by attaching a cell of the form $G \times_K D(W)$, where W is stably equivalent to $\alpha - \epsilon(G/K) + k$, and V is a representation of H stably equivalent to $\alpha - \delta(G/H) + i$ where $i < k$, then any G-map of pairs $G \times_H \bar{D}(V) \to (Y, B)$ is homotopic rel boundary to a map into B. (In the application to the present proposition, $k > n$ and $i \leq n$.) For this we take an ordinary G-triangulation of $G \times_H D(V)$ and show by induction on the cells that we can homotope the map to miss the orbit $G/K \times 0$ in the attached cell. We use the fact that the fixed-set dimensions of V must equal those of $\alpha - \delta(G/H) + i$ and the fixed set dimensions of W equal those of $\alpha - \epsilon(G/K) + k$.

The only simplices that might hit the orbit $G/K \times 0$ have the form $G/J \times \Delta^j$ where $J \leq H$ and J is subconjugate to K; by replacing K with a conjugate we may assume $J \leq K$ also. Because such a simplex is embedded in $G \times_H D(V)$, we have

$$\begin{aligned} j &\leq |[\alpha - \delta(G/H) - \mathscr{L}(H/J)]^J| + i \\ &\leq |[\alpha - \delta(G/H) - \delta(H/J)]^J| + i \\ &= |[\alpha - \delta(G/J)]^J| + i \\ &\leq |[\alpha - \epsilon(G/K)]^J| + i \\ &< |[\alpha - \epsilon(G/K)]^J| + k. \end{aligned}$$

Here we use that

$$|\delta(H/J)^J| \leq |\mathscr{L}(H/J)^J|$$

as in Remarks 1.2.7 and that

$$|\epsilon(G/K)^J| \leq |\delta(G/J)^J|$$

because $\delta \succcurlyeq \epsilon$. Finally, note that $|[\alpha-\epsilon(G/K)]^J|+k$ is the codimension of $(G/K\times 0)^J$ in Y^J, so the desired homotopy exists for dimensional reasons. □

Theorem 1.4.12 (Cellular Approximation of Maps) *Let δ be a familial dimension function and let ϵ be any dimension function such that $\delta \succcurlyeq \epsilon$. Let (X, A) be a relative δ-G-CW(α) complex and let (Y, B) be a relative ϵ-G-CW(α) complex. Let $f: (X, A) \to (Y, B)$ be a G-map and suppose given a subcomplex $(Z, A) \subset (X, A)$ and a G-homotopy h of $f|Z$ to a cellular map. Then h can be extended to a G-homotopy of f to a cellular map.*

Proof This follows by induction on skeleta, using the H.E.L.P. lemma and Proposition 1.4.11 applied to the inclusion $Y^{\alpha+n} \to Y$. □

We digress now to discuss how δ-$(\alpha + n)$-equivalence is related to conditions on fixed sets. We first need the following approximation result.

Lemma 1.4.13 *Let δ be a familial dimension function. If V is an actual representation of H stably equivalent to $\alpha - \delta(G/H) + n$, then the pair $G \times_H \bar{D}(V)$ is G-homotopy equivalent to a pair of δ-G-CW(α) complexes (X, A) where X has dimension $\alpha + n$ and A has dimension $\alpha + n - 1$.*

Proof We first note that, by Proposition 1.4.6, it suffices to show that the H-pair $\bar{D}(V)$ is H-homotopy equivalent to a pair of δ-H-CW(V) complexes of dimensions V and $V - 1$, respectively.

We'll prove a slightly more general result (necessary for an inductive argument): If W is a sub-H-representation of V, then the pair $\bar{D}(V - W) \times D(W)$ is H-homotopy equivalent to a pair of δ-H-CW(V) complexes of dimensions V and $V-1$. We shall prove this for every subgroup H of G and every V and W, by induction on H.

The beginning of the induction is the case where $H = e$. In this case,

$$\bar{D}(V-W)\times D(W) = (D(\mathbb{R}^n), S(\mathbb{R}^{n-m})\times D(\mathbb{R}^m))$$

for some n and m, and $D(\mathbb{R}^n)$ can be given a nonequivariant triangulation of dimension n with $S(\mathbb{R}^{n-m})\times D(\mathbb{R}^m)$ a subcomplex of dimension $n-1$.

Now assume the result for all proper subgroups of H. If $V^H \neq 0$, write $n = |(V-W)^H|$ and $m = |W^H|$, and then we can write

$$\bar{D}(V-W)\times D(W) \cong \bar{D}(V-W-n)\times D(W-m)\times \bar{D}(\mathbb{R}^n)\times D(\mathbb{R}^m).$$

Using a non-equivariant triangulation of $\bar{D}(\mathbb{R}^n)\times D(\mathbb{R}^m)$, we see that it suffices to find the desired cell structure on $\bar{D}(V-W-n)\times D(W-m)$.

This reduces us to the case where V contains no H-trivial summands. It suffices now to show that $(S(V), S(V-W)\times D(W))$ is equivalent to a pair of δ-G-CW(V) complexes of dimension $V-1$, since we then get $D(V)$ by attaching one cell of the form $D(V)$, i.e., the interior. Give $(S(V), S(V-W)\times D(W))$ an ordinary H-triangulation and then take the dual cell structure. A typical dual cell will have the form $H\times_J \bar{D}(V-\mathscr{L}(H/J)-j)$ where $J\subset H$ and $j>0$. Such a cell is H-homotopy equivalent to the pair

$$\begin{aligned}&H\times_J \bar{D}(V-\mathscr{L}(H/J)-j)\times D(\mathscr{L}(H/J)-\delta(H/J))\\&\quad= H\times_J \bar{D}(V-\delta(H/J)-(\mathscr{L}(H/J)-\delta(H/J))-j)\times D(\mathscr{L}(H/J)-\delta(H/J)).\end{aligned}$$

(We use here that $\delta(H/J)\subset \mathscr{L}(H/J)$.) Note that J must be a proper subgroup of H because V has no H-trivial summand, yet its sphere, which contains the cell $H\times_J \bar{D}(V-\mathscr{L}(H/J)-j)$, has a J-fixed point. With $\mathscr{L}(H/J)-\delta(H/J)$ playing the role of W and $V-\delta(H/J)-j$ playing the role of V, we can now apply the inductive hypothesis to say that the J-pair

$$\bar{D}(V-\delta(H/J)-(\mathscr{L}(H/J)-\delta(H/J))-j)\times D(\mathscr{L}(H/J)-\delta(H/J))$$

is J-homotopy equivalent to a pair of δ-J-CW($V-\delta(H/J)$) complexes of dimensions $V-\delta(H/J)-j$ and $V-\delta(H/J)-j-1$. Applying $H\times_J -$ and using Proposition 1.4.6 again, we get the structure we are seeking on $H\times_J\bar{D}(V-\mathscr{L}(H/J)-j)\times D(\mathscr{L}(H/J)-\delta(H/J))$.

Using induction on the dual cell structure of $(S(V), S(V-W)\times D(W))$, we replace each dual cell with an H-homotopy equivalent pair of complexes as above, using a cellular approximation of the attaching map. This constructs a pair of δ-H-CW(V) complexes of dimension $V-1$, H-homotopy equivalent to $(S(V), S(V-W)\times D(W))$, as required. □

Corollary 1.4.14 *If δ is familial, H is a δ-α-admissible subgroup of G, and n is any nonnegative integer, then the pair $G \times_H \bar{D}^n$ is G-homotopy equivalent to a pair of δ-G-CW(α) complexes of dimensions $\alpha - |[\alpha - \delta(G/H)]^H| + n$ and $\alpha - |[\alpha - \delta(G/H)]^H| + n - 1$.*

Proof Because H is admissible, $\alpha - \delta(G/H) + k$ is stably equivalent to an actual representation for some k; in fact, the smallest such k is $-|[\alpha - \delta(G/H)]^H|$, so let V be an actual representation of H stably equivalent to $\alpha - \delta(G/H) - |[\alpha - \delta(G/H)]^H|$. By the preceding lemma (applied to H rather than G), $D(V)$ is H-homotopy equivalent to a δ-H-CW($\alpha - \delta(G/H)$) complex of dimension $\alpha - \delta(G/H) - |[\alpha - \delta(G/H)]^H|$. It follows that

$$G \times_H \bar{D}^n \simeq G \times_H (\bar{D}^n \times D(V))$$

is G-homotopy equivalent to a pair of δ-G-CW(α) complexes of dimensions $\alpha - |[\alpha - \delta(G/H)]^H| + n$ and $\alpha - |[\alpha - \delta(G/H)]^H| + n - 1$ as claimed. □

Lemma 1.4.15 *If δ is familial and $\alpha - \delta(G/H) + n$ is stably equivalent to an actual representation V of H, then the pair $G \times_H \bar{D}(V)$ has an ordinary G-triangulation in which the simplices have the form $G \times_J \Delta^j$ with J δ-α-admissible and $j \le |[\alpha - \delta(G/J)]^J| + n$ [$j \le |[\alpha - \delta(G/J)]^J| + n - 1$ for a cell in $G \times_H S(V)$].*

Proof Take any G-triangulation of the pair $G \times_H \bar{D}(V)$ and let $G \times_J \Delta^j$ be any one of its simplices; we may assume that $J \le H$. Because $H/J \times \Delta^J$ embeds in $D(V)$ we must have

$$\delta(H/J) \subset \mathscr{L}(H/J) \subset V$$

(i.e., $\delta(H/J)$ is isomorphic to a J-subspace of V), hence

$$\delta(G/J) \cong \delta(G/H) + \delta(H/J) \subset V + \delta(G/H),$$

and $V + \delta(G/H)$ is stably equivalent to $\alpha + n$, showing that J is admissible. We also must have

$$\begin{aligned} j &\le |[\alpha - \delta(G/H) - \mathscr{L}(H/J) + n]^J| \\ &\le |[\alpha - \delta(G/H) - \delta(H/J)]^J| + n \\ &= |[\alpha - \delta(G/J)]^J| + n \end{aligned}$$

in general, with the upper limit one lower if the simplex is embedded in the boundary sphere. □

Theorem 1.4.16 *Let δ be a familial dimension function. A map $f: X \to Y$ is a δ-$(\alpha + n)$-equivalence if and only if f^H is a nonequivariant $(|[\alpha - \delta(G/H)]^H| + n)$-*

equivalence for each δ-α-admissible subgroup H. A map f is a δ-weak$_\alpha$ equivalence if and only if f^H is a weak equivalence for every δ-α-admissible subgroup H.

Proof The last statement of the theorem follows directly from the first.

Suppose that f is a δ-$(\alpha+n)$-equivalence, let H be an admissible subgroup, and let $k = |[\alpha-\delta(G/H)]^H|+n$. If $k<0$ then f is a k-equivalence vacuously, so suppose $k \geq 0$. Then, by Corollary 1.4.14, for $0 \leq i \leq k$, $G \times_H \bar{D}^i$ is equivalent to a relative δ-G-CW(α) complex of dimension $\alpha - |[\alpha-\delta(G/H)]^H| + i = \alpha + n - (k-i)$. By the relative Whitehead theorem, it follows that any map from $\bar{D}^i$ to f^H is homotopic to one in which we can find a lift, hence f^H is a k-equivalence.

Conversely, suppose that f has the property that f^H is a nonequivariant $(|[\alpha - \delta(G/H)]^H| + n)$-equivalence for each admissible subgroup H. Consider a pair $G \times_H \bar{D}(V)$ where V is an actual representation of H stably equivalent to $\alpha - \delta(G/H) + i$, where $i \leq n$. By Lemma 1.4.15, this pair has a G-triangulation in which a typical simplex $G/J \times \Delta^j$ has J admissible and $j \leq |[\alpha - \delta(G/J)]^J| + i$. It follows from induction on this cell structure and our assumption on f, that any map from $G \times_H \bar{D}(V)$ to f is homotopic to one in which we can find a lift, hence f is a δ-$(\alpha+n)$-equivalence. □

One reason we need this theorem is to derive the following corollary, which shows that δ-weak$_\alpha$ equivalence deserves the name.

Corollary 1.4.17 (Two-out-of-three Property) *Let δ be familial and consider two G-maps $f: X \to Y$ and $g: Y \to Z$. If two of f, g, and gf are δ-weak$_\alpha$ equivalences, then so is the third.* □

Definition 1.4.18 (See also [42, II.2])

(1) If $\mathscr{F}$ is a collection of subgroups of G, we say that a G-map $f: X \to Y$ is an $\mathscr{F}$*-equivalence* if $f^H: X^H \to Y^H$ is a (non-equivariant) weak equivalence for all $H \in \mathscr{F}$.
(2) Given δ and α, let $\mathscr{F}(\delta,\alpha)$ denote the collection of δ-α-admissible subgroups of G.

Corollary 1.4.19 *Let δ be familial. Then a map f is a δ-weak$_\alpha$ equivalence if and only if it is an $\mathscr{F}(\delta,\alpha)$-equivalence. Further, for a fixed α, there exists an actual representation V such that $\mathscr{F}(\delta,\alpha+V) = \mathscr{F}(\delta)$ and, hence, a map f is a δ-weak$_{\alpha+V}$ equivalence if and only if it is an $\mathscr{F}(\delta)$-equivalence.*

Proof The first claim is a restatement of the last part of Theorem 1.4.16. If V is large enough, then $\alpha + V$ is equivalent to an actual representation containing a copy of $\mathscr{L}(G)$, from which it follows that every $H \in \mathscr{F}(\delta)$ is δ-$(\alpha+V)$-admissible. □

Recall from [42] that, if $\mathscr{F}$ is a family, there is a universal $\mathscr{F}$-space $E\mathscr{F}$, characterized by its fixed sets: $(E\mathscr{F})^H = \emptyset$ if $H \notin \mathscr{F}$ and $(E\mathscr{F})^H$ is contractible if $H \in \mathscr{F}$.

Corollary 1.4.20 *Let δ be a familial dimension function for G.*

(1) *If X is any G-space then the projection $E\mathscr{F}(\delta) \times X \to X$ is a δ-weak$_\alpha$ equivalence.*
(2) *Suppose that $\mathscr{F}(\delta,\alpha)$ is a family (e.g., in the case $\mathscr{F}(\delta,\alpha) = \mathscr{F}(\delta)$). If $f: X \to Y$ is a δ-weak$_\alpha$ equivalence, then $1 \times f: E\mathscr{F}(\delta,\alpha) \times X \to E\mathscr{F}(\delta,\alpha) \times Y$ is a weak G-equivalence.* □

Returning from our digression into weak equivalences, we now discuss approximation of spaces, beginning with the following general result.

Theorem 1.4.21 *Let δ be a familial dimension function, let (A,P) be a relative δ-G-CW(α) complex, let X be a G-space, and let $f: A \to X$ be a map. Then there exists a relative δ-G-CW(α) complex (Y,P), containing (A,P) as a subcomplex, and a δ-weak$_\alpha$ equivalence $g: Y \to X$ extending f.*

Proof We use a variant of the usual technique of killing homotopy groups. We start by letting

$$Y^{\alpha-|\alpha|} = A^{\alpha-|\alpha|} \sqcup \coprod G/K,$$

where the coproduct runs over all subgroups K such that $\delta(G/K) = 0$ and K acts trivially on α, and all maps $G/K \to X$. The map $g: Y^{\alpha-|\alpha|} \to X$ is the one induced by those maps of orbits. $(Y^{\alpha-|\alpha|}, P)$ is then a relative δ-G-CW(α) complex of dimension $\alpha - |\alpha|$, containing $(A^{\alpha-|\alpha|}, P)$, and $Y^{\alpha-|\alpha|} \to X$ is a δ-($\alpha - |\alpha|$)-equivalence.

Inductively, suppose that we have constructed $(Y^{\alpha+n-1}, P)$, a relative δ-G-CW(α) complex of dimension $\alpha + n - 1$ containing $(A^{\alpha+n-1}, P)$ as a subcomplex, and a δ-($\alpha + n - 1$)-equivalence $g: Y^{\alpha+n-1} \to X$ extending f on $A^{\alpha+n-1}$. Let

$$Y^{\alpha+n} = Y^{\alpha+n-1} \cup A^{\alpha+n} \cup \coprod G \times_K D(V),$$

where the coproduct runs over all subgroups K of G and isomorphism classes of representations V stably equivalent to $\alpha - \delta(G/K) + n$, and all diagrams of the form

$$\begin{array}{ccc} G \times_K S(V) & \longrightarrow & Y^{\alpha+n-1} \\ \downarrow & & \downarrow \\ G \times_K D(V) & \longrightarrow & X. \end{array}$$

The union that defines $Y^{\alpha+n}$ is along $A^{\alpha+n-1} \to Y^{\alpha+n-1}$ and the maps $G \times_K S(V) \to Y^{\alpha+n-1}$ displayed above. By construction, $(Y^{\alpha+n}, P)$ is a relative δ-G-CW(α) complex of dimension $\alpha + n$ containing $(A^{\alpha+n}, P)$ as a subcomplex. We let $g: Y^{\alpha+n} \to X$ be the induced map and we claim that g is a δ-($\alpha + n$)-equivalence.

To see this, consider any diagram of the following form, with V stably equivalent to $\alpha - \delta(G/K) + i$ and $i \le n$:

$$\begin{array}{ccc} G \times_K S(V) & \xrightarrow{\zeta} & Y^{\alpha+n} \\ \downarrow & & \downarrow \\ G \times_K D(V) & \xrightarrow[\xi]{} & X \end{array}$$

By Lemma 1.4.13, the sphere $G \times_K S(V)$ is G-homotopy equivalent to a δ-G-CW(α) complex of dimension $\alpha + i - 1$. Now $\alpha + i - 1 < \alpha + n$, so, by cellular approximation of maps, the diagram above is homotopic to one in which ζ maps the sphere into $Y^{\alpha+n-1}$. We can then find a lift of ξ up to homotopy using the inductive hypothesis if $i < n$ or the construction of $Y^{\alpha+n}$ if $i = n$.

Finally, $Y = \operatorname{colim}_n Y^{\alpha+n}$ satisfies the claim of the theorem. □

Theorem 1.4.22 (Approximation by δ-G-CW(α) Complexes) *Let δ be a familial dimension function and let X be a G-space. Then there exists a δ-G-CW(α) complex ΓX and a δ-weak$_\alpha$ equivalence $g: \Gamma X \to X$. If $f: X \to Y$ is a G-map and $g: \Gamma Y \to Y$ is an approximation of Y by a δ-G-CW(α) complex, then there exists a G-map $\Gamma f: \Gamma X \to \Gamma Y$, unique up to G-homotopy, such that the following diagram commutes up to G-homotopy:*

$$\begin{array}{ccc} \Gamma X & \xrightarrow{\Gamma f} & \Gamma Y \\ {\scriptstyle g}\downarrow & & \downarrow{\scriptstyle g} \\ X & \xrightarrow[f]{} & Y \end{array}$$

Proof The existence of $g: \Gamma X \to X$ is the special case of Theorem 1.4.21 in which we take $A = P = \emptyset$. The existence and uniqueness of Γf follows from Whitehead's theorem, which tells us that $\pi G\mathscr{K}(\Gamma X, \Gamma Y) \cong \pi G\mathscr{K}(\Gamma X, Y)$. □

Theorem 1.4.23 (Approximation of Based Spaces) *Let δ be a familial dimension function and let X be a based G-space. Then there exists a based δ-G-CW(α) complex ΓX and a based map $g: \Gamma X \to X$ that is a δ-weak$_\alpha$ equivalence. Further, Γ is functorial up to based homotopy.*

Proof The existence of $g: \Gamma X \to X$ is the special case of Theorem 1.4.21 in which we take $A = P = *$.

Given $f: X \to Y$, the existence and uniqueness of $\Gamma f: \Gamma X \to \Gamma Y$ follows from the relative version of Whitehead's theorem, which tells us that $\pi G\mathscr{K}_*(\Gamma X, \Gamma Y) \cong \pi G\mathscr{K}_*(\Gamma X, Y)$. □

Theorem 1.4.24 (Approximation of Pairs) *Let δ be a familial dimension function and let (X, A) be a pair of G-spaces. Then there exists a pair of δ-G-CW(α) complexes $(\Gamma X, \Gamma A)$ and a pair of δ-weak$_\alpha$ equivalences $g: (\Gamma X, \Gamma A) \to (X, A)$. Further, Γ is functorial on maps of pairs up to homotopy.*

Proof Take any approximation $g: \Gamma A \to A$, then apply Theorem 1.4.21 to $\Gamma A \to X$ (taking $P = \emptyset$ in that theorem) to get ΓX with ΓA as a subcomplex.

Given $f: (X, A) \to (Y, B)$, we first construct $\Gamma f: \Gamma A \to \Gamma B$ using the Whitehead theorem and then extend to $\Gamma X \to \Gamma Y$ using the relative Whitehead theorem (considering the category of spaces under ΓA). □

Finally, we want to show that we can approximate excisive triads by δ-G-CW(α) triads. We need the following pasting result first.

Proposition 1.4.25 *Let δ be a familial dimension function. If $f: (X; A, B) \to (X'; A', B')$ is a map of excisive triads such that $f: A \cap B \to A' \cap B'$, $f: A \to A'$, and $f: B \to B'$ are all δ-weak$_\alpha$ equivalences, then $f: X \to X'$ is a δ-weak$_\alpha$ equivalence.*

Proof This follows from Theorem 1.4.16 and the corresponding nonequivariant result (see [48, §10.7]) applied to f^H for each admissible H. □

Theorem 1.4.26 (Approximation of Triads) *Let δ be a familial dimension function and let $(X; A, B)$ be an excisive triad. Then there exists a δ-G-CW(α) triad $(\Gamma X; \Gamma A, \Gamma B)$ and a map of triads*

$$g: (\Gamma X; \Gamma A, \Gamma B) \to (X; A, B)$$

such that each of the maps $\Gamma A \cap \Gamma B \to A \cap B$, $\Gamma A \to A$, $\Gamma B \to B$, and $\Gamma X \to X$ is a δ-weak$_\alpha$ equivalence. Γ is functorial on maps of excisive triads up to homotopy.

Proof Let $C = A \cap B$. Take a δ-G-CW(α) approximation $g: \Gamma C \to C$. Using Theorem 1.4.21, extend g to approximations $g: (\Gamma A, \Gamma C) \to (A, C)$ and $g: (\Gamma B, \Gamma C) \to (B, C)$. Let $\Gamma X = \Gamma A \cup_{\Gamma C} \Gamma B$. All the statements of the theorem are clear except that the map $g: \Gamma X \to X$ is a δ-weak$_\alpha$ equivalence. This follows from Proposition 1.4.25 using the argument in [48, §10.7]: We can not apply Proposition 1.4.25 directly to g because $(\Gamma X; \Gamma A, \Gamma B)$ is not excisive. However, if we replace ΓX with the double mapping cylinder of the inclusions of ΓC in ΓA and ΓB, we get an equivalent excisive triad to which we can apply that result. □

In general, δ-weak$_\alpha$ equivalence is weaker than weak G-equivalence even if δ is complete. This can happen when the set of admissible subgroups is smaller than the set of all subgroups, by Theorem 1.4.16. Explicit examples where α is not an actual representation are easy to construct. The following gives an example with α being an actual representation.

Example 1.4.27 Let $G = SO(3)$ and let $H = SO(2) < G$. We can identify G/H with the two-sphere S^2, with the action of H being rotation around the z-axis, so $\mathscr{L}(G/H)$ is a two-dimensional representation on which $SO(2)$ acts by rotation in the standard way. Let X be an H-space such that

$$X^K \simeq \begin{cases} S^0 & \text{if } K = H \\ * & \text{if } K < H. \end{cases}$$

Such a space exists by the construction of [25]. Now consider the projection $f: G \times_H X \to G/H$. Clearly, f is not a weak G-equivalence, because f^H is not a weak equivalence. We claim that f is an $\mathscr{L}$-weak$_0$ equivalence (i.e., we take $\alpha = 0$). The only fixed sets we need to check are by the subgroups of H. If $K < H$, then f^K is a weak equivalence. On the other hand, H is not $\mathscr{L}$-0-admissible because $\mathscr{L}(G/H) \not\subset n$ for any integer n, so f^H is not relevant, by Theorem 1.4.16. Thus, f is an $\mathscr{L}$-weak$_0$ equivalence.

The preceding example shows that $G \times_H X$ and G/H cannot both have the G-homotopy type of an $\mathscr{L}$-G-CW(0) complex. In fact, neither does: The cells in an $\mathscr{L}$-G-CW(0) complex have the form $G \times_K D^n$, where K acts trivially on $\mathscr{L}(G/K)$. This implies that $H = SO(2)$ cannot occur as an isotropy group of such a complex. However, both $G \times_H X$ and G/H have $SO(2)$ as a maximal isotropy group, which implies that neither is G-homotopy equivalent to an $\mathscr{L}$-G-CW(0) complex.

It's useful to note that we have the following stability result, which follows from Corollary 1.4.19.

Corollary 1.4.28 *Let δ be a familial dimension function, let α be a virtual representation of G such that every subgroup in $\mathscr{F}(\delta)$ is δ-α-admissible, let $f: X \to Y$ be a δ-weak$_\alpha$ equivalence of well-based G-spaces, and let W be a representation of G. Then*

$$\Sigma^W f: \Sigma^W X \to \Sigma^W Y$$

is a δ-weak$_{\alpha+W}$ equivalence.

Proof Under the assumptions on δ and α, Corollary 1.4.19 shows that δ-weak$_\alpha$ equivalence is the same as $\mathscr{F}(\delta)$-equivalence, as is δ-weak$_{\alpha+W}$ equivalence. That the suspension of an $\mathscr{F}(\delta)$-equivalence of well-based G-spaces is again an $\mathscr{F}(\delta)$-equivalence follows on considering the fixed-point maps f^H and applying the corresponding nonequivariant result. □

The following example shows that the conclusion of the preceding corollary does not hold in general, if α is too small, even if δ is complete.

Example 1.4.29 Recall Example 1.4.27. As there, let $G = SO(3)$, $H = SO(2)$, and $V = 0$. Let $f: G \times_H X \to G/H$ be the map constructed there that is an $\mathscr{L}$-weak$_0$ equivalence but not a weak G-equivalence. Let W be $\mathbb{R}^3$ with the usual action of

$SO(3)$. We claim that

$$\Sigma^W f_+ : \Sigma^W (G \times_H X)_+ \to \Sigma^W G/H_+$$

is not an $\mathscr{L}$-weak$_W$ equivalence. Now $\mathscr{L}(G/H) \subset W$, in fact, $W \cong \mathscr{L}(G/H) \oplus \mathbb{R}$ as a representation of H, so H is $\mathscr{L}$-W-admissible. Consider the fixed set map

$$(\Sigma^W f_+)^H : \Sigma(G \times_H X)^H_+ \to \Sigma(G/H)^H_+.$$

$(G/H)^H$ consists of two points, while $(G \times_H X)^H$ is homotopy equivalent to four points, mapping two-to-one to $(G/H)^H$. It follows that $(\Sigma^W f_+)^H$ is not a weak equivalence, hence, from Theorem 1.4.16, that $\Sigma^W f_+$ is not an $\mathscr{L}$-weak$_W$ equivalence.

1.5 A Brief Introduction to Equivariant Stable Homotopy

Beginning in the next section, we will need to use equivariant stable maps, at least between G-spaces, as well as various suspensions and desuspensions of these. So it is appropriate to discuss such maps in the context of a category of equivariant spectra. We can use any of several available models of equivariant spectra, for example the orthogonal spectra of [44] or the older model expounded in [42]; all models give equivalent stable categories. Readers already familiar with equivariant spectra are free to use their favorite model. For the uninitiated, we briefly outline an approach based on [42], but with the flavor of the more modern models. The "spectra" we use we will call *LMS spectra* in honor of Lewis, May, and Steinberger—these are what, in [42], they called "prespectra."

Definition 1.5.1 A *complete G-universe* $\mathscr{U}$ is a G-inner product space that is the sum of countably many copies of each irreducible representation of G.

Definition 1.5.2 ([42, I.2.1]) Let $\mathscr{U}$ be a complete G-universe. An *LMS G-spectrum D*, indexed on $\mathscr{U}$, is a collection of based G-spaces $D(V)$, one for each finite-dimensional subrepresentation $V \subset \mathscr{U}$, and, for each inclusion $V \subset W$, a structure map

$$\sigma : \Sigma^{W-V} D(V) \to D(W),$$

where $W - V$ is the orthogonal complement of V in W. These structure maps obey two conditions:

(1) $\sigma : \Sigma^0 D(V) \to D(V)$ is the identity for each V; and
(2) for $V \subset W \subset Z$, the following diagram commutes:

$$\begin{array}{ccc}
\Sigma^{Z-W}\Sigma^{W-V}D(V) & \xrightarrow{\Sigma^{Z-W}\sigma} & \Sigma^{Z-W}D(W) \\
\cong\downarrow & & \downarrow\sigma \\
\Sigma^{Z-V}D(V) & \xrightarrow[\sigma]{} & D(Z)
\end{array}$$

A *map* of LMS spectra $f: D \to E$ is a collection of G-maps $f(V): D(V) \to E(V)$ such that, for each $V \subset W$, the following map commutes:

$$\begin{array}{ccc}
\Sigma^{W-V}D(V) & \xrightarrow{\Sigma^{W-V}f(V)} & \Sigma^{W-V}E(V) \\
\sigma\downarrow & & \downarrow\sigma \\
D(W) & \xrightarrow[f(W)]{} & E(W)
\end{array}$$

We let $G\mathscr{P}$ denote the category of LMS spectra. An LMS spectrum E is an Ω-G-*spectrum* if each adjoint map

$$\tilde{\sigma}: E(V) \to \Omega^{W-V}E(W)$$

is a weak equivalence; it is a *strict LMS spectrum* if each $\tilde{\sigma}$ is a homeomorphism.

There is a functor taking LMS spectra to strict LMS spectra (what were called spectra in [42]). It is easily described in the case where the adjoint maps $\tilde{\sigma}$ are inclusions, but not so easy in the general case. See [42] for the details. However, strict LMS spectra play a much reduced role in modern approaches to the stable category, which begin with a definition of homotopy groups and the resulting weak equivalences.

In the remainder of this section, we will shorten "LMS spectrum" to "spectrum."

Definition 1.5.3 Let E be a G-spectrum. If $H \leq G$ and $n \in \mathbb{Z}$, we define

$$\pi_n^H E = \begin{cases} \operatorname{colim}_{V \subset \mathscr{U}} \pi H\mathscr{K}_*(S^{V+n}, E(V)) & \text{if } n \geq 0 \\ \operatorname{colim}_{\mathbb{R}^{|n|} \subset V \subset \mathscr{U}} \pi H\mathscr{K}_*(S^{V-|n|}, E(V)) & \text{if } n < 0. \end{cases}$$

We say that a map of spectra $f: E \to F$ is a *stable equivalence* if it induces an isomorphism

$$f_*: \pi_n^H E \cong \pi_n^H F$$

for all $H \leq G$ and $n \in \mathbb{Z}$.

We can define a model structure on the category of G-spectra whose weak equivalences are the stable equivalences and whose fibrant objects are exactly the Ω-spectra. We can then invert the stable equivalences to get the *stable category* $\mathrm{Ho}\, G\mathscr{P}$. We write $[E, F]_G = \mathrm{Ho}\, G\mathscr{P}(E, F)$.

We have a pair of adjoint functors relating based G-spaces and spectra.

Definition 1.5.4 If X is a G-space, let $\Sigma_G^\infty X = \Sigma^\infty X$ be the *suspension spectrum* of X, the spectrum defined by

$$(\Sigma^\infty X)(V) = \Sigma^V X.$$

If E is a spectrum, let $\Omega_G^\infty E = \Omega^\infty E$ be the G-space $E(0)$.

We then have the adjunction

$$G\mathscr{K}_*(X, \Omega^\infty E) \cong G\mathscr{P}(\Sigma^\infty X, E).$$

This induces an adjunction on homotopy categories,

$$\mathrm{Ho}\, G\mathscr{K}_*(X, \Omega^\infty E) \cong [\Sigma^\infty X, E]_G,$$

but we now have to be careful to interpret $\Omega^\infty E$ to be the zeroth space of a fibrant approximation to E and also to replace X with a G-CW approximation. In this context, we note that a fibrant approximation to $\Sigma^\infty Y$ is given by the Ω-spectrum $R\Sigma^\infty Y$ with

$$R\Sigma^\infty Y(V) = \operatorname*{colim}_{W \supset V} \Omega^{W-V} \Sigma^W Y.$$

It follows that the group of stable maps from a based G-CW space X to a space Y is given by

$$\begin{aligned}[\Sigma^\infty X, \Sigma^\infty Y]_G &\cong \pi G\mathscr{K}_*(X, \Omega^\infty \Sigma^\infty Y) \\ &\cong \pi G\mathscr{K}_*(X, \operatorname*{colim}_{V \subset \mathscr{U}} \Omega^V \Sigma^V Y).\end{aligned}$$

If X is compact, this is isomorphic to

$$\operatorname*{colim}_{V} \pi G\mathscr{K}_*(\Sigma^V X, \Sigma^V Y).$$

We also have sphere objects.

Definition 1.5.5 If $V-W$ is a virtual representation of G, we have a sphere spectrum S^{V-W} defined by

$$S^{V-W}(U) = \begin{cases} S^{V+U-W} & \text{if } W \subset V+U \\ S^0 & \text{otherwise.} \end{cases}$$

(Here, we have to make a particular choice of $W \subset V \oplus \mathscr{U}$.) A fibrant approximation RS^{V-W} is given by the Ω-spectrum

$$RS^{V-W}(U) = \operatorname*{colim}_{Z \supset U} \Omega^{Z-U}\Omega^W\Sigma^Z S^V.$$

In the case when $W = 0$, S^{V-0} is just $\Sigma^\infty S^V$. In general,

$$[S^{V-W}, \Sigma^\infty Y]_G \cong [S^V, \Sigma^\infty\Sigma^W Y]_G \cong \operatorname*{colim}_{Z} \pi G\mathscr{K}_*(\Sigma^Z S^V, \Sigma^Z\Sigma^W Y).$$

If $H \le G$, there is a forgetful homomorphism from G-spectra to H-spectra, defined in the obvious way. There is a left adjoint to the forgetful functor, written $E \mapsto G_+ \wedge_H E$. On suspension spectra, we have

$$G_+ \wedge_H \Sigma_H^\infty X \cong \Sigma_G^\infty(G_+ \wedge_H X).$$

(See, for example, [42, II.4] or [44, V.2]. We use the notation from [44].)

The homotopy groups of spectra can then be expressed as follows: If E is a G-spectrum, $H \le G$, and $n \in \mathbb{Z}$, then

$$\pi_n^H E \cong [S^n, E]_H \cong [G_+ \wedge_H S^n, E]_G.$$

1.6 The Algebra of Mackey Functors

Before we can construct the cellular homology and cohomology theories based on general cell complexes, we need to discuss some algebra.

The *orbit category* $\mathscr{O}_G$ is the topological category whose objects are the orbit spaces G/H and whose morphisms are the G-maps between them. We give $\mathscr{O}_G(G/H, G/K)$ the compact-open topology; it is homeomorphic to $(G/K)^H$. Write $\widehat{\mathscr{O}}_G$ for the *stable orbit category*, the category of G-orbits and stable G-maps between them, so that, using the notation of the preceding section,

$$\widehat{\mathscr{O}}_G(G/H, G/K) = [\Sigma^\infty G/H_+, \Sigma^\infty G/K_+]_G.$$

As in [41] we take a *Mackey functor* to be an additive functor $\widehat{\mathscr{O}}_G \to Ab$ where Ab is the category of abelian groups. (see also [22, 36], and [27]). If we want to specify the group G, we shall refer to a *G-Mackey functor*. Mackey functors can be either covariant or contravariant. When G is finite the variance is essentially irrelevant because $\widehat{\mathscr{O}}_G$ is self-dual ([36, 42]). However, when G is infinite, $\widehat{\mathscr{O}}_G$ is not self-dual and the variance becomes important. We adopt the convention of writing a bar above or below to indicate variance: $\overline{T}$ will denote a contravariant functor, and $\underline{S}$ will denote a covariant one.

More generally, we make the following definition, in which we use the negative sphere spectra defined in the preceding section.

Definition 1.6.1 If δ is a dimension function for G, let $\widehat{\mathscr{O}}_{G,\delta}$ denote the full subcategory of the stable category on the objects $G_+ \wedge_H S^{-\delta(G/H)}$ with H in the domain of δ. To simplify notation, we write $(G/H, \delta)$ for the object $G_+ \wedge_H S^{-\delta(G/H)}$ in $\widehat{\mathscr{O}}_{G,\delta}$. We define a *G-$\delta$-Mackey functor*, or simply a *δ-Mackey functor*, to be an additive functor $\widehat{\mathscr{O}}_{G,\delta} \to Ab$. A δ-Mackey functor can be either covariant or contravariant; we use the convention that $\overline{T}$ denotes a contravariant δ-Mackey functor and $\underline{S}$ denotes a covariant one.

We have the following explicit calculation, which is a special case of Theorem 2.6.4 so we defer the proof until then; the case $\delta = 0$ is a special case of [42, V.9.4].

Proposition 1.6.2 *Let δ be a familial dimension function for G. Then $\widehat{\mathscr{O}}_{G,\delta}((G/H, \delta), (G/K, \delta))$ is the free abelian group generated by the equivalence classes of diagrams of orbits of the form*

$$G/H \xleftarrow{p} G/J \xrightarrow{q} G/K,$$

where $J \le H$, p is the projection, q is a (space-level) G-map with, say, $q(eJ) = gK$, and $\delta(N_H J/J) \cong \mathscr{L}(N_H J/J)$ and $\delta(N_{K^g} J/J) = 0$. Two such diagrams are equivalent if there is a diagram of the following form in which $G/J \to G/J'$ is a G-homeomorphism, the left triangle commutes, and the right triangle commutes up to G-homotopy:

$$\begin{array}{ccccc} & & G/J & & \\ & \swarrow^{p} & \downarrow & \searrow^{q} & \\ G/H & & & & G/K \\ & \nwarrow_{p'} & \downarrow & \nearrow_{q'} & \\ & & G/J' & & \end{array}$$

Composition is given by taking pullbacks and then replacing the resulting manifold with a disjoint union of orbits having the same Euler characteristics on all fixed sets. □

Note the following.

Proposition 1.6.3 *If δ is a dimension function on G and $\mathscr{L} - \delta$ is its dual, then* $\widehat{\mathscr{O}}_{G,\delta} \cong \widehat{\mathscr{O}}^{\mathrm{op}}_{G,\mathscr{L}-\delta}$.

Proof This could be seen as a corollary of the preceding proposition, or proved directly as follows. From [42, II.6.3 & §III.2] we have that the stable dual of an orbit G/H is

$$D(G/H_+) \cong G_+ \wedge_H S^{-\mathscr{L}(G/H)}.$$

From this it follows that

$$D(G_+ \wedge_H S^{-\delta(G/H)}) \cong G_+ \wedge_H S^{-(\mathscr{L}(G/H)-\delta(G/H))},$$

so that

$$\begin{aligned}&[G_+ \wedge_H S^{-\delta(G/H)}, G_+ \wedge_K S^{-\delta(G/K)}]_G \\ &\quad\cong [G_+ \wedge_K S^{-(\mathscr{L}(G/K)-\delta(G/K))}, G_+ \wedge_H S^{-(\mathscr{L}(G/H)-\delta(G/H))}]_G,\end{aligned}$$

which gives the proposition. □

Corollary 1.6.4 *Suppose δ is a dimension function on G. Then the category of contravariant δ-Mackey functors and natural transformations between them is isomorphic to the category of covariant $(\mathscr{L} - \delta)$-Mackey functors.* □

In particular, contravariant ordinary Mackey functors are the same thing as covariant $\mathscr{L}$-Mackey functors, and similarly with the variances reversed.

The definition of a Mackey functor is a special case of the following more general definition. (See [53] or [54] for much more about these objects.)

Definition 1.6.5 Let $\mathscr{A}$ be a small preadditive category (that is, its hom sets are abelian groups and composition is bilinear). Define an *$\mathscr{A}$-module* to be an additive functor from $\mathscr{A}$ to *Ab*, the category of abelian groups. We adopt the convention of writing $\overline{T}$ for a contravariant $\mathscr{A}$-module and $\underline{S}$ for a covariant $\mathscr{A}$-module. We define

$$\mathrm{Hom}_{\mathscr{A}}(\overline{T}, \overline{U}) = \int_{a\in\mathscr{A}} \mathrm{Hom}(\overline{T}(a), \overline{U}(a))$$

to be the group of natural transformations from $\overline{T}$ to $\overline{U}$, and similarly for covariant modules. We define

$$\overline{T} \otimes_{\mathscr{A}} \underline{S} = \int^{a\in\mathscr{A}} \overline{T}(a) \otimes \underline{S}(a).$$

More explicitly,

$$\overline{T} \otimes_{\mathscr{A}} \underline{S} = \left[\bigoplus_a \overline{T}(a) \otimes \underline{S}(a) \right] / \sim,$$

where the sum extends over all objects a in $\mathscr{A}$ and the equivalence relation is generated by

$$(\alpha^* x) \otimes y \sim x \otimes (\alpha_* y)$$

when $\alpha: a \to a'$ is a map in $\mathscr{A}$, $x \in \overline{T}(a')$ and $y \in \underline{S}(a)$.

Let $\mathscr{B}$ be another preadditive category (not necessarily small) and let $F: \mathscr{A} \to \mathscr{B}$ be an additive functor. If $\overline{U}$ is a contravariant $\mathscr{B}$-module and $\underline{S}$ is a covariant $\mathscr{B}$-module, define

$$F^* \overline{U} = \overline{U} \circ F$$

and

$$F^* \underline{S} = \underline{S} \circ F.$$

If $\overline{T}$ is a contravariant $\mathscr{A}$-module, define the contravariant $\mathscr{B}$-module $F_! \overline{T}$ by

$$(F_! \overline{T})(b) = \overline{T} \otimes_{\mathscr{A}} F^* \mathscr{B}(b, -) = \int^{a \in \mathscr{A}} \overline{T}(a) \otimes \mathscr{B}(b, F(a)).$$

Similarly, if $\underline{R}$ is a covariant $\mathscr{A}$-module, define the covariant $\mathscr{B}$-module $F_! \underline{R}$ by

$$(F_! \underline{R})(b) = F^* \mathscr{B}(-, b) \otimes_{\mathscr{A}} \underline{R} = \int^{a \in \mathscr{A}} \mathscr{B}(F(a), b) \otimes \underline{R}(a).$$

Finally, define

$$(F_* \overline{T})(b) = \mathrm{Hom}_{\mathscr{A}}(F^* \mathscr{B}(-, b), \overline{T})$$

and

$$(F_* \underline{R})(b) = \mathrm{Hom}_{\mathscr{A}}(F^* \mathscr{B}(b, -), \underline{R}).$$

Very useful examples are the canonical projective modules: If a is an object of $\mathscr{A}$, let $\overline{A}_a$ be the $\mathscr{A}$-module defined by $\overline{A}_a = \mathscr{A}(-, a)$, and let $\underline{A}^a$ be defined by $\underline{A}^a = \mathscr{A}(a, -)$.

The following facts are standard and straightforward to prove. If $\mathscr{B}$ is not small, claims of isomorphisms below like $\mathrm{Hom}_{\mathscr{B}}(F_! \overline{T}, \overline{U}) \cong \mathrm{Hom}_{\mathscr{A}}(\overline{T}, F^* \overline{U})$ are claims that the large limit or colimit involved on the left exists and is given by the group on the right.

Proposition 1.6.6 *Let $\mathscr{A}$ and $\mathscr{B}$ be preadditive categories with $\mathscr{A}$ small, let $F\colon \mathscr{A} \to \mathscr{B}$ be an additive functor, and let $\overline{T}$, etc., denote $\mathscr{A}$- or $\mathscr{B}$-modules of the appropriate variance below.*

(1) *For any object a of $\mathscr{A}$ we have the following isomorphisms, natural in a as well as in $\overline{T}$ or $\underline{S}$.*

$$\begin{aligned} \mathrm{Hom}_{\mathscr{A}}(\overline{A}_a, \overline{T}) &\cong \overline{T}(a) \\ \overline{T} \otimes_{\mathscr{A}} \underline{A}^a &\cong \overline{T}(a) \\ \mathrm{Hom}_{\mathscr{A}}(\underline{A}^a, \underline{S}) &\cong \underline{S}(a) \\ \overline{A}_a \otimes_{\mathscr{A}} \underline{S} &\cong \underline{S}(a) \end{aligned}$$

(2) *The functor $F_!$ is left adjoint to F^*. That is,*

$$\mathrm{Hom}_{\mathscr{B}}(F_!\overline{T}, \overline{U}) \cong \mathrm{Hom}_{\mathscr{A}}(\overline{T}, F^*\overline{U})$$

and similarly for covariant modules.

(3) *The functor F_* is right adjoint to F^*. That is,*

$$\mathrm{Hom}_{\mathscr{A}}(F^*\overline{U}, \overline{T}) \cong \mathrm{Hom}_{\mathscr{B}}(\overline{U}, F_*\overline{T})$$

and similarly for covariant modules.

(4) *We have isomorphisms*

$$F_!\overline{T} \otimes_{\mathscr{B}} \underline{S} \cong \overline{T} \otimes_{\mathscr{A}} F^*\underline{S}$$

and

$$\overline{U} \otimes_{\mathscr{B}} F_!\underline{R} \cong F^*\overline{U} \otimes_{\mathscr{A}} \underline{R}.$$

(5) *We have the following isomorphisms, natural in a.*

$$F_!\overline{A}_a \cong \overline{A}_{F(a)}$$

and

$$F_!\underline{A}^a \cong \underline{A}^{F(a)}$$

(6) *F^* is exact, $F_!$ is right exact, and F_* is left exact.* □

Remark 1.6.7 Recall that a *zero object* in a preadditive category $\mathscr{A}$ is an object that is both initial and terminal. Such an object z is characterized by the fact that $\mathscr{A}(z,z)$ is the zero group, or by the fact that the identity map on z is the zero element of $\mathscr{A}(z,z)$. Any additive functor on $\mathscr{A}$ must take any zero object to a zero object in the target. In particular, any module on $\mathscr{A}$ must take any zero object to the zero group.

At times it will be convenient to augment the category $\widehat{\mathscr{O}}_{G,\delta}$ with a zero object and there is a natural way to do this: include as an additional object the trivial based space $*$, which acts as a zero object for stable maps. Because a module must take $*$ to 0, modules on $\widehat{\mathscr{O}}_{G,\delta}$ so augmented are equivalent to modules on the original $\widehat{\mathscr{O}}_{G,\delta}$. For most purposes it will not matter which version of $\widehat{\mathscr{O}}_{G,\delta}$ we use; we will point out the places where it is useful to include the zero object. One important example will come while discussing restriction to fixed sets later in this chapter.

Returning to the context of δ-Mackey functors, we shall quite often use the Mackey functors

$$\overline{A}_{G/K,\delta} = \widehat{\mathscr{O}}_{G,\delta}(-, (G/K,\delta))$$

and

$$\underline{A}^{G/K,\delta} = \widehat{\mathscr{O}}_{G,\delta}((G/K,\delta), -).$$

We think of these as *free* Mackey functors because of the isomorphisms

$$\mathrm{Hom}_{\widehat{\mathscr{O}}_{G,\delta}}(\overline{A}_{G/K,\delta}, \overline{T}) \cong \overline{T}(G/K,\delta)$$

and

$$\mathrm{Hom}_{\widehat{\mathscr{O}}_{G,\delta}}(\underline{A}^{G/K,\delta}, \underline{S}) \cong \underline{S}(G/K,\delta)$$

given in the preceding proposition. The isomorphism is given in each case by sending a homomorphism f to $f(1_{G/K,\delta})$. Thus, we can think of $\{1_{G/K,\delta}\}$ as a basis for either $\overline{A}_{G/K,\delta}$ or $\underline{A}^{G/K,\delta}$. In general, we shall use the term *free Mackey functor* for a direct sum of Mackey functors of the form $\overline{A}_{G/K,\delta}$ or $\underline{A}^{G/K,\delta}$. If, for example, $\overline{T} = \bigoplus_i \overline{A}_{G/K_i,\delta}$, we think of the elements $1_{G/K_i,\delta} \in \overline{T}(G/K_i,\delta)$ as forming a basis of $\overline{T}$. As we shall see, the Mackey functors $\overline{A}_{G/K,\delta}$ and $\underline{A}^{G/K,\delta}$ play much the same central role in the theory of equivariant homology and cohomology as does the group $\mathbb{Z}$ nonequivariantly.

We end this section with a general result we'll need later while discussing fixed sets and restriction to subgroups.

Proposition 1.6.8 *Let $\mathscr{A}$, $\mathscr{B}$, $\mathscr{C}$, and $\mathscr{D}$ be small additive categories and assume that we have the following commutative diagram of additive functors:*

$$\begin{array}{ccc} \mathscr{A} & \xrightarrow{\varphi} & \mathscr{B} \\ {\scriptstyle\alpha}\downarrow & & \downarrow{\scriptstyle\beta} \\ \mathscr{C} & \xrightarrow[\psi]{} & \mathscr{D} \end{array}$$

Then, as functors on contravariant $\mathscr{C}$-modules, there is a natural transformation $\xi\colon \varphi_!\alpha^* \to \beta^*\psi_!$, which is an isomorphism if and only if the map

$$\int^{a\in\mathscr{A}} \mathscr{C}(\alpha(a), c) \otimes \mathscr{B}(b, \varphi(a)) \to \mathscr{D}(\beta(b), \psi(c))$$

is an isomorphism for all $b \in \mathscr{B}$ and $c \in \mathscr{C}$. For covariant $\mathscr{C}$-modules we have the dual condition: ξ is an isomorphism if and only if

$$\int^{a\in\mathscr{A}} \mathscr{B}(\varphi(a), b) \otimes \mathscr{C}(c, \alpha(a)) \to \mathscr{D}(\psi(c), \beta(b))$$

is an isomorphism for all $b \in \mathscr{B}$ and $c \in \mathscr{C}$.

Proof The map ξ is the adjoint of the map

$$\alpha^* \xrightarrow{\eta} \alpha^*\psi^*\psi_! \cong \varphi^*\beta^*\psi_!;$$

it is also the adjoint of

$$\beta_!\varphi_!\alpha^* \cong \psi_!\alpha_!\alpha^* \xrightarrow{\epsilon} \psi_!,$$

as can be seen from an appropriate diagram. Explicitly, if $\overline{T}$ is a $\mathscr{C}$-module,

$$\varphi_!\alpha^*\overline{T} = \int^{a\in\mathscr{A}} \overline{T}(\alpha(a)) \otimes \mathscr{B}(-, \varphi(a))$$

while

$$\beta^*\psi_!\overline{T} = \int^{c\in\mathscr{C}} \overline{T}(c) \otimes \mathscr{D}(\beta(-), \psi(c))$$

and the map $\varphi_!\alpha^*\overline{T} \to \beta^*\psi_!\overline{T}$ is given by $t \otimes f \mapsto t \otimes \beta(f)$ for $t \in \overline{T}(\alpha(a))$, setting $c = \alpha(a)$ so $\psi(c) = \beta\varphi(a)$.

Now, if ξ is a natural isomorphism, we can look at the special case $\overline{T} = \mathscr{C}(-, c)$. From above, we have

$$(\varphi_! \alpha^* \mathscr{C}(-, c))(b) = \int^{a \in \mathscr{A}} \overline{\mathscr{C}}(\alpha(a), c) \otimes \mathscr{B}(b, \varphi(a))$$

for each object $b \in \mathscr{B}$, while

$$(\beta^* \psi_! \mathscr{C}(-, c))(b) = \int^{c' \in \mathscr{C}} \mathscr{C}(c', c) \otimes \mathscr{D}(\beta(b), \psi(c')) \cong \mathscr{D}(\beta(b), \psi(c)).$$

Therefore,

$$\int^{a \in \mathscr{A}} \mathscr{C}(\alpha(a), c) \otimes \mathscr{B}(b, \varphi(a)) \cong \mathscr{D}(\beta(b), \psi(c)).$$

For the converse, suppose that this isomorphism holds. Then, for any $\overline{T}$, the following shows that ξ is an isomorphism:

$$\begin{aligned}
(\varphi_! \alpha^* \overline{T})(b) &= \int^{a \in \mathscr{A}} \overline{T}(\alpha(a)) \otimes \mathscr{B}(b, \varphi(a)) \\
&\cong \int^{a \in \mathscr{A}} \int^{c \in \mathscr{C}} \overline{T}(c) \otimes \mathscr{C}(\alpha(a), c) \otimes \mathscr{B}(b, \varphi(a)) \\
&\cong \int^{c \in \mathscr{C}} \overline{T}(c) \otimes \int^{a \in \mathscr{A}} \mathscr{C}(\alpha(a), c) \otimes \mathscr{B}(b, \varphi(a)) \\
&\cong \int^{c \in \mathscr{C}} \overline{T}(c) \otimes \mathscr{D}(\beta(b), \psi(c)) \\
&= (\beta^* \psi_! \overline{T})(b).
\end{aligned}$$

The proof for covariant modules is similar or follows by duality. □

Remark 1.6.9 An appealing way of formulating this proof is to write

$$\varphi_! \alpha^* \overline{T} = \overline{T} \otimes_{\mathscr{A}} \mathscr{B}$$

and

$$\beta^* \psi_! \overline{T} = \overline{T} \otimes_{\mathscr{C}} \mathscr{D}.$$

Using $\overline{T} \cong \overline{T} \otimes_{\mathscr{C}} \mathscr{C}$, the map ξ can be written as

$$\overline{T} \otimes_{\mathscr{A}} \mathscr{B} \cong (\overline{T} \otimes_{\mathscr{C}} \mathscr{C}) \otimes_{\mathscr{A}} \mathscr{B} \cong \overline{T} \otimes_{\mathscr{C}} (\mathscr{C} \otimes_{\mathscr{A}} \mathscr{B}) \to \overline{T} \otimes_{\mathscr{C}} \mathscr{D},$$

which is an isomorphism for all $\overline{T}$ if and only if $\mathscr{C} \otimes_{\mathscr{A}} \mathscr{B} \to \mathscr{D}$ is an isomorphism. When you make sense of this notation you get the explicit description given in the proof.

1.7 Homology and Cohomology of Cell Complexes

We now turn to the construction of cellular equivariant homology and cohomology graded on $RO(G)$.

Let δ be a dimension function for G and let $\alpha = V \ominus W$ be a virtual representation of G. Let (X,A) be a relative δ-G-CW(α) complex with skeleta $X^{\alpha+n}$. Then

$$X^{\alpha+n}/X^{\alpha+n-1} = \bigvee_i G_+ \wedge_{H_i} S^{V_i},$$

where the wedge runs over the $(\alpha+n)$-cells of X and V_i is stably equivalent to $\alpha - \delta(G/H_i) + n$.

To avoid sign ambiguities we need to be careful and rather precise about definitions. We choose a specific sequence of trivial representations

$$\mathbb{R} \subset \mathbb{R}^2 \subset \cdots \subset \mathbb{R}^n \subset \cdots$$

disjoint from V and W, with $\mathbb{R}^{n+1} = \mathbb{R}^n \oplus \mathbb{R}$ specifying how each sits inside the next. We choose another sequence

$$\tilde{\mathbb{R}} \subset \tilde{\mathbb{R}}^2 \subset \cdots \subset \tilde{\mathbb{R}}^n \subset \cdots$$

disjoint from V and W, with $\tilde{\mathbb{R}}^{n+1} = \mathbb{R} \oplus \tilde{\mathbb{R}}^n$. Below, when we write S^n we mean the one-point compactification of this particular choice of $\mathbb{R}^n$, so that we have the identification $S^{n+1} = S^n \wedge S^1$. We write $\tilde{S}^n$ for the one-point compactification of $\tilde{\mathbb{R}}^n$, so that we have $\tilde{S}^{n+1} = S^1 \wedge \tilde{S}^n$.

Definition 1.7.1 Let (X,A) be a relative δ-G-CW(α) complex, where $\alpha = V \ominus W$. The *cellular chain complex of* (X,A), $\overline{C}^{G,\delta}_{\alpha+*}(X,A) = \overline{C}_{\alpha+*}(X,A)$, is the differential graded contravariant δ-Mackey functor specified on orbits by

$$\overline{C}_{\alpha+n}(X,A)(G/H,\delta)$$
$$= \begin{cases} [G_+ \wedge_H S^{-\delta(G/H)} \wedge S^V \wedge S^n, \Sigma^\infty_G X^{\alpha+n}/X^{\alpha+n-1} \wedge S^W]_G & \text{if } n \geq 0 \\ [G_+ \wedge_H S^{-\delta(G/H)} \wedge S^V, \Sigma^\infty_G X^{\alpha+n}/X^{\alpha+n-1} \wedge S^W \wedge \tilde{S}^{-n}]_G & \text{if } n < 0, \end{cases}$$

where $[-,-]_G$ denotes the group of stable G-maps. The differential d is given by the G-map $X^{\alpha+n}/X^{\alpha+n-1} \to \Sigma X^{\alpha+n-1}_+ \to \Sigma X^{\alpha+n-1}/X^{\alpha+n-2}$ as follows: If $n > 0$ we take d to be the composite

$$\begin{aligned}
&\overline{C}_{\alpha+n}(X,A) \\
&= [G_+ \wedge_H S^{-\delta(G/H)} \wedge S^V \wedge S^n, \Sigma_G^\infty X^{\alpha+n}/X^{\alpha+n-1} \wedge S^W]_G \\
&\to [G_+ \wedge_H S^{-\delta(G/H)} \wedge S^V \wedge S^n, \Sigma_G^\infty X^{\alpha+n-1}/X^{\alpha+n-2} \wedge S^1 \wedge S^W]_G \\
&\to [G_+ \wedge_H S^{-\delta(G/H)} \wedge S^V \wedge S^n, \Sigma_G^\infty X^{\alpha+n-1}/X^{\alpha+n-2} \wedge S^W \wedge S^1]_G \\
&\xrightarrow{\Sigma^{-1}} [G_+ \wedge_H S^{-\delta(G/H)} \wedge S^V \wedge S^{n-1}, \Sigma_G^\infty X^{\alpha+n-1}/X^{\alpha+n-2} \wedge S^W]_G \\
&= \overline{C}_{\alpha+n-1}(X,A).
\end{aligned}$$

If $n \leq 0$, d is the composite

$$\begin{aligned}
&\overline{C}_{\alpha+n}(X,A) \\
&= [G_+ \wedge_H S^{-\delta(G/H)} \wedge S^V, \Sigma_G^\infty X^{\alpha+n}/X^{\alpha+n-1} \wedge S^W \wedge \tilde{S}^{-n}]_G \\
&\to [G_+ \wedge_H S^{-\delta(G/H)} \wedge S^V, \Sigma_G^\infty X^{\alpha+n-1}/X^{\alpha+n-2} \wedge S^1 \wedge S^W \wedge \tilde{S}^{-n}]_G \\
&\to [G_+ \wedge_H S^{-\delta(G/H)} \wedge S^V, \Sigma_G^\infty X^{\alpha+n-1}/X^{\alpha+n-2} \wedge S^W \wedge S^1 \wedge \tilde{S}^{-n}]_G \\
&= [G_+ \wedge_H S^{-\delta(G/H)} \wedge S^V, \Sigma_G^\infty X^{\alpha+n-1}/X^{\alpha+n-2} \wedge S^W \wedge \tilde{S}^{-(n-1)}]_G \\
&= \overline{C}_{\alpha+n-1}(X,A).
\end{aligned}$$

For simplicity of notation, we will write

$$\overline{C}_{\alpha+n}(X,A)(G/H,\delta) = [G_+ \wedge_H S^{-\delta(G/H)+\alpha+n}, \Sigma_G^\infty X^{\alpha+n}/X^{\alpha+n-1}]_G,$$

but it is important to keep in mind that the righthand side is shorthand for the precise definition above.

Remarks 1.7.2

(1) If $f\colon (X,A) \to (Y,B)$ is a cellular map of relative δ-G-CW(α) complexes, the maps $X^{\alpha+n} \to Y^{\alpha+n}$ induce a map of chain complexes

$$f_*\colon \overline{C}_{\alpha+*}(X,A) \to \overline{C}_{\alpha+*}(Y,B).$$

This makes $\overline{C}_{\alpha+*}(X,A)$ a covariant functor on the category of relative δ-G-CW(α) complexes and cellular maps.

(2) If $\zeta\colon \alpha \to \alpha'$ is a virtual map, any relative δ-G-CW(α) structure on (X,A) is also a relative δ-G-CW(α') structure, and vice versa. The induced stable map $S^\alpha \to S^{\alpha'}$ (i.e., the pair of homeomorphisms $S^{V\oplus Z_1} \to S^{V'\oplus Z_2}$ and $S^{W\oplus Z_1} \to S^{W'\oplus Z_2}$ given by ζ) then gives a chain isomorphism

$$\zeta^*\colon \overline{C}_{\alpha'+*}(X,A) \to \overline{C}_{\alpha+*}(X,A).$$

This makes $\overline{C}_{\alpha+*}(X,A)$ a contravariant functor on the category of virtual representations of G equivalent to a specified α.

(3) A special case of the preceding remark is the following: If Z is a representation of G, suspension by Z induces a chain isomorphism

$$\overline{C}_{V\ominus W+*}(X,A) \cong \overline{C}_{(V\oplus Z)\ominus(Z\oplus W)+*}(X,A).$$

A reinterpretation of the latter chain complex in the based case gives us a suspension isomorphism

$$\sigma^Z\colon \overline{C}_{V\ominus W+*}(X,*) \cong \overline{C}_{(V\oplus Z)\ominus W+*}(\Sigma^Z X,*)$$

or

$$\sigma^Z\colon \overline{C}_{\alpha+*}(X,*) \cong \overline{C}_{\alpha+Z+*}(\Sigma^Z X,*).$$

Note that we use the fact that, if X is a based δ-G-CW(α) complex, then $\Sigma^Z X$ is a based δ-G-CW($\alpha+Z$) complex with $(\Sigma^Z X)^{\alpha+Z+n} = \Sigma^Z X^{\alpha+n}$.

(4) The functor $\overline{C}_{V\ominus W+n}(X)$ should really be thought of as a functor of the triple (V,W,n). Our notation, although convenient and suggestive, tends to hide this fact. On the other hand, if we're careful, we see that we can shift trivial summands from V or W to n or back. For example, for $n \geq 0$, we can identify $\overline{C}_{(\alpha+1)+n}(X,*)$ with $\overline{C}_{\alpha+(n+1)}(X,*)$ via the isomorphism

$$\begin{aligned}&[G_+ \wedge_H S^{-\delta(G/H)} \wedge S^{V+1} \wedge S^n, \Sigma^\infty_G X^{(\alpha+1)+n}/X^{(\alpha+1)+n-1} \wedge S^W]_G \\ &\quad\cong [G_+ \wedge_H S^{-\delta(G/H)} \wedge S^V \wedge S^{1+n}, \Sigma^\infty_G X^{\alpha+n+1}/X^{\alpha+n} \wedge S^W]_G\end{aligned}$$

where $S^{1+n} = S^1 \wedge S^n$. Taking the smash product in this order makes the identification a chain map. Similarly, if $n < 0$, we identify

$$\begin{aligned}&[G_+ \wedge_H S^{-\delta(G/H)} \wedge S^{V+1}, \Sigma^\infty_G X^{(\alpha+1)+n}/X^{(\alpha+1)+n-1} \wedge S^W \wedge S^{-n}]_G \\ &\quad\cong [G_+ \wedge_H S^{-\delta(G/H)} \wedge S^V, \Sigma^\infty_G X^{\alpha+n+1}/X^{\alpha+n} \wedge S^W \wedge S^{-n-1}]_G\end{aligned}$$

where $S^{-n-1} \wedge S^1 = S^{-n}$.

Now, notice that each $\overline{C}_{\alpha+n}(X,A)$ is a free Mackey functor. Precisely, if

$$X^{\alpha+n}/X^{\alpha+n-1} \cong \bigvee_i G_+ \wedge_{H_i} S^{V_i}$$

as above, where each V_i is stably equivalent to $V - W - \delta(G/H_i) + n$, then

$$\overline{C}_{\alpha+n}(X,A) \cong \bigoplus_i \overline{A}_{G/H_i,\delta}.$$

A basis is given by the set of inclusions

$$G_+ \wedge_{H_i} S^{V_i} \hookrightarrow \bigvee_i G_+ \wedge_{H_i} S^{V_i}.$$

It follows that, if $\underline{S}$ is a covariant δ-Mackey functor and $\overline{T}$ is a contravariant one, then

$$\mathrm{Hom}_{\widehat{\mathscr{O}}_{G,\delta}}(\overline{C}_{\alpha+n}(X,A), \overline{T}) = \prod_i \overline{T}(G/H_i, \delta)$$

and

$$\overline{C}_{\alpha+n}(X,A) \otimes_{\widehat{\mathscr{O}}_{G,\delta}} \underline{S} = \bigoplus_i \underline{S}(G/H_i, \delta).$$

In both cases the induced differential can be understood as coming from the attaching maps of the cells.

Definition 1.7.3 Let $\overline{T}$ be a contravariant δ-Mackey functor and let $\underline{S}$ be a covariant δ-Mackey functor.

(1) Let (X, A) be a relative δ-G-CW(α) complex. We define the $(\alpha + n)$th *cellular homology* of (X, A), with coefficients in $\underline{S}$, to be

$$H_{\alpha+n}^{G,\delta}(X, A; \underline{S}) = H_{\alpha+n}(\overline{C}_{\alpha+*}^{G,\delta}(X, A) \otimes_{\widehat{\mathscr{O}}_{G,\delta}} \underline{S}).$$

and we define the $(\alpha + n)$th *cellular cohomology* of (X, A), with coefficients in $\overline{T}$, to be

$$H_{G,\delta}^{\alpha+n}(X, A; \overline{T}) = H^{\alpha+n}(\mathrm{Hom}_{\widehat{\mathscr{O}}_{G,\delta}}(\overline{C}_{\alpha+*}^{G,\delta}(X, A), \overline{T})).$$

where we introduce a sign in the differential: $(da)(x) = (-1)^{n+1} a(dx)$ if $a \in \mathrm{Hom}_{\widehat{\mathscr{O}}_{G,\delta}}(\overline{C}_{\alpha+n}^{G,\delta}(X, A), \overline{T})$. (The sign is necessary to make evaluation be a chain map.) Homology is covariant in cellular maps of (X, A) while cohomology is contravariant in cellular maps of (X, A).

(2) If X is a δ-G-CW(α) complex, so $(X, \emptyset)$ is a relative δ-G-CW(α) complex, we define

$$H_{\alpha+n}^{G,\delta}(X; \underline{S}) = H_{\alpha+n}^{G,\delta}(X, \emptyset; \underline{S})$$

and

$$H_{G,\delta}^{\alpha+n}(X; \overline{T}) = H_{G,\delta}^{\alpha+n}(X, \emptyset; \overline{T}).$$

(3) If X is a based δ-G-CW(α) complex, so $(X, *)$ is a relative δ-G-CW(α) complex, we define the *reduced* homology and cohomology of X to be

$$\tilde{H}^{G,\delta}_{\alpha+n}(X;\underline{S}) = H^{G,\delta}_{\alpha+n}(X,*;\underline{S})$$

and

$$\tilde{H}_{G,\delta}^{\alpha+n}(X;\overline{T}) = H_{G,\delta}^{\alpha+n}(X,*;\overline{T}).$$

For the moment we concentrate on the reduced theories. In order to have sufficient letters to use, we no longer reserve V and W for the representations defining α.

Theorem 1.7.4 (Reduced Homology and Cohomology of Complexes) *Let δ be a dimension function for G, let α be a virtual representation of G, and let $\underline{S}$ and $\overline{T}$ be respectively a covariant and a contravariant δ-Mackey functor. Then the abelian groups $\tilde{H}^{G,\delta}_{\alpha}(X;\underline{S})$ and $\tilde{H}_{G,\delta}^{\alpha}(X;\overline{T})$ are respectively covariant and contravariant functors on the homotopy category of based δ-G-CW(α) complexes and cellular maps and homotopies. They are also respectively contravariant and covariant functors of α. These functors satisfy the following properties.*

(1) *(Exactness) If A is a based subcomplex of X, then the following sequences are exact:*

$$\tilde{H}^{G,\delta}_{\alpha}(A;\underline{S}) \to \tilde{H}^{G,\delta}_{\alpha}(X;\underline{S}) \to \tilde{H}^{G,\delta}_{\alpha}(X/A;\underline{S})$$

and

$$\tilde{H}_{G,\delta}^{\alpha}(X/A;\overline{T}) \to \tilde{H}_{G,\delta}^{\alpha}(X;\overline{T}) \to \tilde{H}_{G,\delta}^{\alpha}(A;\overline{T}).$$

(2) *(Additivity) If $X = \bigvee_i X_i$ is a wedge of based δ-G-CW(α) complexes, then the inclusions of the wedge summands induce isomorphisms*

$$\bigoplus_i \tilde{H}^{G,\delta}_{\alpha}(X_i;\underline{S}) \cong \tilde{H}^{G,\delta}_{\alpha}(X;\underline{S})$$

and

$$\tilde{H}_{G,\delta}^{\alpha}(X;\overline{T}) \cong \prod_i \tilde{H}_{G,\delta}^{\alpha}(X_i;\overline{T}).$$

(3) *(Suspension) There are suspension isomorphisms*

$$\sigma^V : \tilde{H}^{G,\delta}_{\alpha}(X;\underline{S}) \xrightarrow{\cong} \tilde{H}^{G,\delta}_{\alpha+V}(\Sigma^V X;\underline{S})$$

and

$$\sigma^V: \tilde{H}^{\alpha}_{G,\delta}(X;\overline{T}) \xrightarrow{\cong} \tilde{H}^{\alpha+V}_{G,\delta}(\Sigma^V X;\overline{T}).$$

These isomorphisms satisfy $\sigma^0 = \mathrm{id}$, $\sigma^W \circ \sigma^V = \sigma^{V\oplus W}$, *and the following naturality condition: If* $\zeta: V \to V'$ *is an isomorphism, then the following diagrams commute (cf [47, XIII.1.1]):*

$$\begin{array}{ccc}
\tilde{H}^{G,\delta}_{\alpha}(X;\underline{S}) & \xrightarrow{\sigma^V} & \tilde{H}^{G,\delta}_{\alpha+V}(\Sigma^V X;\underline{S}) \\
\downarrow{\scriptstyle \sigma^{V'}} & & \downarrow{\scriptstyle \tilde{H}_{\mathrm{id}}(\mathrm{id}\wedge\zeta)} \\
\tilde{H}^{G,\delta}_{\alpha+V'}(\Sigma^{V'} X;\underline{S}) & \xrightarrow[\tilde{H}_{\mathrm{id}+\zeta}(\mathrm{id})]{} & \tilde{H}^{G,\delta}_{\alpha+V}(\Sigma^{V'} X;\underline{S})
\end{array}$$

and

$$\begin{array}{ccc}
\tilde{H}^{\alpha}_{G,\delta}(X;\overline{T}) & \xrightarrow{\sigma^V} & \tilde{H}^{\alpha+V}_{G,\delta}(\Sigma^V X;\overline{T}) \\
\downarrow{\scriptstyle \sigma^{V'}} & & \downarrow{\scriptstyle \tilde{H}^{\mathrm{id}+\zeta}(\mathrm{id})} \\
\tilde{H}^{\alpha+V'}_{G,\delta}(\Sigma^{V'} X;\overline{T}) & \xrightarrow[\tilde{H}^{\mathrm{id}}(\mathrm{id}\wedge\zeta)]{} & \tilde{H}^{\alpha+V'}_{G,\delta}(\Sigma^{V} X;\overline{T})
\end{array}$$

(4) *(Dimension Axiom) If* $H \in \mathscr{F}(\delta)$ *and* V *is a representation of* H *so large that* $V - \delta(G/H) + n$ *is an actual representation, then there are natural isomorphisms*

$$\tilde{H}^{G,\delta}_{V+k}(G_+ \wedge_H S^{V-\delta(G/H)+n};\underline{S}) \cong \begin{cases} \underline{S}(G/H,\delta) & \text{if } k = n \\ 0 & \text{if } k \neq n \end{cases}$$

and

$$\tilde{H}^{V+k}_{G,\delta}(G_+ \wedge_H S^{V-\delta(G/H)+n};\overline{T}) \cong \begin{cases} \overline{T}(G/H,\delta) & \text{if } k = n \\ 0 & \text{if } k \neq n. \end{cases}$$

Proof That $\tilde{H}^{G,\delta}_{\alpha}(X;\underline{S})$ and $\tilde{H}^{\alpha}_{G,\delta}(X;\overline{T})$ are functorial on cellular maps and invariant under cellular homotopies is standard. We also need to show that they are functorial in α as stated in the theorem. Let $\zeta: \alpha \to \alpha'$ be a virtual map. As noted in Remark 1.7.2(2), there is an induced chain isomorphism $\zeta^*: \overline{C}_{\alpha'+*}(X,*) \to \overline{C}_{\alpha+*}(X,*)$. This chain map induces the maps in homology and cohomology with the variance claimed.

Exactness: If A is a based subcomplex of X, then $\overline{C}_{\alpha+*}(A,*)$ is a sub-chain complex of $\overline{C}_{\alpha+*}(X,*)$ and $\overline{C}_{\alpha+*}(X/A,*) \cong \overline{C}_{\alpha+*}(X,*)/\overline{C}_{\alpha+*}(A,*)$. The exactness of the homology and cohomology sequences follows from familiar homological algebra.

Additivity: If $X = \bigvee_i X_i$, then $X^{\alpha+n} = \bigvee_i X_i^{\alpha+n}$, hence

$$\overline{C}_{\alpha+*}(X,*) \cong \bigoplus_i \overline{C}_{\alpha+*}(X_i,*).$$

The additivity of homology and cohomology follows from this.

Suspension: The suspension isomorphisms in homology and cohomology are induced by the chain isomorphism σ^V defined in Remark 1.7.2(3). Clearly, $\sigma^0 = \mathrm{id}$ and $\sigma^W \circ \sigma^V = \sigma^{V\oplus W}$.

If $\zeta\colon V \to V'$, then the following diagram commutes for any spectra E and F — around either side it takes a map f to $f \wedge S^\zeta$:

$$\begin{array}{ccc}
[E,F]_G & \xrightarrow{\sigma^V} & [\Sigma^V E, \Sigma^V F]_G \\
\Big\downarrow{\scriptstyle \sigma^{V'}} & & \Big\downarrow{\scriptstyle [\mathrm{id},\mathrm{id}\wedge\zeta]} \\
[\Sigma^{V'} E, \Sigma^{V'} F]_G & \xrightarrow[{[\mathrm{id}\wedge\zeta,\mathrm{id}]}]{} & [\Sigma^V E, \Sigma^{V'} F]_G
\end{array}$$

From this it follows that the following diagram commutes:

$$\begin{array}{ccc}
\overline{C}_{\alpha+*}(X,*) & \xrightarrow{\sigma^V} & \overline{C}_{\alpha+V+*}(\Sigma^V X,*) \\
\Big\downarrow{\scriptstyle \sigma^{V'}} & & \Big\downarrow{\scriptstyle \overline{C}_{\mathrm{id}}(\mathrm{id}\wedge\zeta)} \\
\overline{C}_{\alpha+V'+*}(\Sigma^{V'} X,*) & \xrightarrow[\overline{C}_{\mathrm{id}+\zeta}(\mathrm{id})]{} & \overline{C}_{\alpha+V+*}(\Sigma^{V'} X,*)
\end{array}$$

The naturality diagrams for homology and cohomology now follow.

Dimension Axiom: Let $X = G_+ \wedge_H S^{V-\delta(G/H)+n}$. Then X is a based δ-G-CW(V) complex with a single relative cell of dimension $V+n$. We have then

$$X^{V+n}/X^{V+n-1} \cong G_+ \wedge_H S^{V-\delta(G/H)+n}$$

while the other filtration quotients are trivial, hence

$$\overline{C}_{V+k}(X,*) \cong \begin{cases} \underline{A}^{G/H,\delta} & \text{if } k = n \\ 0 & \text{if } k \neq n. \end{cases}$$

The statements about homology and cohomology follow. □

1.8 Ordinary and Dual Homology and Cohomology

We will shortly extend cellular homology and cohomology to be defined on all G-spaces, but we pause briefly to give names to and discuss the most important specializations to particular choices of the dimension function δ.

Definition 1.8.1 If we set $\delta = 0$ in cellular homology and cohomology, we call the resulting $RO(G)$-graded theories *ordinary homology and cohomology*. We use the notations

$$\tilde{H}^G_*(X; \underline{S}) = \tilde{H}^{G,0}_*(X; \underline{S})$$

and

$$\tilde{H}^*_G(X; \overline{T}) = \tilde{H}^*_{G,0}(X; \overline{T})$$

and similarly for the unreduced theories, where $\underline{S}$ is a covariant ($\widehat{\mathscr{O}}_G$-)Mackey functor and $\overline{T}$ is a contravariant Mackey functor.

The ordinary theories are the ones constructed using ordinary G-CW(α) complexes, in which the orbits do not contribute to dimension.

As we will show later, the ordinary theories are characterized by the dimension axiom, which takes the following form:

$$H^G_k(G/H; \underline{S}) \cong \begin{cases} \underline{S}(G/H) & \text{if } k = 0 \\ 0 & \text{if } k \neq 0 \end{cases}$$

and

$$H^k_G(G/H; \overline{T}) \cong \begin{cases} \overline{T}(G/H) & \text{if } k = 0 \\ 0 & \text{if } k \neq 0 \end{cases}$$

for k an integer. Thus, the ordinary theories constructed here coincide with the theories of the same name in [41] and [47].

Turning to the case $\delta = \mathscr{L}$, recall from Corollary 1.6.4 that a covariant $\mathscr{L}$-Mackey functor is the same thing as a contravariant Mackey functor and vice versa.

Definition 1.8.2 If we set $\delta = \mathscr{L}$ in cellular homology and cohomology, we call the resulting $RO(G)$-graded theories *dual homology and cohomology*. We use the notations

$$\tilde{\mathscr{H}}^G_*(X; \overline{T}) = \tilde{H}^{G,\mathscr{L}}_*(X; \overline{T})$$

and

$$\tilde{\mathscr{H}}_G^*(X;\underline{S}) = \tilde{H}_{G,\mathscr{L}}^*(X;\underline{S})$$

and similarly for the unreduced theories, where $\underline{S}$ is a covariant (0-)Mackey functor (hence a contravariant $\mathscr{L}$-Mackey functor) and $\overline{T}$ is a contravariant Mackey functor.

The dual theories are the ones constructed using dual G-CW(α) complexes, in which an orbit G/H contributes its full geometric dimension, $\mathscr{L}(G/H)$. Thus, the dimension of a cell $G \times_H D(V)$ is $V + \mathscr{L}(G/H)$ in the context of the dual theories.

The reason for the name will become clear when we talk about representing spectra, but is suggested by the following characterization. Recall that the dual of the orbit G/H has the form

$$D(G/H_+) = G_+ \wedge_H S^{-\mathscr{L}(G/H)},$$

so that

$$\Sigma^V D(G/H_+) = \Sigma^\infty(G_+ \wedge_H S^{V-\mathscr{L}(G/H)}),$$

where V is a representation large enough that G/H embeds in it, so $\mathscr{L}(G/H) \subset V$. We introduce the notation

$$\tilde{\mathscr{H}}_\alpha^G(D(G/H_+);\overline{T}) = \tilde{\mathscr{H}}_{V+\alpha}^G(G_+ \wedge_H S^{V-\mathscr{L}(G/H)};\overline{T}),$$

which does not depend on the choice of V, by the suspension isomorphism. The dual theories are then characterized by the following dimension axioms:

$$\tilde{\mathscr{H}}_k^G(D(G/H_+);\overline{T}) \cong \begin{cases} \overline{T}(G/H) & \text{if } k = 0 \\ 0 & \text{if } k \neq 0 \end{cases}$$

and

$$\tilde{\mathscr{H}}_G^k(D(G/H_+);\underline{S}) \cong \begin{cases} \underline{S}(G/H) & \text{if } k = 0 \\ 0 & \text{if } k \neq 0. \end{cases}$$

for k an integer. (Notice that the variances match correctly due to the use of duality.)

1.9 Stable G-CW Approximation of Spaces

We now want to extend the definition of homology and cohomology to arbitrary G-spaces. Our first thought is, given a based G-space X, to take a δ-G-CW(α) approximation $\Gamma_\alpha X \to X$ and then take the homology of $\Gamma_\alpha X$, as defined in

Section 1.7, to be the homology of X. However, the resulting theory may not have a suspension isomorphism. The problem, as illustrated by Example 1.4.29, is that suspension need not preserve δ-weak$_\alpha$ equivalence, so that $\Sigma^W \Gamma_\alpha X \to \Sigma^W X$ need not be a δ-G-CW$(\alpha + W)$ approximation. One way out of this problem that might occur to the reader is given by Corollary 1.4.28: for a fixed δ and α, we can choose a V such that δ-weak$_{\alpha+V}$ equivalence is preserved by further suspension, and calculate the homology of X as an appropriate shift of the homology of $\Gamma_{\alpha+V}\Sigma^V X$. However, this approach does not generalize to the parametrized case we'll be interested in later.

Our approach is to develop a theory of δ-G-CW(α) spectra that can be used to approximate suspensions of G-spaces. Although this theory accomplishes what we need it to, it does not do everything one might hope from a theory of G-CW spectra. In particular, taking homotopy classes of maps between them does not give us the stable category. One reason we don't try to develop such a theory (like the G-CW spectra discussed in [42]) is that we cannot do so in the parametrized case. In fact, there does not appear to be a well-behaved version of even G-CW parametrized spectra in the above sense, a problem discussed in detail in [51, Chap. 24].

If $\Gamma_{\alpha+W}\Sigma^W X \to \Sigma^W X$ is a δ-G-CW$(\alpha + W)$ approximation, then the Whitehead theorem tells us that there is a map $\Sigma^W \Gamma_\alpha X \to \Gamma_{\alpha+W}\Sigma^W X$ over $\Sigma^W X$, unique up to homotopy. In fact, using Theorem 1.4.21, we can arrange that this map is the inclusion of a subcomplex. This suggests taking the colimit over W to define the chains of X. It also suggests using a special, if old-fashioned, kind of spectrum to formalize the process.

Recall from [42] or [51] that, if $\mathscr{U}$ is a G-universe, an *indexing sequence* $\mathscr{V} = \{V_i\}$ in $\mathscr{U}$ is an expanding sequence $V_1 \subset V_2 \subset \cdots$ of finite-dimensional subrepresentations of $\mathscr{U}$ such that $\bigcup_i V_i = \mathscr{U}$.

Definition 1.9.1 (Cf [42, I.2.1]) Let $\mathscr{V}$ be an indexing sequence in a G-universe $\mathscr{U}$. An *(LMS) G-spectrum D indexed on* $\mathscr{V}$ is a collection of G-spaces $D(V_i)$ and based G-maps

$$\sigma\colon \Sigma^{V_i - V_{i-1}} D(V_{i-1}) \to D(V_i).$$

A *map* $f\colon D \to E$ of spectra is a collection of maps $f_i\colon D(V_i) \to E(V_i)$ such that the following diagram commutes:

$$\begin{array}{ccc} \Sigma^{V_i - V_{i-1}} D(V_{i-1}) & \xrightarrow{\Sigma f} & \Sigma^{V_i - V_{i-1}} E(V_{i-1}) \\ \sigma \downarrow & & \downarrow \sigma \\ D(V_i) & \xrightarrow[f]{} & E(V_i) \end{array}$$

We let $G\mathscr{P}\mathscr{V}$ denote the category of G-spectra indexed on $\mathscr{V}$.

One obvious example is the *suspension spectrum* of a based G-space X: $\Sigma_G^\infty X$ is the G-spectrum with

$$(\Sigma_G^\infty X)(V_i) = \Sigma^{V_i} X.$$

We shall use suspension spectra in the next section. Another example is the G-CW approximation we shall define in this section.

Definition 1.9.2 Let $\mathscr{V}$ be an indexing sequence in a universe $\mathscr{U}$.

(1) A G-spectrum D in $G\mathscr{P}\mathscr{V}$ is a *δ-G-CW(α) spectrum* if, for each i, $D(V_i)$ is a based δ-G-CW($\alpha + V_i$) complex and each structure map

$$\sigma\colon \Sigma^{V_i - V_{i-1}} D(V_{i-1}) \to D(V_i)$$

is the inclusion of a subcomplex.

(2) A map $D \to E$ of δ-G-CW(α) spectra is *the inclusion of a subcomplex* if, for each i, the map $D(V_i) \to E(V_i)$ is the inclusion of a subcomplex. We also say simply that D is a subcomplex of E.

(3) A map $D \to E$ of δ-G-CW(α) spectra is *cellular* if, for each i, the map $D(V_i) \to E(V_i)$ is cellular.

(4) A map $f\colon D \to E$ in $G\mathscr{P}\mathscr{V}$ is a *δ-weak$_\alpha$ equivalence* if, for each i, $f_i\colon D(V_i) \to E(V_i)$ is a δ-weak$_{\alpha+V_i}$ equivalence of G-spaces.

(5) If D is a G-spectrum in $G\mathscr{P}\mathscr{V}$, a *δ-G-CW(α) approximation* of D is a δ-G-CW(α) spectrum $\Gamma_\alpha^\delta D$ and a δ-weak$_\alpha$ equivalence $\Gamma_\alpha^\delta D \to D$.

This notion of CW spectrum harks back to the old idea of CW spectra as in [1, 57], and [62]. The notion of δ-weak$_\alpha$ equivalence is similar to the level equivalences discussed in [44] or [51]. We will not need or discuss its relationship to stable equivalence except for the following brief comments: When δ is complete, a δ-weak$_\alpha$ equivalence $D \to E$ implies a weak G-equivalence $D(V_i) \to E(V_i)$ for sufficiently large i, so is a stable equivalence, but this simple argument is not available in the parametrized case, where the relationship appears to be more complicated. Further, we are not attempting to model the stable category and there are good reasons that a CW model of that category is out of reach in the parametrized case, as discussed in [51, Chap. 24].

Our results on G-CW spaces give quick proofs of the following results.

Lemma 1.9.3 (H.E.L.P.) *Let $r\colon E \to F$ be a δ-weak$_\alpha$ equivalence of G-spectra and let D be a δ-G-CW(α) spectrum with subcomplex C. If the following diagram commutes without the dashed arrows, then there exist maps $\tilde g$ and $\tilde h$ making the diagram commute.*

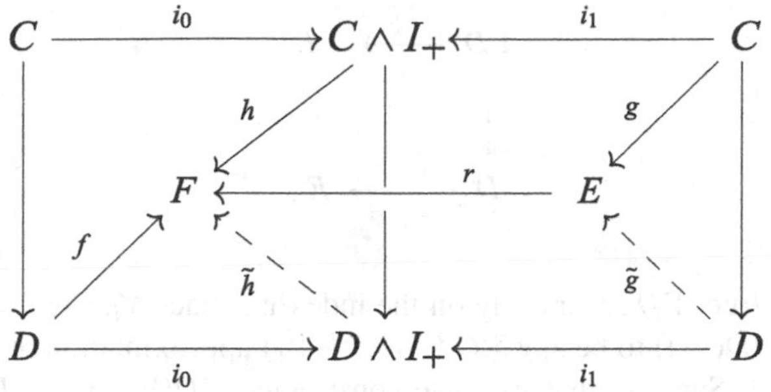

Proof We construct $\tilde{g}$ and $\tilde{h}$ inductively on the indexing space V_i. For $i = 1$ we simply quote the space-level H.E.L.P. lemma to find $\tilde{g}_1$ and $\tilde{h}_1$.

For the inductive step, we assume that we've constructed $\tilde{g}_{i-1}$ and $\tilde{h}_{i-1}$. We then apply the space-level H.E.L.P. lemma with (using the notation of Lemma 1.4.8) $Y = E_i$, $Z = F_i$, $X = D_i$, and $A = \Sigma X_{i-1} \cup C_i$. □

Proposition 1.9.4 (Whitehead) *Suppose that D is a δ-G-CW(α) spectrum and that $f: E \to F$ is a δ-weak$_\alpha$ equivalence. Then*

$$f_*: \pi G\mathscr{P}\mathscr{V}(D, E) \to \pi G\mathscr{P}\mathscr{V}(D, F)$$

is an isomorphism, where $\pi G\mathscr{P}\mathscr{V}(-,-)$ denotes the group of homotopy classes of G-maps. Therefore, any δ-weak$_\alpha$ equivalence of δ-G-CW(α) spectra is a G-homotopy equivalence.

Proof We get surjectivity by applying the H.E.L.P. lemma to D and its subcomplex $*$. We get injectivity by applying it to $D \wedge I_+$ and its subcomplex $D \wedge \partial I_+$. □

Proposition 1.9.5 (Cellular Approximation of Maps) *Let δ be a familial dimension function. Suppose that $f: D \to E$ is a map of δ-G-CW(α) spectra, C is a subcomplex of D, and h is a G-homotopy of $f|C$ to a cellular map. Then h can be extended to a G-homotopy of f to a cellular map.*

Proof This follows by induction on the indexing space V_i. The first case to consider is $f_1: D(V_1) \to E(V_1)$, and we know from the space-level result, Theorem 1.4.12 (which requires that δ be familial), that we can extend h_1 to a G-homotopy k_1 from f_1 to a cellular map g_1. For the inductive step, assume we have a homotopy k_{i-1}, extending h_{i-1}, from f_{i-1} to a cellular map g_{i-1}. Then $\Sigma^{V_i - V_{i-1}} k_{i-1} \cup h_i$ is a homotopy on the subcomplex $\Sigma^{V_i - V_{i-1}} D(V_{i-1}) \cup C_i$ of $D(V_i)$. By the space-level result again, we can extend to a homotopy k_i on $D(V_i)$ from f_i to a cellular map g_i. □

Proposition 1.9.6 (Cellular Approximation of Spectra) *Let δ be a familial dimension function. If D is a G-spectrum in $G\mathscr{P}\mathscr{V}$, then there exists a δ-G-CW(α) approximation $\Gamma D \to D$. If $f: D \to E$ is a map of G-spectra and $\Gamma E \to E$ is an approximation of E, then there exists a cellular map $\Gamma f: \Gamma D \to \Gamma E$, unique up to cellular homotopy, making the following diagram homotopy commute:*

$$\begin{array}{ccc} \Gamma D & \xrightarrow{\Gamma f} & \Gamma E \\ \downarrow & & \downarrow \\ D & \xrightarrow[f]{} & E \end{array}$$

Proof We construct ΓD recursively on the indexing space V_i. For $i = 1$, we take $(\Gamma D)(V_1) = \Gamma(D(V_1))$ to be any δ-G-CW$(\alpha + V_1)$ approximation of $D(V_1)$, using Theorem 1.4.23. Suppose that we have constructed $\Gamma D(V_{i-1}) \to D(V_{i-1})$, a δ-G-CW$(\alpha + V_{i-1})$ approximation. Then $\Sigma^{V_i - V_{i-1}} \Gamma D(V_{i-1})$ is a δ-G-CW$(\alpha + V_i)$ complex and, by Theorem 1.4.21, we can find a δ-G-CW$(\alpha + V_i)$ approximation $\Gamma D(V_i) \to D(V_i)$ making the following diagram commute, in which the map σ at the top is the inclusion of a subcomplex:

$$\begin{array}{ccc} \Sigma^{V_i - V_{i-1}} \Gamma D(V_{i-1}) & \xrightarrow{\sigma} & \Gamma D(V_i) \\ \downarrow & & \downarrow \\ \Sigma^{V_i - V_{i-1}} D(V_{i-1}) & \xrightarrow[\sigma]{} & D(V_i) \end{array}$$

Thus, ΓD is a δ-G-CW(α) spectrum and the map $\Gamma D \to D$ so constructed is a δ-weak$_\alpha$ equivalence.

The existence and uniqueness of Γf follow from the Whitehead theorem and cellular approximation of maps and homotopies. □

As usual, these results imply that we can invert the δ-weak$_\alpha$ equivalences of spectra and that the result is equivalent to the ordinary homotopy category of δ-G-CW(α) spectra. We emphasize that this is not the stable category and our intention is not to model the stable category—this notion of equivalence is much stricter than stable equivalence. These results will be used solely as a mechanism for extending the definition of ordinary homology and cohomology from G-CW complexes to arbitrary G-spaces.

When discussing products we will need to use somewhat more general maps.

Definition 1.9.7 Let D and E be G-spectra in $G\mathscr{P}\mathscr{V}$. A *semistable map*

$$f = (M, \{f_i\}) : D \to E$$

consists of a nonnegative integer M and, for each $i \geq M$, maps $f_i : D(V_i) \to E(V_i)$ compatible in the usual way. Say that two semistable maps $f = (M, \{f_i\})$ and $g = (N, \{g_i\})$ from D to E are *equivalent* if there exists a $P \geq \max(M, N)$ such that $f_i = g_i$ for all $i \geq P$. Equivalence classes of semistable maps can be composed in the obvious way. Let $G\mathscr{P}\mathscr{V}_s$ denote the category of G-spectra indexed on $\mathscr{V}$ and equivalence classes of semistable maps.

1.10   Homology and Cohomology of Spaces

We now want to extend the definition of cellular homology and cohomology to arbitrary G-spaces. To do so, we will replace a based G-space X by a δ-G-CW(α) approximation of its suspension spectrum $\Sigma_G^\infty X$, as in the preceding section.

1.10.1   Cellular Chains of G-Spaces

So, let E be a δ-G-CW(α) spectrum as in Sect. 1.9. For each i we can consider the chain complex $\overline{C}^{G,\delta}_{\alpha+V_i+*}(E(V_i))$. The structure maps induce maps (inclusions, actually)

$$\overline{C}^{G,\delta}_{\alpha+V_{i-1}+*}(E(V_{i-1})) \cong \overline{C}^{G,\delta}_{\alpha+V_i+*}(\Sigma^{V_i-V_{i-1}}E(V_{i-1})) \to \overline{C}^{G,\delta}_{\alpha+V_i+*}(E(V_i)).$$

Definition 1.10.1 Let E be a δ-G-CW(α) spectrum. We define the *cellular chain complex of E* to be the colimit

$$\overline{C}^{G,\delta}_{\alpha+*}(E) = \operatorname*{colim}_i \overline{C}_{\alpha+V_i+*}(E(V_i)).$$

If F is an arbitrary spectrum and δ is familial, we define

$$\overline{C}^{G,\delta}_{\alpha+*}(F) = \overline{C}^{G,\delta}_{\alpha+*}(\Gamma F)$$

where $\Gamma F \to F$ is a δ-G-CW(α) approximation of F. If X is a based G-space and δ is familial, we define

$$\overline{C}^{G,\delta}_{\alpha+*}(X) = \overline{C}^{G,\delta}_{\alpha+*}(\Sigma_G^\infty X) = \overline{C}^{G,\delta}_{\alpha+*}(\Gamma\Sigma_G^\infty X).$$

Our main interest is in the chains of a G-space, so we will concentrate on that case, using more general spectra only as necessary. Because we are not really modeling the stable category by looking at δ-weak$_\alpha$ equivalences, the extra generality of discussing spectra does not seem that useful in itself.

Of course, we want to make these chains functorial, so we make the following definitions.

Definition 1.10.2 Let $f: E \to F$ be a cellular map of δ-G-CW(α) spectra. For each i we have an induced chain map

$$(f_i)_*: \overline{C}_{\alpha+V_i+*}(E(V_i)) \to \overline{C}_{\alpha+V_i+*}(F(V_i))$$

and these maps are compatible under suspension. We define

$$f_* = \operatorname{colim}_i (f_i)_*: \overline{C}^{G,\delta}_{\alpha+*}(E) \to \overline{C}^{G,\delta}_{\alpha+*}(F).$$

If $f: E \to F$ is a map of arbitrary spectra, we define f_* to be the map induced by a cellular approximation $\Gamma f: \Gamma E \to \Gamma F$. If $f: X \to Y$ is a map of based G-spaces, we let

$$f_*: \overline{C}^{G,\delta}_{\alpha+*}(X) \to \overline{C}^{G,\delta}_{\alpha+*}(Y)$$

be the map associated with the suspension $\Sigma^\infty f: \Sigma^\infty_G X \to \Sigma^\infty_G Y$.

It's easy to see that this makes $\overline{C}^{G,\delta}_{\alpha+*}$ functorial on the category of δ-G-CW(α) spectra and cellular maps. Note that we can extend the definition easily to semistable cellular maps. The following is also reassuring and is essentially just the observation that, if X is a based δ-G-CW(α) complex, then $\Sigma^\infty_G X$ is a δ-G-CW(α) spectrum.

Proposition 1.10.3 *Suspension Σ^∞_G defines a functor from the category of based δ-G-CW(α) complexes and cellular maps to the category of δ-G-CW(α) spectra and cellular maps. When X is a based δ-G-CW(α) complex we have a natural isomorphism*

$$\overline{C}^{G,\delta}_{\alpha+*}(X) \cong \overline{C}^{G,\delta}_{\alpha+*}(\Sigma^\infty_G X)$$

where the chains on the left are those of X as a based δ-G-CW(α) complex. □

1.10.2 Definition and Properties of Homology and Cohomology

There is one more technical detail to deal with. Because suspension is guaranteed to respect weak G-equivalences only for well-based G-spaces, we need to recall the following common definitions.

Definition 1.10.4

(1) Let $f: A \to X$ be a map of unbased G-spaces. The *(unreduced) mapping cylinder*, Mf, is the pushout in the following diagram:

$$\begin{array}{ccc} A & \xrightarrow{f} & X \\ {\scriptstyle i_0}\downarrow & & \downarrow \\ A \times I & \longrightarrow & Mf \end{array}$$

The *(unreduced) mapping cone* is $Cf = Mf/(A \times 1)$ with basepoint the image of $A \times 1$.

(2) Let $f: A \to X$ be a map of based G-spaces. The *reduced mapping cylinder*, $\tilde{M}f$, is the pushout in the following diagram:

$$\begin{array}{ccc} A & \xrightarrow{f} & X \\ {\scriptstyle i_0}\downarrow & & \downarrow \\ A \wedge I_+ & \longrightarrow & \tilde{M}f \end{array}$$

The *reduced mapping cone*, $\tilde{C}f$, is the pushout in the following diagram, in which I has basepoint 1:

$$\begin{array}{ccc} A & \xrightarrow{f} & X \\ {\scriptstyle i_0}\downarrow & & \downarrow \\ A \wedge I & \longrightarrow & \tilde{C}f \end{array}$$

In the special case in which $p: * \to X$ is the inclusion of the basepoint, we call Mp the *whiskering construction* and write $X_w = Mp$ for X with a whisker attached.

The point of the mapping cylinder is, of course, that the inclusion $A \to Mf$ induced by $i_1: A \to A \times I$ is always a cofibration. In particular, if we give X_w the basepoint at the end of the whisker, it is well-based. Collapsing the whisker gives an unbased G-homotopy equivalence $X_w \to X$ that is a based G-homotopy equivalence if X is well-based. Hence we can use the whiskering construction as a functorial way of replacing any based G-space with a weakly equivalent well-based space.

Likewise, if $f: A \to X$ is a based map, then the inclusion $A \to \tilde{M}f$ is a based cofibration. If A and X are well-based it is in fact an unbased cofibration as well.

With these definitions and results in place, we can now define the homology and cohomology of arbitrary G-spaces.

Definition 1.10.5 Let δ be a familial dimension function for G, let α be a virtual representation of G, let $\overline{T}$ be a contravariant δ-Mackey functor, and let $\underline{S}$ be a covariant δ-Mackey functor. If X is a based G-space, let

$$\tilde{H}^{G,\delta}_{\alpha+n}(X;\underline{S}) = H_{\alpha+n}(\overline{C}^{G,\delta}_{\alpha+*}(X_w) \otimes_{\widehat{\mathscr{O}}_{G,\delta}} \underline{S})$$

and

$$\tilde{H}^{\alpha+n}_{G,\delta}(X;\overline{T}) = H^{\alpha+n}(\mathrm{Hom}_{\widehat{\mathscr{O}}_{G,\delta}}(\overline{C}^{G,\delta}_{\alpha+*}(X_w), \overline{T})).$$

Note carefully that, unlike the homology of cell complexes, we must always assume that δ is familial when discussing the homology of arbitrary G-spaces.

Proposition 1.10.6 *$\tilde{H}^{G,\delta}_{\alpha+n}(X;\underline{S})$ is a well-defined covariant homotopy functor of X, while $\tilde{H}^{\alpha+n}_{G,\delta}(X;\overline{T})$ is a well-defined contravariant homotopy functor of X. If X is a δ-G-CW(α) complex, these groups are naturally isomorphic to those given by Definition 1.7.3.*

Proof That these groups are well-defined, meaning independent, up to canonical isomorphism, of the choice of universe, indexing sequence, and CW approximation, will be shown in Propositions 1.10.11–1.10.13. That they are homotopy functors of X follows by the usual argument.

If X is a δ-G-CW(α) complex, then $\Sigma^\infty_G X$ is a δ-G-CW(α) spectrum so serves as an approximation of itself. There is then an isomorphism of chains $\overline{C}^{G,\delta}_{\alpha+*}(X) \cong \overline{C}^{G,\delta}_{\alpha+*}(\Sigma^\infty_G X)$ given by the suspension maps. □

Theorem 1.10.7 (Reduced Homology and Cohomology of Spaces) *Let δ be a familial dimension function for G, let α be a virtual representation of G, and let $\underline{S}$ and $\overline{T}$ be respectively a covariant and a contravariant δ-Mackey functor. Then the abelian groups $\tilde{H}^{G,\delta}_\alpha(X;\underline{S})$ and $\tilde{H}^\alpha_{G,\delta}(X;\overline{T})$ are respectively covariant and contravariant functors on the homotopy category of based G-spaces. They are also respectively contravariant and covariant functors of α. These functors satisfy the following properties.*

(1) *(Weak Equivalence) If $f: X \to Y$ is an $\mathscr{F}(\delta)$-equivalence of based G-spaces, then*

$$f_*: \tilde{H}^{G,\delta}_\alpha(X;\underline{S}) \to \tilde{H}^{G,\delta}_\alpha(Y;\underline{S})$$

and

$$f^*: \tilde{H}^\alpha_{G,\delta}(Y;\overline{T}) \to \tilde{H}^\alpha_{G,\delta}(X;\overline{T})$$

are isomorphisms.

(2) *(Exactness) If $A \to X$ is a cofibration, then the following sequences are exact:*

$$\tilde{H}^{G,\delta}_{\alpha}(A;\underline{S}) \to \tilde{H}^{G,\delta}_{\alpha}(X;\underline{S}) \to \tilde{H}^{G,\delta}_{\alpha}(X/A;\underline{S})$$

and

$$\tilde{H}^{\alpha}_{G,\delta}(X/A;\overline{T}) \to \tilde{H}^{\alpha}_{G,\delta}(X;\overline{T}) \to \tilde{H}^{\alpha}_{G,\delta}(A;\overline{T}).$$

(3) *(Additivity) If $X = \bigvee_i X_i$ is a wedge of well-based G-spaces, then the inclusions of the wedge summands induce isomorphisms*

$$\bigoplus_i \tilde{H}^{G,\delta}_{\alpha}(X_i;\underline{S}) \cong \tilde{H}^{G,\delta}_{\alpha}(X;\underline{S})$$

and

$$\tilde{H}^{\alpha}_{G,\delta}(X;\overline{T}) \cong \prod_i \tilde{H}^{\alpha}_{G,\delta}(X_i;\overline{T}).$$

(4) *(Suspension) If X is well-based, there are suspension isomorphisms*

$$\sigma^V : \tilde{H}^{G,\delta}_{\alpha}(X;\underline{S}) \xrightarrow{\cong} \tilde{H}^{G,\delta}_{\alpha+V}(\Sigma^V X;\underline{S})$$

and

$$\sigma^V : \tilde{H}^{\alpha}_{G,\delta}(X;\overline{T}) \xrightarrow{\cong} \tilde{H}^{\alpha+V}_{G,\delta}(\Sigma^V X;\overline{T})$$

These isomorphisms satisfy $\sigma^0 = \mathrm{id}$, $\sigma^W \circ \sigma^V = \sigma^{V\oplus W}$, and the following naturality condition: If $\zeta: V \to V'$ is an isomorphism, then the following diagrams commute (cf [47, XIII.1.1]):

$$\begin{array}{ccc}
\tilde{H}^{G,\delta}_{\alpha}(X;\underline{S}) & \xrightarrow{\sigma^V} & \tilde{H}^{G,\delta}_{\alpha+V}(\Sigma^V X;\underline{S}) \\
\downarrow{\scriptstyle \sigma^{V'}} & & \downarrow{\scriptstyle \tilde{H}_{\mathrm{id}}(\mathrm{id}\wedge\zeta)} \\
\tilde{H}^{G,\delta}_{\alpha+V'}(\Sigma^{V'} X;\underline{S}) & \xrightarrow[\tilde{H}_{\mathrm{id}+\zeta}(\mathrm{id})]{} & \tilde{H}^{G,\delta}_{\alpha+V}(\Sigma^{V'} X;\underline{S})
\end{array}$$

$$\begin{array}{ccc}
\tilde{H}^{\alpha}_{G,\delta}(X;\overline{T}) & \xrightarrow{\sigma^V} & \tilde{H}^{\alpha+V}_{G,\delta}(\Sigma^V X;\overline{T}) \\
\downarrow{\scriptstyle \sigma^{V'}} & & \downarrow{\scriptstyle \tilde{H}^{\mathrm{id}+\zeta}(\mathrm{id})} \\
\tilde{H}^{\alpha+V'}_{G,\delta}(\Sigma^{V'} X;\overline{T}) & \xrightarrow[\tilde{H}^{\mathrm{id}}(\mathrm{id}\wedge\zeta)]{} & \tilde{H}^{\alpha+V'}_{G,\delta}(\Sigma^{V} X;\overline{T})
\end{array}$$

(5) *(Dimension Axiom) If $H \in \mathscr{F}(\delta)$ and V is a representation of H so large that $V-\delta(G/H)+n$ is an actual representation, then there are natural isomorphisms*

$$\tilde{H}^{G,\delta}_{V+k}(G_+ \wedge_H S^{V-\delta(G/H)+n}; \underline{S}) \cong \begin{cases} \underline{S}(G/H, \delta) & \text{if } k = n \\ 0 & \text{if } k \neq n \end{cases}$$

and

$$\tilde{H}_{G,\delta}^{V+k}(G_+ \wedge_H S^{V-\delta(G/H)+n}; \overline{T}) \cong \begin{cases} \overline{T}(G/H, \delta) & \text{if } k = n \\ 0 & \text{if } k \neq n. \end{cases}$$

Proof That homology and cohomology are homotopy functors is Proposition 1.10.6. Functoriality in α on the chain level follows from the same property for spaces shown in the proof of Theorem 1.7.4.

Weak Equivalence: If $f: X \to Y$ is an $\mathscr{F}(\delta)$-equivalence, then any approximation $\Gamma f: \Gamma\Sigma^\infty X_w \to \Gamma\Sigma^\infty Y_w$ is a δ-weak$_\alpha$ equivalence. (Here, we use that X_w and Y_w are well-based to conclude that each $\Sigma^{V_i} X_w \to \Sigma^{V_i} Y_w$ is an $\mathscr{F}(\delta)$-equivalence, hence a δ-weak$_{\alpha+V_i}$-equivalence.) We've shown previously that Γf then has a G-homotopy inverse, hence that f_* and f^* are isomorphisms in homology and cohomology, respectively.

Exactness Let $i: A \to X$ be a cofibration. Using Theorem 1.4.21 we can choose approximations $\Gamma i: \Gamma\Sigma^\infty A_w \to \Gamma\Sigma^\infty X_w$ with Γi the inclusion of a subcomplex. At each level, by Proposition 1.4.25 we have a weak equivalence of mapping cones

$$C(\Gamma i(V_k)) \to C(\Sigma^{V_k} A_w \to \Sigma^{V_k} X_w).$$

We also have homotopy equivalences

$$C(\Gamma i(V_k)) \simeq (\Gamma\Sigma^\infty_{\mathscr{V}} X(V_i))/(\Gamma\Sigma^\infty_{\mathscr{V}} A(V_i))$$

and

$$C(\Sigma^{V_k} A_w \to \Sigma^{V_k} X_w) \simeq \Sigma^{V_k}(X/A)_w$$

We conclude that

$$\Gamma\Sigma^\infty X_w / \Gamma\Sigma^\infty A_w \to \Sigma^\infty (X/A)_w$$

is a CW approximation, from which it follows that

$$\overline{C}^{G,\delta}_{\alpha+*}((X/A)_w) = \overline{C}^{G,\delta}_{\alpha+*}(X_w)/\overline{C}^{G,\delta}_{\alpha+*}(A_w).$$

(That is, we get equality when we choose the CW approximation of $\Sigma^\infty(X/A)_w$ to be $\Gamma\Sigma^\infty X_w/\Gamma\Sigma^\infty A_w$.) The exact sequences now follow from familiar homological algebra.

Additivity This follows from the fact that the wedge of well-based G-spaces preserves δ-weak$_{\alpha+V_i}$ equivalence, so that the wedge of CW approximations is again a CW approximation, and the chains of a wedge can be taken to be the direct sum of the chains of the wedge summands.

Suspension This is where we need the full force of our approximation by CW spectra. We also use the fact, shown in Proposition 1.10.6, that the cellular chains are well-defined, meaning independent, up to canonical chain equivalence, of the choice of universe, indexing sequence, and CW approximation. Assuming that X is well-based, we may use X in place of X_w. To define σ^V, we begin with the case of $V \subset \mathscr{U}$. Let $\mathscr{V}$ be an indexing sequence in $\mathscr{U}$ such that $V \subset V_i$ for all i. Then $\mathscr{V}-V = \{V_i - V\}$ is an indexing sequence in $\mathscr{U}-V$. Let $\Sigma^\infty_{\mathscr{V}}X$ denote the suspension spectrum of X indexed on $\mathscr{V}$. Given a CW approximation $\Gamma\Sigma^\infty_{\mathscr{V}}X$, we can reinterpret each $\Gamma\Sigma^\infty_{\mathscr{V}}X(V_i) \to \Sigma^{V_i}X$ as

$$(\Gamma\Sigma^\infty_{\mathscr{V}-V}\Sigma^V X)(V_i - V) \to \Sigma^{V_i-V}\Sigma^V X.$$

This then gives $\Gamma\Sigma^\infty_{\mathscr{V}-V}\Sigma^V X \to \Sigma^\infty_{\mathscr{V}-V}\Sigma^V X$, a δ-G-CW$(\alpha+V)$ approximation of $\Sigma^V X$ indexed on $\mathscr{V}-V$. From this we get an isomorphism

$$\sigma^V : \overline{C}^{\mathscr{V},\delta}_{\alpha+*}(X) \cong \overline{C}^{\mathscr{V}-V,\delta}_{\alpha+V+*}(\Sigma^V X).$$

For a general V, choose an isomorphism $\xi: V \to V_0$ where $V_0 \subset \mathscr{U}$. Define $\sigma^V : \overline{C}^{\mathscr{V},\delta}_{\alpha+*}(X) \to \overline{C}^{\mathscr{V}-V_0,\delta}_{\alpha+V+*}(\Sigma^V X)$ to be the composite

$$\overline{C}^{\mathscr{V},\delta}_{\alpha+*}(X) \xrightarrow{\sigma^{V_0}} \overline{C}^{\mathscr{V}-V_0,\delta}_{\alpha+V_0+*}(\Sigma^{V_0}X) \xrightarrow{\overline{C}_\xi(\mathrm{id}\wedge\xi^{-1})} \overline{C}^{\mathscr{V}-V_0,\delta}_{\alpha+V+*}(\Sigma^V X).$$

This is a chain isomorphism so induces isomorphisms in homology and cohomology that we also call σ^V. Note that we use the fact that the chains are essentially independent of choice of universe and indexing sequence to justify using chains based on $\mathscr{U}-V_0$ and $\mathscr{V}-V_0$.

Transitivity, $\sigma^W \circ \sigma^V = \sigma^{V\oplus W}$, can be checked directly from this definition.

To show naturality, suppose $\zeta: V \to V'$, choose $\xi': V' \to V_0$ with $V_0 \subset \mathscr{U}$ as above, and let $\xi = \xi' \circ \zeta: V \to V_0$. Then naturality follows from the (chain homotopy) commutativity of the following diagram:

$$
\begin{array}{ccc}
\overline{C}^{\mathscr{V},\delta}_{\alpha+*}(X) & & \\
\Big\downarrow{\scriptstyle \sigma^{V_0}} & & \\
\overline{C}^{\mathscr{V}-V_0,\delta}_{\alpha+V_0+*}(\Sigma^{V_0}X) & \xrightarrow{\overline{C}_{\xi}(\mathrm{id}\wedge\xi^{-1})} & \overline{C}^{\mathscr{V}-V_0,\delta}_{\alpha+V+*}(\Sigma^{V}X) \\
{\scriptstyle \overline{C}_{\xi'}(\mathrm{id}\wedge\xi'^{-1})}\Big\downarrow & & \Big\downarrow{\scriptstyle \overline{C}_{\mathrm{id}}(\mathrm{id}\wedge\zeta)} \\
\overline{C}^{\mathscr{V}-V_0,\delta}_{\alpha+V'+*}(\Sigma^{V'}X) & \xrightarrow[\overline{C}_{\zeta}(\mathrm{id})]{} & \overline{C}^{\mathscr{V}-V_0,\delta}_{\alpha+V+*}(\Sigma^{V'}X)
\end{array}
$$

Dimension Axiom This was already shown in Theorem 1.7.4. (We apply Proposition 1.10.6 to the based complex $G_+ \wedge_H S^{V-\delta(G/H)+n}$.) □

Corollary 1.10.8

(1) *If $f: A \to X$ is a map of well-based G-spaces, then the following sequences are exact:*

$$\tilde{H}^{G,\delta}_{\alpha}(A; \underline{S}) \to \tilde{H}^{G,\delta}_{\alpha}(X; \underline{S}) \to \tilde{H}^{G,\delta}_{\alpha}(\tilde{C}f; \underline{S})$$

and

$$\tilde{H}_{G,\delta}^{\alpha}(\tilde{C}f; \overline{T}) \to \tilde{H}_{G,\delta}^{\alpha}(X; \overline{T}) \to \tilde{H}_{G,\delta}^{\alpha}(A; \overline{T}).$$

(2) *If $f: A \to X$ is any map of G-spaces, then the following sequences are exact, where $f_w: A_w \to X_w$ is the induced map:*

$$\tilde{H}^{G,\delta}_{\alpha}(A; \underline{S}) \to \tilde{H}^{G,\delta}_{\alpha}(X; \underline{S}) \to \tilde{H}^{G,\delta}_{\alpha}(\tilde{C}(f_w); \underline{S})$$

and

$$\tilde{H}_{G,\delta}^{\alpha}(\tilde{C}(f_w); \overline{T}) \to \tilde{H}_{G,\delta}^{\alpha}(X; \overline{T}) \to \tilde{H}_{G,\delta}^{\alpha}(A; \overline{T}).$$

Proof If $f: A \to X$ is a map of well-based G-spaces, then f factors as $A \to \tilde{M}f \to X$, where the first map is a cofibration and the second is an equivalence. Because $\tilde{C}f = \tilde{M}f/A$, the claimed exact sequences in (1) now follow from the preceding theorem.

If $f: A \to X$ is any based map, we apply (1) to f_w to get the exact sequences claimed in (2). □

Finally, we define unreduced homology and cohomology of pairs.

Definition 1.10.9 If (X,A) is a pair of G-spaces, write $i\colon A \to X$ for the inclusion and let

$$H_\alpha^{G,\delta}(X,A;\underline{S}) = \tilde{H}_\alpha^{G,\delta}(Ci;\underline{S})$$

and

$$H^\alpha_{G,\delta}(X,A;\overline{T}) = \tilde{H}^\alpha_{G,\delta}(Ci;\overline{T}).$$

In particular, we write

$$H_\alpha^{G,\delta}(X;\underline{S}) = H_\alpha^{G,\delta}(X,\emptyset;\underline{S}) = \tilde{H}_\alpha^{G,\delta}(X_+;\underline{S})$$

and

$$H^\alpha_{G,\delta}(X;\overline{T}) = H^\alpha_{G,\delta}(X,\emptyset;\overline{T}) = \tilde{H}^\alpha_{G,\delta}(X_+;\overline{T}).$$

Theorem 1.10.10 (Unreduced Homology and Cohomology) *Let δ be a familial dimension function for G, let α be a virtual representation of G, and let $\underline{S}$ and $\overline{T}$ be respectively a covariant and a contravariant δ-Mackey functor. Then the abelian groups $H_\alpha^{G,\delta}(X,A;\underline{S})$ and $H^\alpha_{G,\delta}(X,A;\overline{T})$ are respectively covariant and contravariant functors on the homotopy category of pairs of G-spaces. They are also respectively contravariant and covariant functors of α. These functors satisfy the following properties.*

(1) *(Weak Equivalence) If $f\colon (X,A) \to (Y,B)$ is an $\mathscr{F}(\delta)$-equivalence of pairs of G-spaces, then*

$$f_*\colon H_\alpha^{G,\delta}(X,A;\underline{S}) \to H_\alpha^{G,\delta}(Y,B;\underline{S})$$

and

$$f^*\colon H^\alpha_{G,\delta}(Y,B;\overline{T}) \to H^\alpha_{G,\delta}(X,A;\overline{T})$$

are isomorphisms.

(2) *(Exactness) If (X,A) is a pair of G-spaces, then there are natural homomorphisms*

$$\partial\colon H_{\alpha+n}^{G,\delta}(X,A;\underline{S}) \to H_{\alpha+n-1}^{G,\delta}(A;\underline{S})$$

and

$$d\colon H^{\alpha+n}_{G,\delta}(A;\overline{T}) \to H^{\alpha+n+1}_{G,\delta}(X,A;\overline{T})$$

and long exact sequences

$$\cdots \to H^{G,\delta}_{\alpha+n}(A;\underline{S}) \to H^{G,\delta}_{\alpha+n}(X;\underline{S}) \to H^{G,\delta}_{\alpha+n}(X,A;\underline{S}) \to H^{G,\delta}_{\alpha+n-1}(A;\underline{S}) \to \cdots$$

and

$$\cdots \to H_{G,\delta}^{\alpha+n-1}(A;\overline{T}) \to H_{G,\delta}^{\alpha+n}(X,A;\overline{T}) \to H_{G,\delta}^{\alpha+n}(X;\overline{T}) \to H_{G,\delta}^{\alpha+n}(A;\overline{T}) \to \cdots .$$

(3) *(Excision) If $(X;A,B)$ is an excisive triad, i.e., X is the union of the interiors of A and B, then the inclusion $(A, A\cap B) \to (X,B)$ induces isomorphisms*

$$H^{G,\delta}_{\alpha}(A, A\cap B;\underline{S}) \cong H^{G,\delta}_{\alpha}(X,B;\underline{S})$$

and

$$H_{G,\delta}^{\alpha}(X,B;\overline{T}) \cong H_{G,\delta}^{\alpha}(A, A\cap B;\overline{T}).$$

(4) *(Additivity) If $(X,A) = \coprod_k (X_k, A_k)$ is a disjoint union of pairs of G-spaces, then the inclusions $(X_k, A_k) \to (X,A)$ induce isomorphisms*

$$\bigoplus_k H^{G,\delta}_{\alpha}(X_k, A_k;\underline{S}) \cong H^{G,\delta}_{\alpha}(X,A;\underline{S})$$

and

$$H_{G,\delta}^{\alpha}(X,A;\overline{T}) \cong \prod_k H_{G,\delta}^{\alpha}(X_k, A_k;\overline{T}).$$

(5) *(Suspension) If $A \to X$ is a cofibration, there are suspension isomorphisms*

$$\sigma^V : H^{G,\delta}_{\alpha}(X,A;\underline{S}) \xrightarrow{\cong} H^{G,\delta}_{\alpha+V}((X,A)\times(D(V),S(V));\underline{S})$$

and

$$\sigma^V : H_{G,\delta}^{\alpha}(X,A;\overline{T}) \xrightarrow{\cong} H_{G,\delta}^{\alpha+V}((X,A)\times(D(V),S(V));\overline{T}),$$

where $(X,A)\times(D(V),S(V)) = (X\times D(V), X\times S(V) \cup A\times D(V))$. These isomorphisms satisfy $\sigma^0 = \mathrm{id}$ and $\sigma^W \circ \sigma^V = \sigma^{V\oplus W}$, under the identification $\bar{D}(V)\times\bar{D}(W) \approx \bar{D}(V\oplus W)$ (here we use the notation $\bar{D}(V) = (D(V),S(V))$). They also satisfy the following naturality condition: If $\zeta : V \to V'$ is an isomorphism, then the following diagrams commute:

$$
\begin{array}{ccc}
H_{\alpha}^{G,\delta}(X,A;\underline{S}) & \xrightarrow{\sigma^{V}} & H^{G,\delta}_{\alpha+V}((X,A)\times \bar{D}(V);\underline{S}) \\
\Big\downarrow {\scriptstyle \sigma^{V'}} & & \Big\downarrow {\scriptstyle H^{G}_{\mathrm{id}}(\mathrm{id}\times\zeta)} \\
H^{G,\delta}_{\alpha+V'}((X,A)\times \bar{D}(V');\underline{S}) & \xrightarrow[H^{G}_{\mathrm{id}\oplus\zeta}(\mathrm{id})]{} & H^{G,\delta}_{\alpha+V}((X,A)\times \bar{D}(V');\underline{S})
\end{array}
$$

$$
\begin{array}{ccc}
H^{\alpha}_{G,\delta}(X,A;\overline{T}) & \xrightarrow{\sigma^{V}} & H^{\alpha+V}_{G,\delta}((X,A)\times \bar{D}(V);\overline{T}) \\
\Big\downarrow {\scriptstyle \sigma^{V'}} & & \Big\downarrow {\scriptstyle H_{G}^{\mathrm{id}\oplus\zeta}(\mathrm{id})} \\
H^{\alpha+V'}_{G,\delta}((X,A)\times \bar{D}(V');\overline{T}) & \xrightarrow[H_{G}^{\mathrm{id}}(\mathrm{id}\times\zeta)]{} & H^{\alpha+V'}_{G,\delta}((X,A)\times \bar{D}(V);\overline{T})
\end{array}
$$

(6) *(Dimension Axiom) If $H \in \mathscr{F}(\delta)$ and V is a representation of H so large that $V-\delta(G/H)+n$ is an actual representation, then there are natural isomorphisms*

$$
H^{G,\delta}_{V+k}(G\times_H \bar{D}(V-\delta(G/H)+n);\underline{S}) \cong \begin{cases} \underline{S}(G/H) & \text{if } k=n \\ 0 & \text{if } k\neq n \end{cases}
$$

and

$$
H_{G,\delta}^{V+k}(G\times_H \bar{D}(V-\delta(G/H)+n);\overline{T}) \cong \begin{cases} \overline{T}(G/H) & \text{if } k=n \\ 0 & \text{if } k\neq n. \end{cases}
$$

Proof Let $i\colon A \to X$ be the inclusion. The claims in the first paragraph translate into claims about the reduced homology and cohomology of Ci that we have already proven.

Weak Equivalence If $i\colon A \to X$ and $j\colon B \to Y$ are the inclusions, then $Ci \to Cj$ is an $\mathscr{F}(\delta)$-equivalence by an argument using Proposition 1.4.25. The result follows because reduced homology and cohomology take $\mathscr{F}(\delta)$-equivalences to isomorphisms.

Exactness If $i\colon A \to X$ is the inclusion, we consider the cofiber sequence

$$
A_+ \to X_+ \to Ci \to \Sigma A_+ \to \Sigma X_+ \to \cdots
$$

where we use that $\tilde{C}(i_+) \approx Ci$. As in the nonequivariant case (cf [48, §8.4]), each pair of maps is, up to sign, G-homotopy equivalent to a map followed by the inclusion of its target in its mapping cone. Thus, on applying reduced homology and cohomology, we get exact sequences, which we identify as the long exact sequences claimed, using the suspension isomorphisms. But, note that we must use the *internal* suspension discussed after Remark 1.7.2. The maps ∂ and d are then the natural homomorphisms induced by $Ci \to \Sigma A_+$.

Excision Let $i: A\cap B \to A$ and $j: B \to X$ be the inclusions. Consider the following map of triads: $(Ci; Ci, C(A \cap B)) \to (Cj; Ci, CB)$. Each of these triads is excisive, and the three maps $Ci \to Ci$, $C(A \cap B) \to CB$, and $Ci \cap C(A \cap B) = C(A \cap B) \to Ci \cap CB = C(A \cap B)$ are all weak equivalences. Therefore, by Proposition 1.4.25, the inclusion $Ci \to Cj$ is a weak equivalence. The excision isomorphisms follow.

Additivity Let $i: A \to X$ and $i_k: A_k \to X_k$ be the inclusions. Then $Ci = \bigvee_k Ci_k$, so additivity follows from the reduced case.

Suspension Under the assumption that $i: A \to X$ is a cofibration, $j: X \times S(V) \cup A \times D(V) \to X \times D(V)$ is also a cofibration, and $Ci \simeq X/A$ and $Cj \simeq \Sigma^V X/A$. The result now follows from the reduced suspension isomorphism.

Dimension Axiom If $i: G \times_H S(V - \delta(G/H) + n) \to G \times_H D(V - \delta(G/H) + n)$, then $Ci \approx G_+ \wedge_H S^{V-\delta(G/H)+n}$, so the result follows from the reduced case. □

1.10.3 Independence of Choices

For spectra or spaces that first need to be approximated, there are choices involved. We now show that these choices do not really matter. Throughout this subsection we assume that δ is a familial dimension function.

Proposition 1.10.11

(1) *If E is any G-spectrum and $\Gamma_1 E \to E$ and $\Gamma_2 E \to E$ are two approximations by δ-G-$CW(\alpha)$ spectra, then there is a canonical chain isomorphism $\overline{C}^{G,\delta}_{\alpha+*}(\Gamma_1 E) \cong \overline{C}^{G,\delta}_{\alpha+*}(\Gamma_2 E)$.*

(2) *If $f: E \to F$ is any map of G-spectra and $\Gamma E \to E$ and $\Gamma F \to F$ are chosen approximations, the map $f_*: \overline{C}^{G,\delta}_{\alpha+*}(\Gamma E) \to \overline{C}^{G,\delta}_{\alpha+*}(\Gamma F)$ is well-defined up to chain homotopy.*

Proof For the second part, we note by the last part of Proposition 1.9.6 that the approximating cellular map Γf is well-defined up to cellular homotopy. This implies that the chain map it induces is well-defined up to chain isomorphism. The first part follows from the second applied to the identity map on E. □

Put another way, if we *choose* for each spectrum E an approximation $\Gamma E \to E$, then we get a functor from the category of spectra to the category of chain complexes modulo chain homotopy. Any two collections of choices of approximations lead to canonically naturally isomorphic functors. Here, we can interpret "canonically" as meaning that, if we consider the category (groupoid, actually) whose objects are collections of choices of approximations and in which there is a unique morphism from any collection to any other, then the natural isomorphisms are functorial on this groupoid.

Now, if X is a based G-space, $\overline{C}^{G,\delta}_{\alpha+*}(X)$ appears to depend also on the choice of universe $\mathscr{U}$ and indexing sequence $\mathscr{V}$, but we now argue that, up to canonical natural chain homotopy equivalence, it really does not. We first show that the chains are independent of the choice of $\mathscr{V}$. To emphasize the possible dependence on $\mathscr{V}$, we write $\overline{C}^{\mathscr{V},\delta}_{\alpha+*}(X)$ for chains defined using $\mathscr{V}$ as the indexing sequence.

Proposition 1.10.12 *Let $\mathscr{V}$ and $\mathscr{W}$ be two indexing sequences in the same universe $\mathscr{U}$. Then the chain complexes $\overline{C}^{\mathscr{V},\delta}_{\alpha+*}(X)$ and $\overline{C}^{\mathscr{W},\delta}_{\alpha+*}(X)$ are canonically naturally chain homotopy equivalent.*

Proof As in the preceding subsection, let $\Sigma^\infty_{\mathscr{V}}X$ denote the suspension spectrum of X indexed on $\mathscr{V}$ and let $\Sigma^\infty_{\mathscr{W}}X$ denote the one indexed on $\mathscr{W}$. Choose δ-G-CW(α) approximations $\Gamma\Sigma^\infty_{\mathscr{V}}X \to \Sigma^\infty_{\mathscr{V}}X$ and $\Gamma\Sigma^\infty_{\mathscr{W}}X \to \Sigma^\infty_{\mathscr{W}}X$, so that, for example, $\Gamma\Sigma^\infty_{\mathscr{V}}X(V_i) \to \Sigma^{V_i}X$ is an approximation by a δ-G-CW($\alpha + V_i$) complex. Because $\mathscr{W}$ is an indexing sequence, we can find a strictly increasing sequence $\{j(i)\}$ such that $V_i \subset W_{j(i)}$ for each i. Using the relative Whitehead theorem, we choose cellular maps

$$\zeta_i\colon \Sigma^{W_{j(i)}-V_i}\Gamma\Sigma^\infty_{\mathscr{V}}X(V_i) \to \Gamma\Sigma^\infty_{\mathscr{W}}X(W_{j(i)})$$

over $\Sigma^{W_{j(i)}}X$ inductively so that the following diagram commutes for each $i > 1$:

$$\begin{array}{ccc}
\Sigma^{W_{j(i)}-V_{i-1}}\Gamma\Sigma^\infty_{\mathscr{V}}X(V_{i-1}) & \xrightarrow{\Sigma\zeta_{i-1}} & \Sigma^{W_{j(i)}-W_{j(i-1)}}\Gamma\Sigma^\infty_{\mathscr{W}}X(W_{j(i-1)}) \\
{\scriptstyle\Sigma\sigma}\downarrow & & \downarrow{\scriptstyle\sigma} \\
\Sigma^{W_{j(i)}-V_i}\Gamma\Sigma^\infty_{\mathscr{V}}X(V_i) & \xrightarrow[\zeta_i]{} & \Gamma\Sigma^\infty_{\mathscr{W}}X(W_{j(i)})
\end{array}$$

This determines a chain map $\zeta\colon \overline{C}^{\mathscr{V},\delta}_{\alpha+*}(X) \to \overline{C}^{\mathscr{W},\delta}_{\alpha+*}(X)$, unique up to chain homotopy—on each term in the colimit it is the composite

$$\begin{aligned}
\overline{C}_{\alpha+V_i+*}(\Gamma\Sigma^\infty_{\mathscr{V}}X(V_i)) &\xrightarrow{\Sigma} \overline{C}_{\alpha+W_{j(i)}+*}(\Sigma^{W_{j(i)}-V_i}\Gamma\Sigma^\infty_{\mathscr{V}}X(V_i)) \\
&\xrightarrow{(\zeta_i)_*} \overline{C}_{\alpha+W_{j(i)}+*}(\Gamma\Sigma^\infty_{\mathscr{W}}X(W_{j(i)}))
\end{aligned}$$

When we say that these chain maps are canonical, we mean that the composite of the map $\overline{C}^{\mathscr{V},\delta}_{\alpha+*}(X) \to \overline{C}^{\mathscr{W},\delta}_{\alpha+*}(X)$ with the map $\overline{C}^{\mathscr{W},\delta}_{\alpha+*}(X) \to \overline{C}^{\mathscr{Z},\delta}_{\alpha+*}(X)$ agrees, up to chain homotopy, with the map $\overline{C}^{\mathscr{V},\delta}_{\alpha+*}(X) \to \overline{C}^{\mathscr{Z},\delta}_{\alpha+*}(X)$. This is a straightforward, if tedious, check.

In particular, the composite

$$\overline{C}^{\mathscr{V},\delta}_{\alpha+*}(X) \to \overline{C}^{\mathscr{W},\delta}_{\alpha+*}(X) \to \overline{C}^{\mathscr{V},\delta}_{\alpha+*}(X)$$

is chain homotopic to the identity, and similarly when $\mathscr{V}$ and $\mathscr{W}$ are reversed, so the maps we've constructed are chain homotopy equivalences. □

Now suppose that $\mathscr{U}$ and $\mathscr{U}'$ are two complete G-universes. (The following line of argument leads to a result similar in concept to the more general [42, II.1.7], but the latter result deals with the stable category, so we give a separate argument here for our case.) Because they are both complete, they are isomorphic; choose a linear isometric isomorphism $f: \mathscr{U} \to \mathscr{U}'$. If $\mathscr{V}$ is an indexing sequence in $\mathscr{U}$, let $V_i' = f(V_i)$ for all i and let $\mathscr{V}' = f(\mathscr{V}) = \{V_i'\}$, which is an indexing sequence in $\mathscr{U}'$. If $\Gamma\Sigma^\infty_{\mathscr{V}}X$ is a CW approximation to $\Sigma^\infty_{\mathscr{V}}X$, we get a CW approximation indexed on $\mathscr{V}'$ by taking the same spaces with the composite maps

$$\Gamma\Sigma^\infty_{\mathscr{V}}X(V_i) \to \Sigma^{V_i}X \xrightarrow{f} \Sigma^{V_i'}X.$$

We freely use the fact that a δ-G-CW$(\alpha + V_i)$ structure is also a δ-G-CW$(\alpha + V_i')$ structure. In what follows, any CW approximation $\Gamma\Sigma^\infty_{\mathscr{V}'}X$ can be used in place of the one just given and we do not assume any particular choice. In any case, the relative Whitehead theorem allows us to inductively define cellular maps $\bar{f}$ making the top square in the following diagram commute and the bottom square commute up to homotopy:

$$\begin{array}{ccc}
\Sigma^{V_i-V_{i-1}}\Gamma\Sigma^\infty_{\mathscr{V}}X(V_{i-1}) & \xrightarrow{\bar{f}} & \Sigma^{V_i'-V_{i-1}'}\Gamma\Sigma^\infty_{\mathscr{V}'}X(V_{i-1}') \\
\downarrow & & \downarrow \\
\Gamma\Sigma^\infty_{\mathscr{V}}X(V_i) & \xrightarrow{\bar{f}} & \Gamma\Sigma^\infty_{\mathscr{V}'}X(V_i') \\
\downarrow & & \downarrow \\
\Sigma^{V_i}X & \xrightarrow{f} & \Sigma^{V_i'}X
\end{array}$$

(If we simply make the right hand side a copy of the left, the map commutes on the nose.) The maps $\bar{f}$ then induce a chain map $f_*: \overline{C}^{\mathscr{V},\delta}_{\alpha+*}(X) \to \overline{C}^{\mathscr{V}',\delta}_{\alpha+*}(X)$, unique up to chain homotopy — on each term in the colimit it is given by

$$\overline{C}_{\alpha+V_i+*}(\Gamma\Sigma^\infty_{\mathscr{V}}X(V_i)) \xrightarrow{\bar{f}_*} \overline{C}_{\alpha+V_i'+*}(\Gamma\Sigma^\infty_{\mathscr{V}'}X(V_i'))$$

The similar map $(f^{-1})_*$ is clearly a chain homotopy inverse.

To what extent does the chain homotopy equivalence $\overline{C}^{\mathscr{V},\delta}_{\alpha+*}(X) \to \overline{C}^{\mathscr{V}',\delta}_{\alpha+*}(X)$ depend on the choice of f? Hardly at all, in the following sense.

Proposition 1.10.13 *Let f and g be two linear isometric isomorphisms $\mathscr{U} \to \mathscr{U}'$. Let $\mathscr{V}$ be an indexing sequence in $\mathscr{U}$, let $\mathscr{V}' = f(\mathscr{V})$, and let $\mathscr{W}' = g(\mathscr{V})$.*

Let $\zeta\colon \overline{C}^{\mathscr{V}',\delta}_{\alpha+*}(X) \to \overline{C}^{\mathscr{W}',\delta}_{\alpha+*}(X)$ *be the canonical chain homotopy equivalence of Proposition 1.10.12. Then the following diagram of chain homotopy equivalences commutes up to chain homotopy:*

$$\begin{array}{ccc} & \overline{C}^{\mathscr{V},\delta}_{\alpha+*}(X) & \\ {}^{f_*}\swarrow & & \searrow{}^{g_*} \\ \overline{C}^{\mathscr{V}',\delta}_{\alpha+*}(X) & \xrightarrow[\zeta]{} & \overline{C}^{\mathscr{W}',\delta}_{\alpha+*}(X) \end{array}$$

Proof Choose a strictly increasing sequence $j(i)$ such that $V'_i \subset W'_{j(i)}$ and $j(i) \geq i$. The map ζf_* is represented by the composite $(\zeta_i)_*\Sigma\bar{f}_*$ down the left side of the following commutative diagram, hence by the composite across the top and down the right side.

$$\begin{array}{ccc} \overline{C}_{\alpha+V_i+*}(\Gamma\Sigma^\infty_{\mathscr{V}}X(V_i)) & \xrightarrow{\Sigma} & \overline{C}_{\alpha+V_i+(W'_{j(i)}-V'_i)+*}(\Sigma^{W'_{j(i)}-V'_i}\Gamma\Sigma^\infty_{\mathscr{V}}X(V_i)) \\ \bar{f}_*\downarrow & & \downarrow(f+1)_* \\ \overline{C}_{\alpha+V'_i+*}(\Gamma\Sigma^\infty_{\mathscr{V}'}X(V'_i)) & \xrightarrow{\Sigma} & \overline{C}_{\alpha+W'_{j(i)}+*}(\Sigma^{W'_{j(i)}-V'_i}\Gamma\Sigma^\infty_{\mathscr{V}'}X(V'_i)) \\ & & \downarrow(\zeta_i)_* \\ & & \overline{C}_{\alpha+W'_{j(i)}+*}(\Gamma\Sigma^\infty_{\mathscr{W}'}X(W'_{j(i)})) \end{array}$$

In order to compare this with g_*, we can assume that $W'_{j(i)}$ is so large that there is a linear isometry $h\colon W'_{j(i)} - V'_i \to W'_{j(i)} - W'_i$ such that the map

$$V_i \oplus (W'_{j(i)} - V'_i) \xrightarrow{g+h} W'_i \oplus (W'_{j(i)} - W'_i) = W'_{j(i)}$$

is homotopic through linear isometries to

$$V_i \oplus (W'_{j(i)} - V'_i) \xrightarrow{f+1} V'_i \oplus (W'_{j(i)} - V'_i) = W'_{j(i)}.$$

Now consider the following cube:

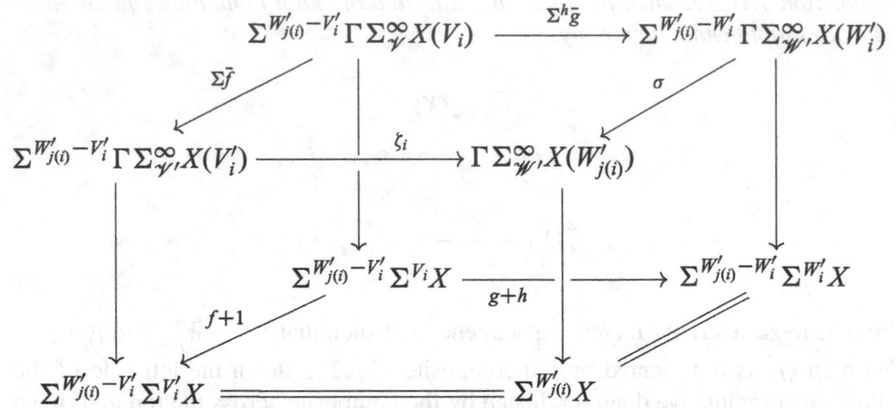

Because the four vertical squares and the bottom square commute up to homotopy, the Whitehead theorem implies that the top square also commutes up to homotopy. Moreover, we can choose a cellular homotopy. Finally, we now have the following chain homotopy commutative diagram:

$$
\begin{array}{ccc}
\overline{C}_{\alpha+V_i+*}(\Gamma\Sigma^\infty_{\mathscr{V}}X(V_i)) & \xrightarrow{\tilde{g}_*} & \overline{C}_{\alpha+W'_i+*}(\Gamma\Sigma^\infty_{\mathscr{W}'}X(W'_i)) \\
\Sigma\big\downarrow & & \big\downarrow\Sigma \\
\overline{C}_{\alpha+V_i+(W'_{j(i)}-V'_i)+*}(\Sigma^{W'_{j(i)}-V'_i}\Gamma\Sigma^\infty_{\mathscr{V}}X(V_i)) & \xrightarrow{\tilde{g}_*} & \overline{C}_{\alpha+W'_i+(W'_{j(i)}-V'_i)+*}(\Sigma^{W'_{j(i)}-V'_i}\Gamma\Sigma^\infty_{\mathscr{W}'}X(W'_i)) \\
 & & \big\downarrow\Sigma^h \\
(\overline{f+1})_*\big\downarrow & & \overline{C}_{\alpha+W'_i+(W'_{j(i)}-W'_i)+*}(\Sigma^{W'_{j(i)}-W'_i}\Gamma\Sigma^\infty_{\mathscr{W}'}X(W'_i)) \\
 & & \big\downarrow\sigma_* \\
\overline{C}_{\alpha+V'_i+(W'_{j(i)}-V'_i)+*}(\Sigma^{W'_{j(i)}-V'_i}\Gamma\Sigma^\infty_{\mathscr{V}'}X(V'_i)) & \xrightarrow[(\zeta_i)_*]{} & \overline{C}_{\alpha+W'_{j(i)}+*}(\Gamma\Sigma^\infty_{\mathscr{W}'}X(W'_{j(i)}))
\end{array}
$$

The composite down the left side represents $\zeta f_*(u)$ while the composite down the right represents $g_*(u)$. Here, we use that the composite $\Sigma^h \circ \Sigma$ on the right coincides with the simple suspension by $W'_{j(i)} - W'_i$, which follows by inspecting the definition of the chain complexes. Hence, ζf_* is cellularly homotopic to g_* as claimed. □

1.11 Atiyah-Hirzebruch Spectral Sequences and Uniqueness

Definition 1.11.1 Recall from [71] or [47, XIII.1] that a *(generalized) reduced RO(G)-graded homology or cohomology theory* is a collection of functors satisfying all of Theorem 1.10.7 except that it is required to take weak G-equivalences (not necessarily all $\mathscr{F}(\delta)$-equivalences) to isomorphisms and it need not obey the dimension axiom. Given a reduced $RO(G)$-graded homology theory $\tilde{h}^G_*(-)$ and a dimension function δ for G, we can regard $\tilde{h}^G_*(-)$ as G-δ-Mackey functor-valued by defining the covariant functor $\underline{h}^{G,\delta}_\alpha(X)$, for a well-based G-space X, by

$$\underline{h}^{G,\delta}_\alpha(X)(G/H,\delta) = \tilde{h}^G_{\alpha+V}(X \wedge G_+ \wedge_H S^{V-\delta(G/H)}),$$

where V is a representation of G so large that $\delta(G/H) \subset V$. In particular, we write $\underline{h}^{G,\delta}_* = \underline{h}^{G,\delta}_*(S^0)$ for the *coefficient system* of $\tilde{h}^G_*(-)$. Similarly, if $\tilde{h}^*_G(-)$ is a reduced $RO(G)$-graded cohomology theory, we write $\overline{h}^\alpha_{G,\delta}(X)$ for the contravariant δ-Mackey functor defined by

$$\overline{h}^\alpha_{G,\delta}(X)(G/H,\delta) = \tilde{h}^{\alpha+V}_G(X \wedge G_+ \wedge_H S^{V-\delta(G/H)})$$

and we write $\overline{h}^*_{G,\delta} = \overline{h}^*_{G,\delta}(S^0)$.

In these terms, the dimension axioms take the form of the following statements about the coefficient systems of cellular homology and cohomology:

$$\underline{H}^{G,\delta}_n(S^0;\underline{S}) \cong \begin{cases} \underline{S} & \text{if } n = 0 \\ 0 & \text{if } n \neq 0 \end{cases}$$

and

$$\overline{H}^n_{G,\delta}(S^0;\overline{T}) \cong \begin{cases} \overline{T} & \text{if } n = 0 \\ 0 & \text{if } n \neq 0. \end{cases}$$

The Atiyah-Hirzebruch spectral sequence generalizes as follows.

Theorem 1.11.2 (Atiyah-Hirzebruch Spectral Sequence) *Suppose that $\tilde{h}^G_*(-)$ is a reduced $RO(G)$-graded homology theory, let δ be a familial dimension function for G, and let α and β be virtual representations of G. Assume that $\tilde{h}^G_*(-)$ takes $\mathscr{F}(\delta)$-equivalences to isomorphisms. Then there is a strongly convergent spectral sequence*

$$E^2_{p,q} = \tilde{H}^{G,\delta}_{\alpha+p}(X;\underline{h}^{G,\delta}_{\beta+q}) \Rightarrow \tilde{h}^G_{\alpha+\beta+p+q}(X).$$

Similarly, if $\tilde{h}^*_G(-)$ is a reduced $RO(G)$-graded cohomology theory taking $\mathscr{F}(\delta)$-equivalences to isomorphisms, there is a conditionally convergent spectral sequence

$$E_2^{p,q} = \tilde{H}^{\alpha+p}_{G,\delta}(X; \overline{h}^{\beta+q}_{G,\delta}) \Rightarrow \tilde{h}^{\alpha+\beta+p+q}_G(X).$$

Proof These spectral sequences arise in the usual way. (See, for example, [2, §12].) Recall from Corollary 1.4.19 that, for sufficiently large V, δ-weak$_{\alpha+V}$ equivalence coincides with $\mathscr{F}(\delta)$-equivalence; by taking suspensions, it suffices to consider the case where δ-weak$_\alpha$ equivalence coincides with $\mathscr{F}(\delta)$-equivalence. By taking δ-G-CW(α) approximation, using that $\tilde{h}^G_*(-)$ takes $\mathscr{F}(\delta)$-equivalences to isomorphisms, it suffices to consider the case where X is a based δ-G-CW(α) complex. The skeletal filtration $\{X^{\alpha+*}\}$ leads to an exact couple on applying either $\tilde{h}^G_{\alpha+\beta+*}(-)$ or $\tilde{h}_G^{\alpha+\beta+*}(-)$. We identify the E^1 and E_1 terms using the natural isomorphisms

$$E^1_{p,q} = \tilde{h}^G_{\alpha+\beta+p+q}(X^{\alpha+p}/X^{\alpha+p-1}) \cong \overline{C}^{G,\delta}_{\alpha+p}(X,*) \otimes_{\widehat{\mathscr{O}}_{G,\delta}} \underline{h}^{G,\delta}_{\beta+q}$$

and

$$E_1^{p,q} = \tilde{h}_G^{\alpha+\beta+p+q}(X^{\alpha+p}/X^{\alpha+p-1}) \cong \mathrm{Hom}_{\widehat{\mathscr{O}}_{G,\delta}}(\overline{C}^{G,\delta}_{\alpha+p}(X,*), \overline{h}^{\beta+q}_{G,\delta}),$$

and the spectral sequences follow.

It will be useful later to show explicitly how the natural isomorphisms displayed above arise. In the case of homology, recall that

$$\begin{aligned}
&\overline{C}^{G,\delta}_{\alpha+p}(X,*) \otimes_{\widehat{\mathscr{O}}_{G,\delta}} \underline{h}^{G,\delta}_{\beta+q} \\
&\quad\cong \int^{(G/H,\delta)} [G_+ \wedge_H S^{-\delta(G/H)+\alpha+p}, \Sigma^\infty_G X^{\alpha+p}/X^{\alpha+p-1}]_G \\
&\qquad\qquad \otimes \tilde{h}^G_{\beta+q}(G_+ \wedge_H S^{-\delta(G/H)}) \\
&\quad\cong \int^{(G/H,\delta)} [G_+ \wedge_H S^{-\delta(G/H)+\alpha+p}, \Sigma^\infty_G X^{\alpha+p}/X^{\alpha+p-1}]_G \\
&\qquad\qquad \otimes \tilde{h}^G_{\alpha+\beta+p+q}(G_+ \wedge_H S^{-\delta(G/H)+\alpha+p}).
\end{aligned}$$

Evaluation defines a natural map

$$\begin{aligned}
&[G_+ \wedge_H S^{-\delta(G/H)+\alpha+p}, \Sigma^\infty_G X^{\alpha+p}/X^{\alpha+p-1}]_G \\
&\quad \otimes \tilde{h}^G_{\alpha+\beta+p+q}(G_+ \wedge_H S^{-\delta(G/H)+\alpha+p}) \to \tilde{h}^G_{\alpha+\beta+p+q}(X^{\alpha+p}/X^{\alpha+p-1}).
\end{aligned}$$

Note that, because G/H is compact, any stable map from $G_+ \wedge_H S^{-\delta(G/H)+\alpha+p}$ into $\Sigma^\infty_G X^{\alpha+p}/X^{\alpha+p-1}$ is represented by a map of G-spaces

$$G_+ \wedge_H S^{V-\delta(G/H)+\alpha+p} \to \Sigma^V_G X^{\alpha+p}/X^{\alpha+p-1}$$

for some V, which we need because we are assuming only that $\tilde{h}_*^G$ is defined on spaces. The evaluation maps are compatible as G/H varies, so define a map

$$\overline{C}_{\alpha+p}^{G,\delta}(X,*) \otimes_{\widehat{\mathscr{O}}_{G,\delta}} \underline{h}_{\beta+q}^{G,\delta} \to \tilde{h}_{\alpha+\beta+p+q}^{G}(X^{\alpha+p}/X^{\alpha+p-1})$$

which is an isomorphism because $X^{\alpha+p}/X^{\alpha+p-1}$ is a wedge of spheres.

In cohomology, we start with the evaluation map

$$\begin{aligned}
&[G_+ \wedge_H S^{-\delta(G/H)+\alpha+p}, \Sigma_G^\infty X^{\alpha+p}/X^{\alpha+p-1}]_G \otimes \tilde{h}_G^{\alpha+\beta+p+q}(X^{\alpha+p}/X^{\alpha+p-1}) \\
&\to \tilde{h}_G^{\alpha+\beta+p+q}(G_+ \wedge_H S^{-\delta(G/H)+\alpha+p}) \\
&\cong \tilde{h}_G^{\beta+q}(G_+ \wedge_H S^{-\delta(G/H)}) \\
&= \overline{h}_{G,\delta}^{\beta+q}(G/H,\delta).
\end{aligned}$$

In adjoint form this gives a homomorphism

$$\tilde{h}_G^{\alpha+\beta+p+q}(X^{\alpha+p}/X^{\alpha+p-1}) \to \mathrm{Hom}(\overline{C}_{\alpha+p}^{G,\delta}(X,*)(G/H,\delta), \overline{h}_{G,\delta}^{\beta+q}(G/H,\delta))$$

for each G/H. These homomorphisms are compatible as G/H varies, giving a map

$$\tilde{h}_G^{\alpha+\beta+p+q}(X^{\alpha+p}/X^{\alpha+p-1}) \to \mathrm{Hom}_{\widehat{\mathscr{O}}_{G,\delta}}(\overline{C}_{\alpha+p}^{G,\delta}(X,*), \overline{h}_{G,\delta}^{\beta+q})$$

which is again an isomorphism because $X^{\alpha+p}/X^{\alpha+p-1}$ is a wedge of spheres. □

Corollary 1.11.3 (Uniqueness of Cellular *RO*(*G*)-Graded Homology and Cohomology) *Let δ be a familial dimension function for G. Let $\tilde{h}_*^G(-)$ be a reduced $RO(G)$-graded homology theory that takes $\mathscr{F}(\delta)$-equivalences to isomorphisms and obeys a δ-dimension axiom, in the sense that*

$$\underline{h}_n^{G,\delta} = 0 \quad \textit{for integers } n \neq 0.$$

Then there is a natural isomorphism

$$\tilde{h}_*^G(-) \cong \tilde{H}_*^{G,\delta}(-;\underline{h}_0^{G,\delta})$$

of $RO(G)$-graded homology theories. Similarly, if $\tilde{h}_G^*(-)$ is a reduced $RO(G)$-graded cohomology theory that takes $\mathscr{F}(\delta)$-equivalences to isomorphisms and satisfies

$$\overline{h}_{G,\delta}^n = 0 \quad \textit{for integers } n \neq 0,$$

then there is a natural isomorphism

$$\tilde{h}_G^*(-) \cong \tilde{H}_{G,\delta}^*(-;\overline{h}_{G,\delta}^0)$$

of $RO(G)$-graded cohomology theories.

Proof There is an Atiyah-Hirezebruch spectral sequence

$$E^2_{p,q} = \tilde{H}^{G,\delta}_{\alpha+p}(X; \underline{h}^{G,\delta}_q) \Rightarrow \tilde{h}^G_{\alpha+p+q}(X).$$

If $\tilde{h}^G_*(-)$ obeys a δ-dimension axiom, then the E^2 term collapses to the line $q = 0$, giving the isomorphism. (Note that $\underline{h}^{G,\delta}_*$ involves only those subgroups in $\mathscr{F}(\delta)$.) The argument for cohomology is the same. □

We also have a universal coefficients spectral sequence. The universal coefficients in this case are the canonical projectives $\underline{A}^{G/K,\delta}$. Write $\overline{H}^{G,\delta}_\alpha(X)$ for the contravariant Mackey functor given by

$$\overline{H}^{G,\delta}_\alpha(X)(G/K,\delta) = \tilde{H}^{G,\delta}_\alpha(X; \underline{A}^{G/K,\delta}).$$

With this notation, we have

$$\overline{C}^{G,\delta}_{\alpha+n}(X,*) \cong \overline{H}^{G,\delta}_{\alpha+n}(X^{\alpha+n}/X^{\alpha+n-1})$$

for any based δ-G-CW(α) complex X. In the following, $\mathrm{Tor}^{\widehat{\mathscr{O}}_{G,\delta}}_*$ and $\mathrm{Ext}^*_{\widehat{\mathscr{O}}_{G,\delta}}$ are the derived functors of $\otimes_{\widehat{\mathscr{O}}_{G,\delta}}$ and $\mathrm{Hom}_{\widehat{\mathscr{O}}_{G,\delta}}$, respectively.

Theorem 1.11.4 (Universal Coefficients Spectral Sequence) *Let δ be a familial dimension function. If $\underline{S}$ is a covariant δ-Mackey functor and $\overline{T}$ is a contravariant δ-Mackey functor, there are spectral sequences*

$$E^2_{p,q} = \mathrm{Tor}^{\widehat{\mathscr{O}}_{G,\delta}}_p(\overline{H}^{G,\delta}_{\alpha+q}(X), \underline{S}) \Rightarrow \tilde{H}^{G,\delta}_{\alpha+p+q}(X; \underline{S})$$

and

$$E^{p,q}_2 = \mathrm{Ext}^p_{\widehat{\mathscr{O}}_{G,\delta}}(\overline{H}^{G,\delta}_{\alpha+q}(X), \overline{T}) \Rightarrow \tilde{H}^{\alpha+p+q}_{G,\delta}(X; \overline{T})$$

Proof These spectral sequences are constructed in the usual way by taking resolutions of $\underline{S}$ and $\overline{T}$ and then taking the spectral sequences of the bigraded complexes obtained from the chain complex of (an approximation to) X and the resolutions. □

1.12 The Representing Spectra

In this section we discuss the equivariant spectra representing cellular homology and cohomology of spaces. As in [47], we say that a G-spectrum E represents an $RO(G)$-graded cohomology theory $\tilde{h}^*_G(-)$ defined on based G-spaces if there is a natural isomorphism of reduced theories

$$\tilde{h}^\alpha_G(X) \cong [\Sigma^\infty_G X, \Sigma^\alpha_G E]_G,$$

where Σ^α_G is suspension by the virtual representation α (for which the usual cautions apply, see Sect. 1.3). Similarly, E represents an $RO(G)$-grade homology theory $\tilde{h}^G_*(-)$ if there is a natural isomorphism

$$\tilde{h}^G_\alpha(X) \cong [S^\alpha, E \wedge X]_G.$$

(Here and elsewhere, S^α or S^V denotes either a space or a spectrum, as appropriate from the context.) Given any $RO(G)$-graded cohomology theory $\tilde{h}^*_G(-)$, there exists a G-spectrum representing $\tilde{h}^*_G(-)$, which is unique up to non-unique stable equivalence, and similarly for homology ([47, XIII.3]).

In particular, for each familial dimension function δ and each contravariant δ-Mackey functor $\overline{T}$, there is a G-spectrum $H_\delta\overline{T}$ representing cellular cohomology, $\tilde{H}^*_{G,\delta}(-;\overline{T})$. Similarly, if $\underline{S}$ is a covariant δ-Mackey functor, there is a G-spectrum $H^\delta\underline{S}$ representing cellular homology, $\tilde{H}^{G,\delta}_*(-;\underline{S})$. (This establishes only that $H_\delta\overline{T}$ is unique up to non-unique stable equivalence. In fact, the stable equivalence is unique as well. This was shown in [47] for $H_0\overline{T}$ and we will return to this point in §1.19.) Explicit constructions of the spectrum $H\overline{T} = H_0\overline{T}$ have previously been announced or given in several places, including [41] (announcement), [13] (for finite G only), [27, 38], and [47]. Below we shall give constructions of $H_\delta\overline{T}$ and $H^\delta\underline{S}$, but first we discuss how these spectra are characterized.

By Corollary 1.11.3 and the uniqueness of representing spectra, $H_\delta\overline{T}$ is characterized up to stable equivalence by its stable homotopy, in the following sense. For any G-spectrum E, we write $\overline{\pi}^{G,\delta}_n(E)$ for the δ-Mackey functor defined by

$$\overline{\pi}^{G,\delta}_n(E)(G/H,\delta) = [G_+ \wedge_H S^{-\delta(G/H)+n}, E]_G.$$

With this notation, $H_\delta\overline{T}$ is characterized by

$$\overline{\pi}^{G,\delta}_n(H_\delta\overline{T}) \cong \begin{cases} \overline{T} & \text{if } n = 0 \\ 0 & \text{if } n \neq 0 \end{cases}$$

for integers n, together with the fact that the cohomology theory it represents takes $\mathscr{F}(\delta)$-equivalences to isomorphisms. The latter property can be expressed as the fact that the map $H_\delta\overline{T} \to F(E\mathscr{F}(\delta)_+, H_\delta\overline{T})$ is a stable equivalence, where $F(-,-)$ denotes the mapping spectrum. Note the strength of this characterization: Given any two G-spectra with these properties, there exists a stable equivalence between them. We shall call such a spectrum a *δ-Eilenberg-Mac Lane spectrum* of type $\overline{T}$.

There is a similar characterization of $H^\delta\underline{S}$, which we call a *δ-Mac Lane-Eilenberg spectrum* of type $\underline{S}$, but with some crucial changes. By Corollary 1.11.3, the relevant homotopy groups are

$$\begin{aligned}[S^n, H^\delta\underline{S} \wedge G_+ \wedge_H S^{-\delta(G/H)}]_G &\cong [D(G_+ \wedge_H S^{-\delta(G/H)}) \wedge S^n, H^\delta\underline{S}]_G \\ &\cong [G_+ \wedge_H S^{-(\mathscr{L}(G/H)-\delta(G/H))+n}, H^\delta\underline{S}]_G \\ &\cong \overline{\pi}^{G,\mathscr{L}-\delta}_n(H^\delta\underline{S})(G/H, \mathscr{L}-\delta),\end{aligned}$$

where $D(-)$ denotes the stable dual and we use that the stable dual of an orbit G/H_+ is $G_+ \wedge_H S^{-\mathscr{L}(G/H)}$ [42, II.6.3]. Thus, we require that

$$\overline{\pi}_n^{G,\mathscr{L}-\delta}(H^\delta \underline{S}) \cong \begin{cases} \underline{S} & \text{if } n = 0 \\ 0 & \text{if } n \neq 0 \end{cases}$$

(using that $\underline{S}$ may be considered a contravariant $(\mathscr{L}-\delta)$-Mackey functor). The other requirement, corresponding to the fact that the represented homology theory must take $\mathscr{F}(\delta)$-equivalences to isomorphisms, becomes the fact that $H^\delta \underline{S} \wedge E\mathscr{F}(\delta)_+ \to H^\delta \underline{S}$ is a stable equivalence.

For finite groups, there is only one possible complete dimension function, the 0 function, and, in that case, there is no distinction between Eilenberg-MacLane spectra and Mac Lane-Eilenberg spectra. In fact, we have the following.

Proposition 1.12.1 *Let G be finite and let $\overline{T}$ be a Mackey functor (defined on all orbits), which, by the self-duality of $\widehat{\mathscr{O}}_G$, we can consider as either covariant or contravariant. Then the same Eilenberg-MacLane spectrum $H\overline{T}$ represents both $H_*^G(-;\overline{T})$ and $H_G^*(-;\overline{T})$.*

Proof This follows from the self-duality of the stable orbit category and the characterizations above of the spectra representing homology and cohomology. □

When the dimension function is not complete, Example 1.12.4 will show that homology and cohomology need not be dual for finite group actions, i.e., they are not represented by the same spectrum. Further, as has been noted before, in [47, XIII.4] for example, for infinite groups it does not even make sense to expect the theories to be dual, because they use different kinds of coefficient systems—the stable orbit category is not self dual. The characterizations above establish the following result, which shows how the various spectra are related.

Theorem 1.12.2 *Let δ be a familial dimension function for G with dual $\mathscr{L}-\delta$. Let $\overline{T}$ be a contravariant δ-Mackey functor, which we may also consider as a covariant $(\mathscr{L}-\delta)$-Mackey functor. Then there is a G-equivalence*

$$H_\delta\overline{T} \wedge E\mathscr{F}(\delta)_+ \simeq H^{\mathscr{L}-\delta}\overline{T}$$

and a G-equivalence

$$H_\delta\overline{T} \simeq F(E\mathscr{F}(\delta)_+, H^{\mathscr{L}-\delta}\overline{T}).$$

□

Thus, when δ is complete (so $E\mathscr{F}(\delta)_+ \simeq S^0$), we have $H_\delta\overline{T} \simeq H^{\mathscr{L}-\delta}\overline{T}$, so $\tilde{H}_{G,\delta}^*(-;\overline{T})$ and $\tilde{H}_*^{G,\mathscr{L}-\delta}(-;\overline{T})$ are dual theories, in the sense that they are represented by the same spectrum. In particular, we have the following.

Corollary 1.12.3 *The following are pairs of dual theories.*

(1) $\tilde{H}^*_G(-;\overline{T})$ *and* $\tilde{\mathscr{H}}^G_*(-;\overline{T})$ *are both represented by* $H\overline{T} = H_0\overline{T} \simeq H^{\mathscr{L}}\overline{T}$.
(2) $\tilde{H}^G_*(-;\underline{S})$ *and* $\tilde{\mathscr{H}}^*_G(-;\underline{S})$ *are both represented by* $H\underline{S} = H_{\mathscr{L}}\underline{S} \simeq H^0\underline{S}$. □

Example 1.12.4 If δ is not complete, then, in general, $H_\delta\overline{T}$ and $H^{\mathscr{L}-\delta}\overline{T}$ need not be equivalent. Consider a simple example: Let $G = \mathbb{Z}/2$ and let δ be the dimension function that is 0 (as it must be) on $\mathscr{F}(\delta) = \{e\}$. Let $\overline{T}$ be the δ-Mackey functor that is constant at $\mathbb{Z}/2$. Then, in integer dimensions we have

$$H^*_{G,\delta}(G/G;\overline{T}) \cong H^*_{G,\delta}(EG;\overline{T}) \cong H^*(B\mathbb{Z}/2;\mathbb{Z}/2)$$

and

$$H^k_{G,\delta}(G/e;\overline{T}) \cong \begin{cases} \mathbb{Z}/2 & \text{if } k=0 \\ 0 & \text{otherwise.} \end{cases}$$

A similar calculation shows that

$$H^{G,\delta}_*(G/G;\overline{T}) \cong H_*(B\mathbb{Z}/2;\mathbb{Z}/2)$$

and

$$H^{G,\delta}_k(G/e;\overline{T}) \cong \begin{cases} \mathbb{Z}/2 & \text{if } k=0 \\ 0 & \text{otherwise,} \end{cases}$$

where we can consider $\overline{T}$ as both a contravariant and a covariant δ-Mackey functor because G is finite. Looking at the values at G/G these two theories are clearly not dual. (Note that they are, in fact, Borel homology and cohomology with $\mathbb{Z}/2$ coefficients.) This is related to the observation in [42, V.8] that there are two distinct ways to restrict cohomology theories to (pairs of) families. In the present case, $\overline{T}$ extends to a G-Mackey functor with an associated Eilenberg-MacLane spectrum $H\overline{T}$ (for example, $\overline{T} = \overline{A}_{G/G} \otimes \mathbb{Z}/2$). The spectrum representing $H^{G,\delta}_*(-;\overline{T})$ is then $H\overline{T} \wedge E\mathbb{Z}/2_+$ while the spectrum representing $H^*_{G,\delta}(-;\overline{T})$ is the function spectrum $F(E\mathbb{Z}/2_+, H\overline{T})$.

Finally, we give an elementary, functorial, explicit construction of $H_\delta\overline{T}$. This particular structure will be of use later.

Construction 1.12.5 Let δ be a familial dimension function for G and let $\overline{T}$ be a contravariant δ-Mackey functor. Define

$$F_\delta\overline{T} = \bigvee_{\overline{T}(G/H,\delta)} G_+ \wedge_H S^{-\delta(G/H)},$$

where the wedge runs over all objects $(G/H, \delta)$ in $\widehat{\mathscr{O}}_{G,\delta}$ and all elements in the group $\overline{T}(G/H, \delta)$. Then F_δ is a functor taking δ-G-Mackey functors to G-spectra and we have a natural epimorphism

$$\epsilon: \overline{\pi}_0^{G,\delta} F_\delta \overline{T} \to \overline{T}.$$

Let

$$K = \{\kappa: G_+ \wedge_{H_\kappa} S^{-\delta(G/H_\kappa)} \to F_\delta \overline{T} \mid \epsilon(\kappa) = 0\},$$

and let

$$R_\delta \overline{T} = \bigvee_{\kappa \in K} G_+ \wedge_{H_\kappa} S^{-\delta(G/H_\kappa)}.$$

Then R_δ is also a functor, there is a natural transformation $R_\delta \overline{T} \to F_\delta \overline{T}$ (given by the maps κ), and we have an exact sequence

$$\overline{\pi}_0^{G,\delta} R_\delta \overline{T} \to \overline{\pi}_0^{G,\delta} F_\delta \overline{T} \to \overline{T} \to 0.$$

It follows that, if we let $C_\delta \overline{T}$ be the cofiber of $R_\delta \overline{T} \to F_\delta \overline{T}$, then $\overline{\pi}_0^{G,\delta} C_\delta \overline{T} \cong \overline{T}$ and $\overline{\pi}_n^{G,\delta} C_\delta \overline{T} = 0$ for $n < 0$. The vanishing of the homotopy groups for $n < 0$ depends on the calculation

$$[G_+ \wedge_H S^{-\delta(G/H)+n}, G_+ \wedge_K S^{-\delta(G/K)}]_G = 0$$

for $n < 0$. This calculation follows from our being able to write this stable mapping group as a colimit of maps of spaces under suspension (using compactness) and then using the same argument as was used in the proof of Proposition 1.4.11.

We can then functorially kill all the homotopy $\overline{\pi}_n^{G,\delta} C_\delta \overline{T}$ for $n > 0$, obtaining a functor P_δ with

$$\overline{\pi}_n^{G,\delta} P_\delta \overline{T} = \begin{cases} \overline{T} & \text{if } n = 0 \\ 0 & \text{if } n \neq 0. \end{cases}$$

Finally, we let

$$H_\delta \overline{T} = F(E\mathscr{F}(\delta)_+, P_\delta \overline{T})$$

and

$$H^\delta \underline{S} = P_{\mathscr{L}-\delta} \underline{S} \wedge E\mathscr{F}(\delta)_+$$

if $\underline{S}$ is a covariant δ-Mackey functor. Then H_δ and H^δ are functors and produce spectra that satisfy the characterizations of the spectra representing cohomology and homology, respectively. □

1.13 Change of Groups

We now discuss various change-of-groups maps. We first discuss restriction to subgroups, where the underlying result is the Wirthmüller isomorphism.

1.13.1 Subgroups

Given a G-homology theory $\tilde{h}^G_*(-)$ and a subgroup K of G, $\tilde{h}^G_*(G_+ \wedge_K -)$ defines a homology theory on K-spaces, and similarly for cohomology. When we apply this construction to our cellular theories, the Wirthmüller isomorphisms identify the resulting theories as again cellular. This can be shown by calculating $\tilde{H}^*_G(G_+ \wedge_K K/L; \overline{T})$ and using the dimension axiom, but it is useful to see that the isomorphism starts at the chain level. Later we shall look at how it is represented on the spectrum level.

Continue to let K be a subgroup of G. Recall from Sect. 1.5 that there is a forgetful functor $E \mapsto E|K$ taking G-spectra to K-spectra, which has a left adjoint denoted $G_+ \wedge_K -$. This adjunction passes to the stable category to give the adjunction

$$[D, E|K]_K \cong [G_+ \wedge_K D, E]_G$$

for K-spectra D and G-spectra E. When the meaning is clear from context, we shall simply write E for $E|K$, E regarded as a K-spectrum. As mentioned in Sect. 1.5, we also have that $G_+ \wedge_K \Sigma^\infty_K X = \Sigma^\infty_G (G_+ \wedge_K X)$, relating the spectrum- and space-level functors.

On the algebraic side, suppose that δ is a dimension function for G and that $K \in \mathscr{F}(\delta)$. Write δ again for $\delta|K$ as defined in Definition 1.4.5.

Definition 1.13.1 Suppose $K \in \mathscr{F}(\delta)$.

(1) Let $i^G_K\colon \widehat{\mathscr{O}}_{K,\delta} \to \widehat{\mathscr{O}}_{G,\delta}$ be the functor $G_+ \wedge_K \Sigma^{-\delta(G/K)}(-)$. Note that

$$G_+ \wedge_K \Sigma^{-\delta(G/K)}(K_+ \wedge_L S^{-\delta(K/L)}) \cong G_+ \wedge_L S^{-\delta(G/L)},$$

using that $\delta(G/K) \oplus \delta(K/L) = \delta(G/L)$.

(2) If $\overline{T}$ is a contravariant G-δ-Mackey functor, let

$$\overline{T}|K = (i^G_K)^*\overline{T} = \overline{T} \circ i^G_K,$$

a contravariant K-δ-Mackey functor. If $\underline{S}$ is a covariant G-δ-Mackey functor, let $\underline{S}|K = (i_K^G)^*\underline{S}$ similarly. We call this operation on Mackey functors *restriction to a subgroup*.

(3) If $\overline{C}$ is a contravariant K-δ-Mackey functor, let

$$G \times_K \overline{C} = (i_K^G)_!\overline{C}$$

as in Definition 1.6.5, a contravariant G-δ-Mackey functor. If $\underline{D}$ is a covariant K-δ-Mackey functor, define $G \times_K \underline{D} = (i_K^G)_!\underline{D}$ similarly. We call this operation *induction from a subgroup*.

It follows from Proposition 1.6.6 that, if $\overline{C}$ is a contravariant K-δ-Mackey functor and $\overline{T}$ is a contravariant G-δ-Mackey functor, then

$$\mathrm{Hom}_{\widehat{\mathscr{O}}_{G,\delta}}(G \times_K \overline{C}, \overline{T}) \cong \mathrm{Hom}_{\widehat{\mathscr{O}}_{K,\delta}}(\overline{C}, \overline{T}|K).$$

If $\underline{S}$ is a covariant G-δ-Mackey functor and $\underline{D}$ is a covariant K-δ-Mackey functor, we also have the isomorphism

$$(G \times_K \overline{C}) \otimes_{\widehat{\mathscr{O}}_{G,\delta}} \underline{S} \cong \overline{C} \otimes_{\widehat{\mathscr{O}}_{K,\delta}} (\underline{S}|K).$$

Proposition 1.6.6 gives us the calculation

$$G \times_K \overline{A}_{K/J,\delta} \cong \overline{A}_{G/J,\delta}.$$

The restrictions of the functors $\overline{A}_{G/J,\delta}$ are complicated in general, but we do have the following simple special cases:

$$\overline{A}_{G/G,0}|K \cong \overline{A}_{K/K,0}$$

and

$$\underline{A}^{G/G,\mathscr{L}}|K \cong \underline{A}^{K/K,\mathscr{L}}.$$

The following is the main calculation that leads to the Wirthmüller isomorphisms.

Proposition 1.13.2 *Let δ be a dimension function for G, let $K \in \mathscr{F}(\delta)$, and let $\alpha \in RO(G)$. For notational simplicity, write α again for $\alpha|K$. Let X be a based δ-K-CW($\alpha - \delta(G/K)$) complex and give $G_+ \wedge_K X$ the δ-G-CW(α) structure from Proposition 1.4.6. Then*

$$G \times_K \overline{C}^{K,\delta}_{\alpha-\delta(G/K)+*}(X,*) \cong \overline{C}^{G,\delta}_{\alpha+*}(G_+ \wedge_K X,*).$$

This isomorphism respects suspension in the sense that, if W is a representation of G, then the following diagram commutes:

$$\begin{array}{ccc} G \times_K \overline{C}^{K,\delta}_{\alpha-\delta(G/K)+*}(X,*) & \xrightarrow{\cong} & \overline{C}^{G,\delta}_{\alpha+*}(G_+ \wedge_K X,*) \\ \downarrow{\scriptstyle \sigma^W} & & \downarrow{\scriptstyle \sigma^W} \\ G \times_K \overline{C}^{K,\delta}_{\alpha-\delta(G/K)+W+*}(\Sigma^W X,*) & \xrightarrow{\cong} & \overline{C}^{G,\delta}_{\alpha+W+*}(G_+ \wedge_K \Sigma^W X,*) \end{array}$$

Proof Geometrically, it's clear that we have a one-to-one correspondence between the cells in the two cell complexes. To see that the algebra tracks the geometry, we define a map

$$G \times_K \overline{C}^{K,\delta}_{\alpha-\delta(G/K)+*}(X,*) \to \overline{C}^{G,\delta}_{\alpha+*}(G_+ \wedge_K X,*)$$

as follows. By Definition 1.6.5,

$$\begin{aligned}&(G \times_K \overline{C}^{K,\delta}_{\alpha-\delta(G/K)+*}(X,*))(G/J,\delta)\\ &= \int^{(K/L,\delta)} \overline{C}^{K,\delta}_{\alpha-\delta(G/K)+*}(X,*)(K/L) \otimes \widehat{\mathscr{O}}_{G,\delta}(G/J, G \times_K K/L).\end{aligned}$$

We have the extension and evaluation map

$$\begin{aligned}&\overline{C}^{K,\delta}_{\alpha-\delta(G/K)+n}(X,*)(K/L) \otimes \widehat{\mathscr{O}}_{G,\delta}(G/J, G \times_K K/L)\\ &\quad \to \overline{C}^{G,\delta}_{\alpha+n}(G_+ \wedge_K X,*)(G/J)\end{aligned}$$

defined as follows. Let $c \in \overline{C}^{K,\delta}_{\alpha-\delta(G/K)+n}(X,*)(K/L)$, so c is a stable map

$$\begin{aligned}c\colon K_+ \wedge_L S^{\alpha-\delta(G/K)-\delta(K/L)+n} &= K_+ \wedge_L S^{\alpha-\delta(G/L)+n}\\ &\to \Sigma^\infty_K X^{\alpha-\delta(G/K)+n}/X^{\alpha-\delta(G/K)+n-1},\end{aligned}$$

and let $d \in \widehat{\mathscr{O}}_{G,\delta}(G/J, G \times_K K/L)$, so d is a stable map

$$d\colon G_+ \wedge_J S^{-\delta(G/J)} \to G_+ \wedge_K K_+ \wedge_L S^{-\delta(G/L)}.$$

Then the extension and evaluation map takes $c \otimes d$ to the stable map

$$\begin{aligned}&(G_+ \wedge_K c) \circ d\colon G_+ \wedge_J S^{\alpha-\delta(G/J)+n}\\ &\quad \to \Sigma^\infty_G (G_+ \wedge_K X^{\alpha-\delta(G/K)+n})/(G_+ \wedge_K X^{\alpha-\delta(G/K)+n-1}).\end{aligned}$$

As mentioned above, we can take $G_+ \wedge_K X^{\alpha-\delta(G/K)+n}$ as the $(\alpha+n)$-skeleton of a G-δ-CW(α) structure on $G_+ \wedge_K X$, i.e.,

$$G_+ \wedge_K X^{\alpha-\delta(G/K)+n} = (G_+ \wedge_K X)^{\alpha+n}.$$

Thus, $(G_+ \wedge_K c) \circ d$ defines an element of $\overline{C}^{G,\delta}_{\alpha+n}(G_+ \wedge_K X, *)(G/J)$ as claimed. The extension and evaluation maps are compatible as K/L varies, defining a map out of the coend. The isomorphism $G \times_K \overline{A}_{K/J,\delta} \cong \overline{A}_{G/J,\delta}$ then implies that the map so defined is an isomorphism

$$G \times_K \overline{C}^{K,\delta}_{\alpha-\delta(G/K)+n}(X, *) \cong \overline{C}^{G,\delta}_{\alpha+n}(G_+ \wedge_K X, *).$$

That $G_+ \wedge_K \Sigma^W X \cong \Sigma^W G_+ \wedge_K X$ implies that the isomorphism respects suspension. □

Theorem 1.13.3 (Wirthmüller Isomorphisms) *Let δ be a dimension function for G, let $K \in \mathscr{F}(\delta)$, and let $\alpha \in RO(G)$. For notational simplicity, write α again for $\alpha|K$. Then, for X in $K\mathscr{W}_*^{\delta,\alpha-\delta(G/K)}$, or for δ familial and X in $K\mathscr{K}_*$ and well-based, there are natural isomorphisms*

$$\tilde{H}^{G,\delta}_\alpha(G_+ \wedge_K X; \underline{S}) \cong \tilde{H}^{K,\delta}_{\alpha-\delta(G/K)}(X; \underline{S}|K)$$

and

$$\tilde{H}^\alpha_{G,\delta}(G_+ \wedge_K X; \overline{T}) \cong \tilde{H}^{\alpha-\delta(G/K)}_{K,\delta}(X; \overline{T}|K).$$

These isomorphisms respect suspension in the sense that, if W is a representation of G, then the following diagram commutes:

$$\begin{array}{ccc}
\tilde{H}^{G,\delta}_\alpha(G_+ \wedge_K X; \underline{S}) & \xrightarrow{\cong} & \tilde{H}^{K,\delta}_{\alpha-\delta(G/K)}(X; \underline{S}|K) \\
\sigma^W \downarrow & & \downarrow \sigma^W \\
\tilde{H}^{G,\delta}_{\alpha+W}(G_+ \wedge_K \Sigma^W X; \underline{S}) & \xrightarrow{\cong} & \tilde{H}^{K,\delta}_{\alpha+W-\delta(G/K)}(\Sigma^W X; \underline{S}|K)
\end{array}$$

and similarly for cohomology.

Proof Consider first the case where X is a K-δ-CW($\alpha - \delta(G/K)$) complex. Using the isomorphism of the preceding proposition and isomorphisms we noted earlier coming from Proposition 1.6.6, we see that there are chain isomorphisms

$$\overline{C}^{G,\delta}_{\alpha+*}(G_+ \wedge_K X, *) \otimes_{\widehat{\mathscr{O}}_{G,\delta}} \underline{S} \cong \overline{C}^{K,\delta}_{\alpha-\delta(G/K)+*}(X, *) \otimes_{\widehat{\mathscr{O}}_{K,\delta}} (\underline{S}|K)$$

and

$$\mathrm{Hom}_{\widehat{\mathcal{O}}_{G,\delta}}(\overline{C}^{G,\delta}_{\alpha+*}(G_+ \wedge_K X, *), \overline{T}) \cong \mathrm{Hom}_{\widehat{\mathcal{O}}_{K,\delta}}(\overline{C}^{K,\delta}_{\alpha-\delta(G/K)+*}(X, *), \overline{T}|K).$$

These isomorphisms induce the Wirthmüller isomorphisms. That the Wirthmüller isomorphisms respect suspension follows from the similar statement in the preceding proposition.

Now consider a general well-based K-space X. Let $\mathscr{U}$ be a complete G-universe, which will also be complete when considered as a K-universe. Let $\mathscr{V}$ be an indexing sequence in $\mathscr{U}$. Let $\Gamma\Sigma^\infty_K X \to \Sigma^\infty_K X$ be a δ-K-CW$(\alpha - \delta(G/K))$ approximation indexed on $\mathscr{V}$. We can then construct a δ-G-CW(α) approximation $\Gamma\Sigma^\infty_G X \to \Sigma^\infty_G X$ indexed on $\mathscr{V}$ such that there are inclusions

$$G_+ \wedge_K (\Gamma\Sigma^\infty_K X)(V_i) \to (\Gamma\Sigma^\infty_G(G_+ \wedge_K X))(V_i)$$

over $\Sigma^{V_i}(G_+ \wedge X)$ and compatible with the structure maps. For sufficiently large i, $(\Gamma\Sigma^\infty_K X)(V_i) \to \Sigma^{V_i}X$ is an $\mathscr{F}(\delta)$-weak equivalence of K-spaces. It follows that $G_+ \wedge_K (\Gamma\Sigma^\infty_K X)(V_i) \to \Sigma^{V_i}(G_+ \wedge_K X)$ is an $\mathscr{F}(\delta)$-weak equivalence of G-spaces for such i, hence a δ-weak$_{\alpha+V_i}$ equivalence. Using the preceding proposition, we then get a chain homotopy equivalence

$$G \times_K \overline{C}^{\mathscr{V},\delta}_{\alpha-\delta(G/K)+*}(X) \simeq \overline{C}^{\mathscr{V},\delta}_{\alpha+*}(G_+ \wedge_K X).$$

The theorem now follows as it did in the CW case. □

Recall from Definition 1.11.1 that we can consider any $RO(G)$-graded theory as G-δ-Mackey functor-valued. Applying that definition to cellular homology and cohomology we get, for $K \in \mathscr{F}(\delta)$,

$$\begin{aligned} \underline{H}^{G,\delta}_\alpha(X; \underline{S})(G/K, \delta) &= \tilde{H}^{G,\delta}_\alpha(X \wedge G_+ \wedge_K S^{-\delta(G/K)}; \underline{S}) \\ &= \tilde{H}^{G,\delta}_{\alpha+V}(X \wedge G_+ \wedge_K S^{V-\delta(G/K)}; \underline{S}) \end{aligned}$$

for V so large that $\delta(G/K) \subset V$, and

$$\overline{H}^\alpha_{G,\delta}(X; \overline{T})(G/K, \delta) = \tilde{H}^\alpha_{G,\delta}(X \wedge G_+ \wedge_K S^{-\delta(G/K)}; \overline{T})$$

similarly. We can use the Wirthmüller isomorphisms to interpret the components of these functors.

Corollary 1.13.4 *If X is a based G-space, δ is a familial dimension function for G, $\underline{S}$ is a covariant G-δ-Mackey functor, and $\overline{T}$ is a contravariant G-δ-Mackey functor, then, for $K \in \mathscr{F}(\delta)$, we have*

$$\underline{H}^{G,\delta}_\alpha(X; \underline{S})(G/K, \delta) \cong \tilde{H}^{K,\delta}_\alpha(X; \underline{S}|K)$$

and

$$\overline{H}^{\alpha}_{G,\delta}(X;\overline{T})(G/K,\delta) \cong \tilde{H}^{\alpha}_{K,\delta}(X;\overline{T}|K).$$

Proof For X well-based, this is simply a restatement of the Wirthmüller isomorphisms. Explicitly,

$$\begin{aligned} \underline{H}^{G,\delta}_{\alpha}(X;\underline{S})(G/K,\delta) &= \tilde{H}^{G,\delta}_{\alpha}(X \wedge G_+ \wedge_K S^{-\delta(G/K)};\overline{S}) \\ &= \tilde{H}^{G,\delta}_{\alpha+V}(X \wedge G_+ \wedge_K S^{V-\delta(G/K)};\overline{S}) \\ &\cong \tilde{H}^{K,\delta}_{\alpha+V-\delta(G/K)}(X \wedge S^{V-\delta(G/K)};\overline{S}|K) \\ &\cong \tilde{H}^{K,\delta}_{\alpha}(X;\overline{S}|K) \end{aligned}$$

and similarly for cohomology. If X is an arbitrary based G-space, the statement follows because we define the homology and cohomology of X to be those X_w, and the whisker construction commutes with the functor $X \mapsto X|K$. □

Note that this is *not* the structure used in Theorem 1.11.4, the universal coefficients theorem.

Another application of the Wirthmüller isomorphism, or inspection of the preceding corollary, gives the following.

Corollary 1.13.5 *Let δ be a familial dimension function for G and let $K \in \mathscr{F}(\delta)$. If X is a based G-space, $\underline{S}$ is a covariant G-δ-Mackey functor, and $\overline{T}$ is a contravariant G-δ-Mackey functor, then we have*

$$\underline{H}^{G,\delta}_{\alpha}(X;\underline{S})|K \cong \underline{H}^{K,\delta}_{\alpha}(X;\underline{S}|K)$$

and

$$\overline{H}^{\alpha}_{G,\delta}(X;\overline{T})|K \cong \overline{H}^{\alpha}_{K,\delta}(X;\overline{T}|K).$$

□

We now look at how the Wirthmüller isomorphisms are represented. The main result we need is the following.

Proposition 1.13.6 *Let δ be a familial dimension function for G, let $H_\delta\overline{T}$ be a G-δ-Eilenberg-Mac Lane spectrum of type $\overline{T}$, and let $K \in \mathscr{F}(\delta)$. Then*

$$(H_\delta\overline{T})|K \simeq \Sigma_K^{-\delta(G/K)} H_\delta(\overline{T}|K).$$

Similarly, for a covariant $\underline{S}$, we have

$$(H^\delta\underline{S})|K \simeq \Sigma_K^{-(\mathscr{L}(G/K)-\delta(G/K))} H^\delta(\underline{S}|K).$$

Proof We verify that $\Sigma_K^{\delta(G/K)}(H_\delta\overline{T})|K$ has the homotopy that characterizes $H_\delta(\overline{T}|K)$. If L is a subgroup of K we have

$$\begin{aligned}
\pi_n^{K,\delta}(\Sigma_K^{\delta(G/K)}(H_\delta\overline{T})|K)(K/L,\delta) &= [K_+ \wedge_L S^{-\delta(K/L)+n}, \Sigma_K^{\delta(G/K)}(H_\delta\overline{T})|K]_K \\
&\cong [K_+ \wedge_L S^{-\delta(K/L)-\delta(G/K)+n}, (H_\delta\overline{T})|K]_K \\
&\cong [K_+ \wedge_L S^{-\delta(G/L)+n}, (H_\delta\overline{T})|K]_K \\
&\cong [G_+ \wedge_L S^{-\delta(G/L)+n}, H_\delta\overline{T}]_G \\
&\cong \pi_n^{G,\delta}(H_\delta\overline{T})(G/L,\delta) \\
&\cong \begin{cases} \overline{T}(G/L,\delta) & \text{if } n=0 \\ 0 & \text{if } n\neq 0. \end{cases}
\end{aligned}$$

Further, $\mathscr{F}(\delta|K)$ contains all the subgroups of K because δ is familial, so the homotopy above does characterize $H_\delta(\overline{T}|K)$.

The argument for $H^\delta \underline{S}$ is similar. □

The cohomology Wirthmüller isomorphism is then the adjunction

$$\begin{aligned}
[G_+ \wedge_K \Sigma_K^\infty X, \Sigma_G^\alpha H_\delta\overline{T}]_G &\cong [\Sigma_K^\infty X, \Sigma_K^\alpha (H_\delta\overline{T})|K]_K \\
&\cong [\Sigma_K^\infty X, \Sigma_K^{\alpha-\delta(G/K)} H_\delta(\overline{T}|K)]_K,
\end{aligned}$$

using the equivalence $(H_\delta\overline{T})|K \cong \Sigma_K^{-\delta(G/K)} H_\delta(\overline{T}|K)$ shown above. For the homology isomorphism, we have

$$\begin{aligned}
[S^\alpha, H^\delta\underline{S} \wedge (G_+ \wedge_K X)]_G &\cong [S^\alpha, G_+ \wedge_K ((H^\delta\underline{S})|K \wedge X)]_G \\
&\cong [S^\alpha, \Sigma_K^{\mathscr{L}(G/K)}(H^\delta\underline{S})|K \wedge X]_K \\
&\cong [S^\alpha, \Sigma_K^{\mathscr{L}(G/K)}\Sigma_K^{-(\mathscr{L}(G/K)-\delta(G/K))} H^\delta(\underline{S}|K) \wedge X]_K \\
&\cong [S^\alpha, \Sigma_K^{\delta(G/K)} H^\delta(\underline{S}|K) \wedge X]_K \\
&\cong [S^{\alpha-\delta(G/K)}, H^\delta(\underline{S}|K) \wedge X]_K.
\end{aligned}$$

Here, the second isomorphism is shown in [42, II.6.5] and the third is again from Proposition 1.13.6. That these isomorphisms agree with the ones constructed on the chain level above follows by comparing their behavior on the filtration quotients $X^{\alpha-\delta(G/K)+n}/X^{\alpha-\delta(G/K)+n-1}$.

We now distinguish a particular map from G homology to K-homology. The definition of a "restriction to subgroups" map is clear on the represented level. Let δ be a familial dimension function for G and let $\overline{T}$ be a contravariant δ-Mackey functor. Then restriction of cohomology from G to a subgroup $K \in \mathscr{F}(\delta)$ should be the following map:

$$\begin{aligned}\tilde{H}^{\alpha}_{G,\delta}(X;\overline{T}) &\cong [\Sigma^{\infty}_G X, \Sigma^{\alpha}_G H_\delta \overline{T}]_G \\ &\to [\Sigma^{\infty}_K X, \Sigma^{\alpha}_K (H_\delta \overline{T})|K]_K \\ &\cong [\Sigma^{\infty}_K X, \Sigma^{\alpha}_K \Sigma^{-\delta(G/K)}_K H_\delta(\overline{T}|K)]_K \\ &\cong \tilde{H}^{\alpha-\delta(G/K)}_{K,\delta}(X;\overline{T}|K).\end{aligned}$$

The arrow above is restriction of G-stable maps to K-stable maps and we use Proposition 1.13.6 to identify the restriction of the Eilenberg-Mac Lane spectrum. For homology, we have the following map:

$$\begin{aligned}\tilde{H}^{G,\delta}_{\alpha}(X;\underline{S}) &\cong [S^\alpha, H^\delta \underline{S} \wedge X]_G \\ &\to [S^\alpha, (H^\delta \underline{S})|K \wedge X]_K \\ &\cong [S^\alpha, \Sigma_K^{-(\mathscr{L}(G/K)-\delta(G/K))} H^\delta(\underline{S}|K) \wedge X]_K \\ &\cong [S^{\alpha+\mathscr{L}(G/K)-\delta(G/K)}, H^\delta(\underline{S}|K) \wedge X]_K \\ &\cong \tilde{H}^{K,\delta}_{\alpha+\mathscr{L}(G/K)-\delta(G/K)}(X;\underline{S}|K).\end{aligned}$$

We write these maps as $a \mapsto a|K$ for a an element of homology or cohomology.

We can also describe restriction to subgroups entirely in terms of space-level maps. The key to doing so is noticing that the restriction map $[E,F]_G \to [E,F]_K$ factors as

$$[E,F]_G \to [G/K_+ \wedge E, F]_G \cong [E,F]_K,$$

where the first map is induced by the projection $G/K \to *$; using duality we see that it also factors as

$$[E,F]_G \to [E, D(G/K_+) \wedge F]_G \cong [E,F]_K$$

where the first map is induced by the stable map $S \to D(G/K_+)$ dual to the projection and the isomorphism is the one shown in [42, II.6.5]. (Note that, if $G \notin \mathscr{F}(\delta)$ or $\delta(G/K) \neq 0$, the projection $G/K \to *$ will not be a map in $\widehat{\mathscr{O}}_{G,\delta}$, so the restriction map we're now considering may not be one of the maps in the Mackey functor structure on $\overline{H}^*_{G,\delta}$.) With the first factorization in mind, we can see that the restriction in cohomology is given by

$$\begin{aligned}\tilde{H}^{\alpha}_{G,\delta}(X;\overline{T}) &\to \tilde{H}^{\alpha}_{G,\delta}(G/K_+ \wedge X;\overline{T}) \\ &\cong \tilde{H}^{\alpha-\delta(G/K)}_{K,\delta}(X;\overline{T}|K)\end{aligned}$$

where the first map is induced by the projection and the second map is the Wirthmüller isomorphism. Note that, if X is a based δ-G-CW(α) complex, there is, in general, no canonical cell structure on $G/K_+ \wedge X$, so we do not attempt to describe restriction on the chain level. (Put another way, if X is a based δ-G-CW(α) complex, there is no canonical δ-K-CW($\alpha - \delta(G/K)$) structure on X.)

To describe the map in homology, we need a space-level map representing the stable map $S \to D(G/K_+)$ dual to the projection $G/K \to *$. Here's the classical construction: Let V be a representation large enough that there is an embedding $G/K \subset V$. A tubular neighborhood has the form $G \times_K D(V - \mathscr{L}(G/K))$, so there is a collapse map

$$c\colon S^V \to G_+ \wedge_K S^{V-\mathscr{L}(G/K)}.$$

This represents $S \to D(G/K_+)$. The restriction in homology can now be described as follows:

$$\begin{aligned}
\tilde{H}^{G,\delta}_{\alpha}(X;\underline{S}) &\cong \tilde{H}^{G,\delta}_{\alpha+V}(\Sigma^V_G X;\underline{S}) \\
&\xrightarrow{c_*} \tilde{H}^{G,\delta}_{\alpha+V}(G_+ \wedge_K \Sigma^{V-\mathscr{L}(G/K)} X;\underline{S}) \\
&\cong \tilde{H}^{K,\delta}_{\alpha+V-\delta(G/K)}(\Sigma^{V-\mathscr{L}(G/K)} X;\underline{S}|K) \\
&\cong \tilde{H}^{K,\delta}_{\alpha+\mathscr{L}(G/K)-\delta(G/K)}(X;\underline{S}|K).
\end{aligned}$$

This again uses the Wirthmüller isomorphism.

Remark 1.13.7 The specializations of these maps to the ordinary and dual theories give us the following maps, each of which we denote by $a \mapsto a|K$:

$$\tilde{H}^G_{\alpha}(X;\underline{S}) \to \tilde{H}^K_{\alpha+\mathscr{L}(G/K)}(X;\underline{S}|K) \qquad \tilde{\mathscr{H}}^{\alpha}_G(X;\underline{S}) \to \tilde{\mathscr{H}}^{\alpha-\mathscr{L}(G/K)}_K(X;\underline{S}|K)$$

$$\tilde{H}^{\alpha}_G(X;\overline{T}) \to \tilde{H}^{\alpha}_K(X;\overline{T}|K) \qquad \tilde{\mathscr{H}}^G_{\alpha}(X;\overline{T}) \to \tilde{\mathscr{H}}^K_{\alpha}(X;\overline{T}|K)$$

1.13.2 Quotient Groups

Let N be a normal subgroup of G and let $\epsilon\colon G \to G/N$ denote the quotient map. The results of this section are essentially an elaboration of two observations: First, that, if Y is a G/N-CW complex, then it can be considered a G-CW complex via ϵ. Second, that, if X is a G-CW complex, then X^N has a natural structure as a G/N-CW complex.

We use repeatedly the fact that

$$(G/H)^N = \begin{cases} G/H & \text{if } N \le H \\ \emptyset & \text{if } N \not\le H, \end{cases}$$

which can be seen by observing that G, hence G/N, acts transitively on $(G/H)^N$, hence $(G/H)^N$ must be either empty or an orbit of G/N. In the case that G/H is fixed by N, N will act trivially on $\mathscr{L}(G/H)$, hence on $\delta(G/H)$ for any dimension function δ for G.

This calculation allows us to make the following definition.

Definition 1.13.8 Let δ be a dimension function for G and let N be a normal subgroup of G. We write $\delta|G/N$ for the dimension function on G/N defined by

$$\mathscr{F}(\delta|G/N) = \{H/N \mid N \le H \text{ and } H \in \mathscr{F}(\delta)\}$$

and

$$(\delta|G/N)((G/N)/(H/N)) = \delta(G/H).$$

We will usually write δ again for $\delta|G/N$, for simplicity of notation.

The following two results record the observations we mentioned at the beginning of the section. If Y is a G/N-space, we write $\epsilon^* Y$ for Y considered as a G-space via ϵ. (We use this notation only when we want to be very explicit; when the meaning is understood we will write Y for $\epsilon^* Y$.)

Proposition 1.13.9 *Let δ be a dimension function for G and let N be a normal subgroup of G. Then, for α a virtual representation of G/N, ϵ^* defines functors*

$$\epsilon^*\colon (G/N)\mathscr{W}^{\delta,\alpha} \to G\mathscr{W}^{\delta,\alpha}$$

and

$$\epsilon^*\colon (G/N)\mathscr{W}_*^{\delta,\alpha} \to G\mathscr{W}_*^{\delta,\alpha}.$$

Proof This follows from the fact that a G/N-δ-α-cell of the form

$$G/N \times_{H/N} \bar{D}(\alpha - \delta((G/N)/(H/N)) + n)$$

can be considered a G-δ-α-cell of the form

$$G \times_H \bar{D}(\alpha - \delta(G/H) + n).$$

If f is a cellular G/N-map, it remains cellular when considered as a G-map. □

Proposition 1.13.10 *Let δ be a dimension function for G and let N be a normal subgroup of G. Then, for α a virtual representation of G, $(-)^N$ defines functors*

$$(-)^N\colon G\mathscr{W}^{\delta,\alpha} \to (G/N)\mathscr{W}^{\delta,\alpha^N}$$

and

$$(-)^N\colon G\mathscr{W}_*^{\delta,\alpha} \to (G/N)\mathscr{W}_*^{\delta,\alpha^N}.$$

Proof Let X be a based δ-G-CW(α) complex. The fixed set of a cell in X of the form $G \times_H \bar{D}(\alpha - \delta(G/H) + n)$ with $N \leq H$ is a cell in X^N of the form $G/N \times_{H/N} \bar{D}(\alpha^N - \delta(G/H) + n)$. Any such cell in X with $N \not\leq H$ will not be fixed by N hence not appear in X^N. Thus, X^N has a canonical structure as a δ-G/N-CW(α^N) complex. If a map f is cellular, then so is f^N. □

To describe the algebra involved on chains, we describe two related functors,

$$\theta\colon \widehat{\mathscr{O}}_{G/N,\delta} \to \widehat{\mathscr{O}}_{G,\delta}$$

and

$$\Phi^N\colon \widehat{\mathscr{O}}_{G,\delta} \to \widehat{\mathscr{O}}_{G/N,\delta}.$$

Recall from [42] or [44] that there is a functor $\epsilon^\sharp$ taking G/N-spectra to G-spectra, obtained by letting G act via ϵ and then extending to a complete G-universe. This functor has a right adjoint $(-)^N$, called the *categorical* fixed point functor, and there results a stable adjunction

$$[\epsilon^\sharp D, E]_G \cong [D, E^N]_{G/N}.$$

If Y is a well-based G/N-space, then

$$\epsilon^\sharp \Sigma^\infty_{G/N} Y = \Sigma^\infty_G \epsilon^* Y.$$

However, the similar statement for N-fixed points does not hold: $(\Sigma^\infty_G X)^N \not\simeq \Sigma^\infty_{G/N} X^N$ in general in the stable category. We also have $(E \wedge D)^N \not\simeq E^N \wedge D^N$ in general, but we do have the following useful fact, not appearing elsewhere to our knowledge.

Proposition 1.13.11 *If E is a G-spectrum and A is a G/N-spectrum, then there is an equivalence*

$$E^N \wedge A \simeq (E \wedge \epsilon^\sharp A)^N$$

in the stable category.

Proof We begin by noticing that there is a stable equivalence

$$\epsilon^\sharp(A \wedge B) \simeq \epsilon^\sharp A \wedge \epsilon^\sharp B$$

for G/N-spectra A and B, which follows from the definitions and the fact that both sides preserve cofibrations and acyclic cofibrations. By adjunction, we also get an equivalence

$$F(A, E^N) \simeq F(\epsilon^\sharp A, E)^N.$$

For every spectrum X, there is a stable map $\epsilon^\sharp F(X,B) \to F(\epsilon^\sharp X, \epsilon^\sharp B)$, adjoint to the composite

$$\epsilon^\sharp F(X,B) \wedge \epsilon^\sharp X \simeq \epsilon^\sharp(F(X,B) \wedge X) \to \epsilon^\sharp B,$$

that need not be an equivalence in general. However, if X is finite (i.e., dualizable) and $B = S$, then $\epsilon^\sharp F(X,S) \simeq F(\epsilon^\sharp X, S)$ (note that $\epsilon^\sharp S = S$) because the dual of $\epsilon^\sharp X$ is equivalent to $\epsilon^\sharp DX$.

We can now define $E^N \wedge A \to (E \wedge \epsilon^\sharp A)^N$ as the composite

$$\begin{aligned} E^N \wedge A &\to [\epsilon^\sharp(E^N \wedge A)]^N \\ &\simeq [\epsilon^\sharp(E^N) \wedge \epsilon^\sharp A]^N \\ &\to (E \wedge \epsilon^\sharp A)^N, \end{aligned}$$

where the first map is the unit of the $\epsilon^\sharp$-$(-)^N$ adjunction and the last map is induced by the counit. Writing down an appropriate diagram shows that, when $A = F(X,S)$ with X (hence A) finite, this composite agrees with the composite

$$\begin{aligned} E^N \wedge F(X,S) &\simeq F(X, E^N) \\ &\simeq F(\epsilon^\sharp X, E)^N \\ &\simeq [E \wedge F(\epsilon^\sharp X, S)]^N \\ &\simeq [E \wedge \epsilon^\sharp F(X,S)]^N, \end{aligned}$$

where the first and third maps are equivalences by duality. Therefore, $E^N \wedge A \to (E \wedge \epsilon^\sharp A)^N$ is an equivalence when A is finite, so, in particular, when A is an orbit. Because both sides preserve wedges and cofibration sequences, it follows that the map is an equivalence for all cell complexes, thence for all spectra. (Put another way, the stable homotopy groups of $E^N \wedge A$ and $(E \wedge \epsilon^\sharp A)^N$ define G/N-homology theories in A that agree on orbits, hence are isomorphic.) □

Better than the categorical fixed point functor for many purposes is the so-called *geometric* fixed-point construction discussed in [42, II.9], [47, XVI.3], and [44, V.4]. The following is the quickest definition.

Definition 1.13.12 If N is a normal subgroup of G and E is a G-spectrum, we define the *geometric fixed point spectrum* of E to be the G/N-spectrum given by

$$\Phi^N(E) = (E \wedge \tilde{E}\mathscr{F}[N])^N,$$

where $\tilde{E}\mathscr{F}[N]$ is the based G-space characterized by

$$\tilde{E}\mathscr{F}[N]^L \simeq \begin{cases} S^0 & \text{if } N \leq L \\ * & \text{if } N \not\leq L. \end{cases}$$

There is another useful definition of Φ^N—we recall the definition for LMS G-spectra, the details being somewhat different for the orthogonal spectra of [44]. Consider then a G-spectrum D indexed on a sequence $\mathscr{V} = \{V_i\}$ in a complete G-universe $\mathscr{U}$. Then $\mathscr{V}^N = \{V_i^N\}$ is an indexing sequence in the complete G/N-universe $\mathscr{U}^N$. We define

$$(\Phi^N D)(V_i^N) = D(V_i)^N$$

and define the structure maps to be

$$\Sigma^{V_{i+1}^N - V_i^N} D(V_i)^N = [\Sigma^{V_{i+1}-V_i} D(V_i)]^N \to D(V_i)^N,$$

the restriction to the N-fixed points of the original structure map. The equivalence of this to Definition 1.13.12 is shown in [42, II.9.8].

The main facts we need to know about Φ^N are that we have the following natural stable equivalences for well-based G-spaces X and G-spectra E and E':

$$\Phi^N(\Sigma_G^\infty X) \simeq \Sigma_{G/N}^\infty X^N$$
$$\Phi^N(E \wedge X) \simeq \Phi^N(E) \wedge X^N$$
$$\Phi^N(E \wedge E') \simeq \Phi^N(E) \wedge \Phi^N(E')$$

In particular, Φ^N respects suspension in the sense that $\Phi^N(\Sigma^W E) \simeq \Sigma^{W^N} \Phi^N(E)$. (See [42, II.9] and [44, V.4].) Not shown in those references, but following directly from the alternate definition of Φ^N, we have the following equivalence if $N \leq L \leq G$:

$$\Phi^N(G_+ \wedge_L E) \simeq (G/N)_+ \wedge_{L/N} \Phi^N E.$$

We also have the following relationship between Φ^N and $\epsilon^\sharp$, which is [42, II.9.10]:

$$\Phi^N \epsilon^\sharp D \simeq D,$$

which can also be written as

$$(\epsilon^\sharp D \wedge \tilde{E}\mathscr{F}[N])^N \simeq D.$$

We have the following behavior of these functors on spheres:

$$\epsilon^\sharp(G/N_+ \wedge_{H/N} S^{-\delta((G/N)/(G/H))}) = G_+ \wedge_H S^{-\delta(G/H)}$$

for $N \le H \le G$, and

$$\Phi^N(G_+ \wedge_H S^{-\delta(G/H)}) \simeq \begin{cases} (G/N)_+ \wedge_{H/N} S^{-\delta(G/H)} & \text{if } N \le H \\ * & \text{if } N \not\le H. \end{cases}$$

These allow us to make the following definitions.

Definition 1.13.13 Let N be a normal subgroup of G. We write

$$\theta \colon \widehat{\mathscr{O}}_{G/N,\delta} \to \widehat{\mathscr{O}}_{G,\delta}$$

for the restriction of $\epsilon^\sharp$. As in Remark 1.6.7, we can consider $\widehat{\mathscr{O}}_{G,\delta}$ and $\widehat{\mathscr{O}}_{G/N,\delta}$ as augmented with a zero object $*$ given by the trivial spectrum. We write

$$\Phi^N \colon \widehat{\mathscr{O}}_{G,\delta} \to \widehat{\mathscr{O}}_{G/N,\delta}$$

for the restriction of Φ^N to the augmented δ-orbit category. In this context we write $((G/H)^N, \delta) \in \widehat{\mathscr{O}}_{G/N}$ for $\Phi^N(G/H, \delta)$. Note that $((G/H)^N, \delta)$ is $*$ if $N \not\le H$ and is $((G/N)/(H/N), \delta)$ if $N \le H$.

We use these functors to define operations on Mackey functors. We have the operations θ^* on G-δ-Mackey functors and $\theta_!$ on G/N-δ-Mackey functors. We give special names to the functors associated with Φ^N:

Definition 1.13.14 If $\underline{S}$ is a covariant and $\overline{T}$ a contravariant G-δ-Mackey functor, let

$$\underline{S}^N = \Phi^N_! \underline{S} \qquad \text{and}$$
$$\overline{T}^N = \Phi^N_! \overline{T}.$$

We call this the *N-fixed point functor.* If $\underline{U}$ is a covariant and $\overline{V}$ is a contravariant G/N-δ-Mackey functor, let

$$\begin{aligned}\operatorname{Inf}^G_{G/N}\underline{U} &= (\Phi^N)^*\underline{U} \qquad \text{and}\\ \operatorname{Inf}^G_{G/N}\overline{V} &= (\Phi^N)^*\overline{V}.\end{aligned}$$

We call this the *inflation functor.*

The fixed point and inflation constructions are, of course, adjoint:

$$\operatorname{Hom}_{\widehat{\mathscr{O}}_{G/N,\delta}}(\underline{S}^N, \underline{U}) \cong \operatorname{Hom}_{\widehat{\mathscr{O}}_{G,\delta}}(\underline{S}, \operatorname{Inf}^G_{G/N}\underline{U})$$

and similarly for contravariant functors.

Thévenaz and Webb gave essentially this definition of the inflation functors for finite groups in [60, §5]. They defined the left adjoint, which they wrote as $\underline{S}^+$ rather than $\underline{S}^N$, in a different way than we did above but, given the uniqueness of adjuncts, their $\underline{S}^+$ must agree with $\Phi^N_!\underline{S}$.

Note that we have $\Phi^N\theta = 1$, which gives

$$\theta^* \operatorname{Inf}^G_{G/N}\underline{U} \cong \underline{U}$$

and

$$(\theta_!\underline{S})^N \cong \underline{S},$$

and similarly for contravariant Mackey functors.

We next record the effects of $\theta_!$ and $(-)^N$ on the canonical projectives. To make explicit the varying underlying group, we write

$$\underline{A}_G^{G/H,\delta} = \widehat{\mathscr{O}}_{G,\delta}((G/H,\delta),-)$$

and

$$\overline{A}^G_{G/H,\delta} = \widehat{\mathscr{O}}_G(-,(G/H,\delta)).$$

Proposition 1.13.15 *If N is normal in G, then*

$$(\underline{A}_G^{G/H,\delta})^N \cong \begin{cases}\underline{A}_{G/N}^{G/H,\delta} & \text{if } N \le H\\ 0 & \text{if } N \not\le H\end{cases}$$

and

$$(\overline{A}^G_{G/H,\delta})^N \cong \begin{cases}\overline{A}^{G/N}_{G/H,\delta} & \text{if } N \le H\\ 0 & \text{if } N \not\le H\end{cases}$$

for any subgroup H of G. For $N \le L \le G$ we have

$$\theta_! \underline{A}^{G/L,\delta}_{G/N} \cong \underline{A}^{G/L,\delta}_{G}$$

and

$$\theta_! \overline{A}^{G/N}_{G/L,\delta} \cong \overline{A}^{G}_{G/L,\delta}.$$

Proof These are special cases of Proposition 1.6.6(5).

Our main calculations are then the following two results.

Proposition 1.13.16 *Let N be a normal subgroup of G, let δ be a dimension function for G, and let α be a virtual representation of G/N. Let Y be a based δ-G/N-CW(α) complex and give $\epsilon^* Y$ the corresponding δ-G-CW(α) structure as in Proposition 1.13.9. Then we have a natural chain isomorphism*

$$\theta_! \overline{C}^{G/N,\delta}_{\alpha+*}(Y, *) \cong \overline{C}^{G,\delta}_{\alpha+*}(Y, *),$$

where we write Y for $\epsilon^* Y$ on the right. This isomorphism respects suspension in the sense that, if W is a representation of G/N, then the following diagram commutes:

$$\begin{array}{ccc}
\theta_! \overline{C}^{G/N,\delta}_{\alpha+*}(Y, *) & \xrightarrow{\cong} & \overline{C}^{G,\delta}_{\alpha+*}(Y, *) \\
\sigma^W \downarrow & & \downarrow \sigma^W \\
\theta_! \overline{C}^{G/N,\delta}_{\alpha+W+*}(\Sigma^W Y, *) & \xrightarrow[\cong]{} & \overline{C}^{G,\delta}_{\alpha+W+*}(\Sigma^W Y, *).
\end{array}$$

Proof The isomorphism is the chain map adjoint to the map

$$\overline{C}^{G/N,\delta}_{\alpha+*}(Y, *) \to \theta^* \overline{C}^{G,\delta}_{\alpha+*}(Y, *)$$

given by

$$\begin{aligned}
\epsilon^\sharp \colon & [G/N_+ \wedge_{H/N} S^{\alpha-\delta((G/N)/(H/N))+n}, \Sigma^\infty_{G/N} Y^{\alpha+n}/Y^{\alpha+n-1}]_{G/N} \\
& \to [G_+ \wedge_H S^{\alpha-\delta(G/H)+n}, \Sigma^\infty_G Y^{\alpha+n}/Y^{\alpha+n-1}]_G
\end{aligned}$$

when $N \le H \le G$. That the adjoint is an isomorphism follows from Proposition 1.13.15. That the isomorphism respects suspension follows from the fact that $\epsilon^\sharp$ does. □

Proposition 1.13.17 *Let N be a normal subgroup of G, let δ be a dimension function for G, and let α be a virtual representation of G. Let X be a based δ-G-CW(α)*

complex and give X^N the δ-G/N-CW(α^N) structure from Proposition 1.13.10. With this structure, we have a natural chain isomorphism

$$\overline{C}^{G,\delta}_{\alpha+*}(X,*)^N \cong \overline{C}^{G/N,\delta}_{\alpha^N+*}(X^N,*).$$

This isomorphism respects suspension in the sense that, if W is a representation G, then the following diagram commutes:

$$\begin{array}{ccc} \overline{C}^{G,\delta}_{\alpha+*}(X,*)^N & \xrightarrow{\cong} & \overline{C}^{G/N,\delta}_{\alpha^N+*}(X^N,*) \\ {\scriptstyle (\sigma^W)^N}\downarrow & & \downarrow{\scriptstyle \sigma^{W^N}} \\ \overline{C}^{G,\delta}_{\alpha+W+*}(\Sigma^W X,*)^N & \xrightarrow[\cong]{} & \overline{C}^{G/N,\delta}_{\alpha^N+W^N+*}(\Sigma^{W^N} X^N,*) \end{array}$$

Proof To see the chain isomorphism, note that, for each subgroup L of G containing N, we have the map

$$\begin{aligned} \Phi^N\colon & [G_+ \wedge_L S^{\alpha-\delta(G/L)+n}, \Sigma^\infty_G X^{\alpha+n}/X^{\alpha+n-1}]_G \\ & \to [G/N_+ \wedge_{L/N} S^{\alpha^N-\delta(G/L)+n}, \Sigma^\infty_{G/N}(X^{\alpha+n}/X^{\alpha+n-1})^N]_{G/N}; \end{aligned}$$

if L does not contain N then the result of applying Φ^N is 0. This defines (the adjoint of) the map $\overline{C}^{G,\delta}_{\alpha+n}(X,*)^N \to \overline{C}^{G/N,\delta}_{\alpha^N+n}(X^N,*)$, which we see is an isomorphism using Proposition 1.13.15. That this isomorphism respects suspension follows from the fact that Φ^N does. □

Before stating the consequences for homology and cohomology we insert the following definition and lemma.

Definition 1.13.18 Let N be a normal subgroup of G and let δ be a dimension function for G. We say that δ is *N-closed* if, whenever $K \in \mathscr{F}(\delta)$, we also have $KN \in \mathscr{F}(\delta)$.

Lemma 1.13.19 *Let N be a normal subgroup of G, let δ be an N-closed familial dimension function for G, and let α be a virtual representation of G/N. If $f\colon X \to Y$ is a $(\delta|G/N)$-weak$_\alpha$ equivalence of G/N-spaces, then it is a δ-weak$_\alpha$ equivalence of G-spaces.*

Further, if all subgroups in $\mathscr{F}(\delta|G/N)$ are admissible and X and Y are well-based, then $\Sigma^W_G f$ is a δ-weak$_\alpha$ equivalence of G-spaces for any representation W of G.

Proof Suppose K is a δ-α-admissible subgroup of G. Because X and Y are fixed by N we have $f^K = f^{KN}$. To say that K is admissible is to say that $\alpha - \delta(G/K) + n$ is equivalent to an actual representation for some n, which implies that $\alpha - \delta(G/KN) + n \cong (\alpha - \delta(G/K) + n) + \delta(KN/K)$ is also equivalent to an actual representation.

Hence, KN is also admissible. By Theorem 1.4.16 we have that f^{KN} is a weak equivalence, hence f^K is, too. Applying Theorem 1.4.16 again, we conclude that f is a δ-weak$_\alpha$ equivalence of G-spaces.

If all subgroups in $\mathscr{F}(\delta|G/N)$ are δ-α-admissible, then the argument above shows that all subgroups in $\mathscr{F}(\delta)$ are δ-α-admissible as well. Thus, in this case, f^K is a weak equivalence for all $K \in \mathscr{F}(\delta)$, so $(\Sigma^W_{G}f)^K$ is a weak equivalence for all $K \in \mathscr{F}(\delta)$, hence $\Sigma^W_{G}f$ is again a δ-weak$_\alpha$ equivalence. □

Theorem 1.13.20 *Let N be a normal subgroup of G, let δ be a dimension function for G, let α be a virtual representation of G/N, let $\underline{S}$ be a covariant G-δ-Mackey functor, and let $\overline{T}$ be a contravariant G-δ-Mackey functor. Then, for Y in $(G/N)\mathscr{W}_*^{\delta,\alpha}$, or for δ N-closed and familial and Y in $(G/N)\mathscr{K}_*$, we have natural isomorphisms*

$$\tilde{H}^{G,\delta}_\alpha(Y;\underline{S}) \cong \tilde{H}^{G/N,\delta}_\alpha(Y;\theta^*\underline{S}) \qquad \text{and}$$
$$\tilde{H}_{G,\delta}^\alpha(Y;\overline{T}) \cong \tilde{H}_{G/N,\delta}^\alpha(Y;\theta^*\overline{T}).$$

These isomorphisms respect suspension in the sense that, if W is a representation of G/N and Y is well-based, then the following diagram commutes:

$$\begin{array}{ccc}
\tilde{H}^{G,\delta}_\alpha(Y;\underline{S}) & \xrightarrow{\cong} & \tilde{H}^{G/N,\delta}_\alpha(Y;\theta^*\underline{S}) \\
\downarrow{\scriptstyle\sigma^W} & & \downarrow{\scriptstyle\sigma^W} \\
\tilde{H}^{G,\delta}_{\alpha+W}(\Sigma^W Y;\underline{S}) & \xrightarrow[\cong]{} & \tilde{H}^{G/N,\delta}_{\alpha+W}(\Sigma^W Y;\theta^*\underline{S})
\end{array}$$

and similarly for cohomology.

Proof For Y in $(G/N)\mathscr{W}_*^{\delta,\alpha}$, the theorem follows from Propositions 1.13.16 and 1.6.6, which give

$$\begin{aligned}
\overline{C}^{G,\delta}_{\alpha+*}(Y) \otimes_{\widehat{\mathscr{O}}_{G,\delta}} \underline{S} &\cong \theta_!\overline{C}^{G/N,\delta}_{\alpha+*}(Y) \otimes_{\widehat{\mathscr{O}}_{G,\delta}} \underline{S} \\
&\cong \overline{C}^{G/N,\delta}_{\alpha+*}(Y) \otimes_{\widehat{\mathscr{O}}_{G/N,\delta}} \theta^*\underline{S}
\end{aligned}$$

and

$$\begin{aligned}
\mathrm{Hom}_{\widehat{\mathscr{O}}_{G,\delta}}(\overline{C}^{G,\delta}_{\alpha+*}(Y),\overline{T}) &\cong \mathrm{Hom}_{\widehat{\mathscr{O}}_{G,\delta}}(\theta_!\overline{C}^{G/N,\delta}_{\alpha+*}(Y),\overline{T}) \\
&\cong \mathrm{Hom}_{\widehat{\mathscr{O}}_{G/N,\delta}}(\overline{C}^{G/N,\delta}_{\alpha+*}(Y),\theta^*\overline{T}).
\end{aligned}$$

That the isomorphisms respect suspension follows from the similar statement in Proposition 1.13.16.

Now consider a general based G/N-space Y, which we may assume is well-based. Let $\mathscr{U}$ be a complete G-universe so that $\mathscr{U}^N$ is a complete G/N-universe. Let $\mathscr{V}$ be an indexing sequence in $\mathscr{U}$ so that $\mathscr{V}^N$ is an indexing sequence in $\mathscr{U}^N$. Let $\Gamma\Sigma^\infty_{G/N}Y \to \Sigma^\infty_{G/N}Y$ be a δ-G/N-CW(α) approximation indexed on $\mathscr{V}^N$. We can then construct a δ-G-CW(α) approximation $\Gamma\Sigma^\infty_G Y \to \Sigma^\infty_G Y$ indexed on $\mathscr{V}$ such that there are inclusions

$$\Sigma_G^{V_i-V_i^N}(\Gamma\Sigma^\infty_{G/N}Y)(V_i^N) \to (\Gamma\Sigma^\infty_G Y)(V_i)$$

over $\Sigma_G^{V_i}Y$ and compatible with the structure maps. For sufficiently large i, all subgroups in $\mathscr{F}(\delta|G/N)$ are δ-$(\alpha + V_i^N)$-admissible. It follows from the preceding lemma that

$$\Sigma_G^{V_i-V_i^N}(\Gamma\Sigma^\infty_{G/N}Y)(V_i^N) \to \Sigma_G^{V_i}Y$$

is a δ-weak$_{\alpha+V_i}$ equivalence. Using Proposition 1.13.16, we have a chain homotopy equivalence

$$\theta_! \overline{C}^{\mathscr{V}^N,\delta}_{\alpha+*}(Y) \simeq \overline{C}^{\mathscr{V},\delta}_{\alpha+*}(Y).$$

The theorem now follows as it did in the CW case. □

This theorem leads to a statement about Mackey functor-valued homology and cohomology.

Corollary 1.13.21 *Let N be a normal subgroup of G, let δ be an N-closed familial dimension function for G, let α be a virtual representation of G/N, let $\underline{S}$ be a covariant G-δ-Mackey functor, and let $\overline{T}$ be a contravariant G-δ-Mackey functor. Then, for any based G/N-space Y, we have*

$$\theta^* \underline{H}^{G,\delta}_\alpha(Y;\underline{S}) \cong \underline{H}^{G/N,\delta}_\alpha(Y;\theta^*\underline{S})$$

and

$$\theta^* \overline{H}^\alpha_{G,\delta}(Y;\overline{T}) \cong \overline{H}^\alpha_{G/N,\delta}(Y;\theta^*\overline{T}).$$

Proof By replacing Y by Y_w, we may assume that Y is well-based. If $N \le L \le G$, we have the following commutative diagram:

$$\begin{array}{ccc}
\widehat{\mathscr{O}}_{L/N,\delta} & \xrightarrow{\theta} & \widehat{\mathscr{O}}_{L,\delta} \\
{\scriptstyle i_{L/N}^{G/N}}\downarrow & & \downarrow{\scriptstyle i_L^G} \\
\widehat{\mathscr{O}}_{G/N,\delta} & \xrightarrow[\theta]{} & \widehat{\mathscr{O}}_{G,\delta}
\end{array}$$

From this we get that $\theta^*(i_L^G)^* \cong (i_{L/N}^{G/N})^*\theta^*$, so $\theta^*(\underline{S}|L) \cong (\theta^*\underline{S})|L$ and the following diagram commutes, in which we write the spectrum $S^{-\delta(G/L)}$ but implicitly take suspensions by a G/N-representation containing $\delta(G/L)$ to work entirely with spaces:

$$\begin{array}{ccc}
\tilde{H}_\alpha^{G,\delta}(Y \wedge G_+ \wedge_L S^{-\delta(G/L)}; \underline{S}) & \xrightarrow{\cong} & \tilde{H}_\alpha^{L,\delta}(Y; \underline{S}|L) \\
\cong\downarrow & & \downarrow\cong \\
\tilde{H}_\alpha^{G/N,\delta}(Y \wedge G_+ \wedge_L S^{-\delta(G/L)}; \theta^*\underline{S}) & \xrightarrow[\cong]{} & \tilde{H}_\alpha^{L/N,\delta}(Y; (\theta^*\underline{S})|L).
\end{array}$$

This shows that $\theta^*\underline{H}_\alpha^{G,\delta}(Y;\underline{S}) \cong \underline{H}_\alpha^{G/N,\delta}(Y;\theta^*\underline{S})$ and the argument for cohomology is the same. □

Theorem 1.13.22 *Let δ be a dimension function for G, let N be a normal subgroup of G, let $\underline{S}$ be a covariant G/N-δ-Mackey functor, and let $\overline{T}$ be a contravariant G/N-δ-Mackey functor. Then, for X in $G\mathscr{W}_*^{\delta,\alpha}$, or for δ familial and X in $G\mathscr{K}_*$, we have natural isomorphisms*

$$\tilde{H}_\alpha^{G,\delta}(X; \mathrm{Inf}_{G/N}^G \underline{S}) \cong \tilde{H}_{\alpha^N}^{G/N,\delta}(X^N; \underline{S}) \qquad \textit{and}$$
$$\tilde{H}^\alpha_{G,\delta}(X; \mathrm{Inf}_{G/N}^G \overline{T}) \cong \tilde{H}^{\alpha^N}_{G/N,\delta}(X^N; \overline{T}).$$

These isomorphisms respect suspension in the sense that, if W is a representation of G and X is well-based, then the following diagram commutes:

$$\begin{array}{ccc}
\tilde{H}_\alpha^{G,\delta}(X; \mathrm{Inf}_{G/N}^G \underline{S}) & \xrightarrow{\cong} & \tilde{H}_{\alpha^N}^{G/N,\delta}(X^N; \underline{S}) \\
{\scriptstyle \sigma^W}\downarrow & & \downarrow{\scriptstyle \sigma^{W^N}} \\
\tilde{H}_{\alpha+W}^{G,\delta}(\Sigma^W X; \mathrm{Inf}_{G/N}^G \underline{S}) & \xrightarrow[\cong]{} & \tilde{H}_{\alpha^N+W^N}^{G/N,\delta}(\Sigma^{W^N} X^N; \underline{S})
\end{array}$$

and similarly for cohomology.

Proof For X in $G\mathscr{W}_*^{\delta,\alpha}$, the theorem follows from the following isomorphisms from Propositions 1.6.6 and 1.13.17:

$$\begin{aligned}\overline{C}^{G,\delta}_{\alpha+*}(X)\otimes_{\widehat{\mathscr{O}}_{G,\delta}}\mathrm{Inf}^G_{G/N}\underline{S} &\cong \overline{C}^{G,\delta}_{\alpha+*}(X)^N\otimes_{\widehat{\mathscr{O}}_{G/N,\delta}}\underline{S}\\ &\cong \overline{C}^{G/N,\delta}_{\alpha^N+*}(X^N)\otimes_{\widehat{\mathscr{O}}_{G/N,\delta}}\underline{S}\end{aligned}$$

and

$$\begin{aligned}\mathrm{Hom}_{\widehat{\mathscr{O}}_{G,\delta}}(\overline{C}^{G,\delta}_{\alpha+*}(X),\mathrm{Inf}^G_{G/N}\overline{T}) &\cong \mathrm{Hom}_{\widehat{\mathscr{O}}_{G/N,\delta}}(\overline{C}^{G,\delta}_{\alpha+*}(X)^N,\overline{T})\\ &\cong \mathrm{Hom}_{\widehat{\mathscr{O}}_{G/N,\delta}}(\overline{C}^{G/N,\delta}_{\alpha^N+*}(X^N),\overline{T}).\end{aligned}$$

That the isomorphisms respect suspension follows from the similar statement in Proposition 1.13.17.

Now consider a general based G-space X, which we may assume is well-based. Let $\mathscr{V}$ be an indexing sequence in a complete G-universe and let $\Gamma\Sigma^\infty_G X\to\Sigma^\infty_G X$ be a δ-G-CW(α) approximation. Recall that $\mathscr{V}^N$ is an indexing sequence in the complete G/N-universe $\mathscr{U}^N$. We can therefore form the δ-G/N-CW(α^N) spectrum $\Phi^N(\Gamma\Sigma^\infty_G X)$ indexed on $\mathscr{V}^N$ given by

$$\Phi^N(\Gamma\Sigma^\infty_G X)(V_i^N)=[(\Gamma\Sigma^\infty_G X)(V_i)]^N,$$

which comes with a map $\Phi^N(\Gamma\Sigma^\infty_G X)\to\Sigma^\infty_{G/N}X^N$. This map may not be a δ-G-CW(α) approximation because taking N-fixed points does not necessarily preserve δ-weak$_\alpha$ equivalence. However, for fixed α, $(-)^N$ does preserve δ-weak$_{\alpha+V}$ equivalence for sufficiently large V. Therefore, if $\Gamma\Sigma^\infty_{G/N}X^N\to\Sigma^\infty_{G/N}X^N$ is a δ-G/N-CW(α^N) approximation, the induced map $\Phi^N(\Gamma\Sigma^\infty_G X)\to\Gamma\Sigma^\infty_{G/N}X^N$ is a δ-weak$_{\alpha+V_i^N}$ equivalence at level i for sufficiently large i. Using Proposition 1.13.17, we see that the map induces a chain homotopy equivalence

$$\overline{C}^{\mathscr{V},\delta}_{\alpha+*}(X)^N\simeq\overline{C}^{\mathscr{V}^N,\delta}_{\alpha^N+*}(X^N).$$

The theorem now follows as it did in the CW case. □

Again, we have a corresponding statement about Mackey functor-valued homology and cohomology.

Corollary 1.13.23 *Let N be a normal subgroup of G, let δ be a familial dimension function for G, let α be a virtual representation of G, let $\underline{S}$ be a covariant G/N-δ-Mackey functor, and let $\overline{T}$ be a contravariant G/N-δ-Mackey functor. Then, for X any based G-space, we have*

$$\mathrm{Inf}^G_{G/N}\underline{H}^{G/N,\delta}_{\alpha^N}(X^N;\underline{S})\cong\underline{H}^{G,\delta}_{\alpha}(X;\mathrm{Inf}^G_{G/N}\underline{S})$$

and

$$\mathrm{Inf}^G_{G/N}\,\overline{H}^{\alpha^N}_{G/N,\delta}(X^N;\overline{T}) \cong \overline{H}^{\alpha}_{G,\delta}(X;\mathrm{Inf}^G_{G/N}\,\overline{T}).$$

Proof We may assume that X is well-based. If $N \le L \le G$, we have the following commutative diagram:

$$\begin{array}{ccc} \widehat{\mathscr{O}}_{L,\delta} & \xrightarrow{\Phi^N} & \widehat{\mathscr{O}}_{L/N,\delta} \\ {\scriptstyle i_L^G}\big\downarrow & & \big\downarrow{\scriptstyle i_{L/N}^{G/N}} \\ \widehat{\mathscr{O}}_{G,\delta} & \xrightarrow[\Phi^N]{} & \widehat{\mathscr{O}}_{G/N,\delta} \end{array}$$

From this we get that $(\Phi^N)^*(i^{G/N}_{L/N})^* \cong (i^G_L)^*(\Phi^N)^*$, so

$$\mathrm{Inf}^L_{L/N}(\underline{S}|L/N) \cong (\mathrm{Inf}^G_{G/N}\,\underline{S})|L$$

and the following diagram commutes:

$$\begin{array}{ccc} \tilde{H}^{G/N,\delta}_{\alpha^N}(X^N \wedge G_+ \wedge_L S^{-\delta(G/L)};\underline{S}) & \xrightarrow{\cong} & \tilde{H}^{L/N,\delta}_{\alpha^N}(X^B;\underline{S}|L/N) \\ \cong\big\downarrow & & \big\downarrow\cong \\ \tilde{H}^{G,\delta}_{\alpha}(X \wedge G_+ \wedge_L S^{-\delta(G/L)};\mathrm{Inf}^G_{G/N}\underline{S}) & \xrightarrow[\cong]{} & \tilde{H}^{L,\delta}_{\alpha}(X;(\mathrm{Inf}^G_{G/N}\underline{S})|L). \end{array}$$

This shows that $\mathrm{Inf}^G_{G/N}\,\underline{H}^{G/N,\delta}_{\alpha^N}(X^N;\underline{S}) \cong \underline{H}^{G,\delta}_{\alpha}(X;\mathrm{Inf}^G_{G/N}\,\underline{S})$ and the argument for cohomology is the same. □

Remarks 1.13.24

(1) The isomorphisms in Theorems 1.13.20 and 1.13.22 are compatible, in the sense that the composite of the following isomorphisms is the identity if X is a G/N space, α is a representation of G/N, and $\overline{T}$ is a G/N-δ-Mackey functor:

$$\begin{aligned} \tilde{H}^{G/N,\delta}_{\alpha}(X;\overline{T}) &= \tilde{H}^{G/N,\delta}_{\alpha^N}(X^N;\overline{T}) \\ &\cong \tilde{H}^{G,\delta}_{\alpha}(X;\mathrm{Inf}^G_{G/N}\,\overline{T}) \\ &\cong \tilde{H}^{G/N,\delta}_{\alpha}(X;\theta^*\,\mathrm{Inf}^G_{G/N}\,\overline{T}) \\ &\cong \tilde{H}^{G/N,\delta}_{\alpha}(X;\overline{T}). \end{aligned}$$

That the composite is the identity follows from considering the chain-level isomorphisms. On that level we are simply composing the $(\Phi^N)_!$-$(\Phi^N)^*$ adjunction and the $\theta_!$-θ^* adjunction, which gives the $(\Phi^N\theta)_!$-$(\Phi^N\theta)^*$ adjunction, which is the identity. The similar statement for homology is also true.

(2) The two isomorphisms combine to give a third isomorphism, if X is a G-space and $\overline{T}$ is a G/N-δ-Mackey functor:

$$\begin{aligned}\tilde{H}^G_\alpha(X; \operatorname{Inf}^G_{G/N}\overline{T}) &\cong \tilde{H}^{G/N,\delta}_{\alpha^N}(X^N; \overline{T})\\ &\cong \tilde{H}^{G/N,\delta}_{\alpha^N}(X^N; \theta^* \operatorname{Inf}^G_{G/N}\overline{T})\\ &\cong \tilde{H}^{G,\delta}_{\alpha^N}(X^N; \operatorname{Inf}^G_{G/N}\overline{T}).\end{aligned}$$

We get a similar isomorphism in homology.

Using Theorems 1.13.20 and 1.13.22 we can define induction from G/N to G and restriction to fixed sets.

Definition 1.13.25 Let δ be an N-closed familial dimension function for G, let α be a virtual representation of G/N, let $\underline{S}$ be a covariant G/N-δ-Mackey functor, and let $\overline{T}$ be a contravariant G/N-δ-Mackey functor. If Y is a based G/N-space, we define *induction from G/N to G* to be the composites

$$\epsilon^*: \tilde{H}^{G/N,\delta}_\alpha(Y; \underline{S}) \to \tilde{H}^{G/N,\delta}_\alpha(Y; \theta^*\theta_!\underline{S}) \cong \tilde{H}^{G,\delta}_\alpha(Y; \theta_!\underline{S})$$

and

$$\epsilon^*: \tilde{H}_{G/N,\delta}^\alpha(Y; \overline{T}) \to \tilde{H}_{G/N,\delta}^\alpha(Y; \theta^*\theta_!\overline{T}) \cong \tilde{H}_{G,\delta}^\alpha(Y; \theta_!\overline{T}).$$

The first map in each case is induced by the unit of the $\theta_!$-θ^* adjunction.

Definition 1.13.26 Let δ be a familial dimension function for G, let α be a virtual representation of G, let $\underline{S}$ be a covariant G-δ-Mackey functor, and let $\overline{T}$ be a contravariant G-δ-Mackey functor. If X is a based G-space, define *restriction to the N-fixed set* to be the composites

$$(-)^N: \tilde{H}^{G,\delta}_\alpha(X; \underline{S}) \to \tilde{H}^{G,\delta}_\alpha(X; \operatorname{Inf}^G_{G/N}\underline{S}^N) \xrightarrow{\cong} \tilde{H}^{G/N,\delta}_{\alpha^N}(X^N; \underline{S}^N)$$

and

$$(-)^N: \tilde{H}_{G,\delta}^\alpha(X; \overline{T}) \to \tilde{H}_{G,\delta}^\alpha(X; \operatorname{Inf}^G_{G/N}\overline{T}^N) \xrightarrow{\cong} \tilde{H}_{G/N,\delta}^{\alpha^N}(X^N; \overline{T}^N).$$

The first map in each case is induced by the unit of the $(-)^N$-$\operatorname{Inf}^G_{G/N}$ adjunction.

The reader should check that if, for example, $\underline{S} = \operatorname{Inf}^G_{G/N}\underline{U}$, then the composite

$$\tilde{H}^{G,\delta}_\alpha(X; \operatorname{Inf}^G_{G/N}\underline{U}) \to \tilde{H}^{G/N,\delta}_{\alpha^N}(X^N; (\operatorname{Inf}^G_{G/N}\underline{U})^N) \to \tilde{H}^{G/N,\delta}_{\alpha^N}(X^N; \underline{U})$$

agrees with the isomorphism of Theorem 1.13.22.

It follows from Theorem 1.13.20 that induction respects suspension, in the sense that, if X is well-based and W is a representation of G/N, then the following diagram commutes, as does the similar one for cohomology:

$$\begin{array}{ccc} \tilde{H}^{G/N,\delta}_{\alpha}(X;\underline{S}) & \xrightarrow{\epsilon^*} & \tilde{H}^{G,\delta}_{\alpha}(X;\theta_!\underline{S}) \\ {\scriptstyle\sigma^W}\big\downarrow & & \big\downarrow{\scriptstyle\sigma^W} \\ \tilde{H}^{G/N,\delta}_{\alpha+W}(\Sigma^W X;\underline{S}) & \xrightarrow{\epsilon^*} & \tilde{H}^{G,\delta}_{\alpha+W}(\Sigma^W X;\theta_!\underline{S}) \end{array}$$

It follows from Theorem 1.13.22 that restriction to fixed sets respects suspension, in the sense that, if X is well-based, the following diagram commutes, as does the similar one for cohomology:

$$\begin{array}{ccc} \tilde{H}^{G,\delta}_{\alpha}(X;\underline{S}) & \xrightarrow{(-)^N} & \tilde{H}^{G/N,\delta}_{\alpha^N}(X^N;\underline{S}^N) \\ {\scriptstyle\sigma^W}\big\downarrow & & \big\downarrow{\scriptstyle\sigma^{W^N}} \\ \tilde{H}^{G,\delta}_{\alpha+W}(\Sigma^W X;\underline{S}) & \xrightarrow{(-)^N} & \tilde{H}^{G/N,\delta}_{\alpha^N+W^N}(\Sigma^{W^N} X^N;\underline{S}^N) \end{array}$$

We also have the following relationship between induction and restriction to fixed sets.

Proposition 1.13.27 *Let δ be an N-closed familial dimension function for G, let α be a virtual representation of G/N, let $\underline{S}$ be a covariant G/N-δ-Mackey functor, and let $\overline{T}$ be a contravariant G/N-δ-Mackey functor. Then, if Y is a G/N-space, each of the following composites is the identity:*

$$\tilde{H}^{G/N,\delta}_{\alpha}(Y;\underline{S}) \xrightarrow{\epsilon^*} \tilde{H}^{G,\delta}_{\alpha}(Y;\theta_!\underline{S}) \xrightarrow{(-)^N} \tilde{H}^{G/N,\delta}_{\alpha}(Y;(\theta_!\underline{S})^N) = \tilde{H}^{G/N,\delta}_{\alpha}(Y;\underline{S})$$

and

$$\tilde{H}_{G/N,\delta}^{\alpha}(Y;\overline{T}) \xrightarrow{\epsilon^*} \tilde{H}_{G,\delta}^{\alpha}(Y;\theta_!\overline{T}) \xrightarrow{(-)^N} \tilde{H}_{G/N,\delta}^{\alpha}(Y;(\theta_!\overline{T})^N) = \tilde{H}_{G/N,\delta}^{\alpha}(Y;\overline{T})$$

Proof As in Remarks 1.13.24, this comes down to the fact that the $(\Phi^N\theta)_!$-$(\Phi^N\theta)^*$ adjunction is the identity. □

Now we look at how the isomorphisms of Theorems 1.13.20 and 1.13.22 are represented on the spectrum level.

Proposition 1.13.28 *Let N be a normal subgroup of G, let δ be an N-closed familial dimension function for G, let $\underline{S}$ be a covariant G-δ-Mackey functor and let $\overline{T}$ be a contravariant G-δ-Mackey functor. Then*

$$(H^{\delta}\underline{S})^N \simeq H^{\delta}(\theta^*\underline{S})$$

and

$$(H_\delta\overline{T})^N \simeq H_\delta(\theta^*\overline{T}).$$

Proof Consider $(H^\delta\underline{S})^N$ first. For $K/N \in \mathscr{F}(\delta|G/N)$ we have

$$\begin{aligned}
&\pi_n^{G/N,\mathscr{L}-\delta}(G/K, \mathscr{L}-\delta)((H^\delta\underline{S})^N) \\
&= [G_+ \wedge_K S^{-(\mathscr{L}(G/K)-\delta(G/K))+n}, (H^\delta\underline{S})^N]_{G/N} \\
&\cong [G_+ \wedge_K S^{-(\mathscr{L}(G/K)-\delta(G/K))+n}, H^\delta\underline{S}]_G \\
&\cong \begin{cases} \underline{S}(G/K,\delta) & \text{if } n=0 \\ 0 & \text{if } n \neq 0 \end{cases}
\end{aligned}$$

Thus, we have

$$\pi_n^{G/N,\mathscr{L}-\delta}((H^\delta\underline{S})^N) \cong \begin{cases} \theta^*\underline{S} & \text{if } n=0 \\ 0 & \text{if } n\neq 0. \end{cases}$$

We also have the equivalences

$$\begin{aligned}
(H^\delta\underline{S})^N \wedge E\mathscr{F}(\delta|G/N)_+ &\simeq (H^\delta\underline{S} \wedge \epsilon^* E\mathscr{F}(\delta|G/N)_+)^N \\
&\simeq (H^\delta\underline{S} \wedge E\mathscr{F}(\delta)_+ \wedge \epsilon^* E\mathscr{F}(\delta|G/N)_+)^N \\
&\simeq (H^\delta\underline{S} \wedge E\mathscr{F}(\delta)_+)^N \\
&\simeq (H^\delta\underline{S})^N.
\end{aligned}$$

The G-homotopy equivalence $E\mathscr{F}(\delta)_+ \wedge \epsilon^* E\mathscr{F}(\delta|G/N)_+ \simeq E\mathscr{F}(\delta)_+$ follows because, if $K \in \mathscr{F}(\delta)$, then

$$(\epsilon^* E\mathscr{F}(\delta|G/N)_+)^K = (\epsilon^* E\mathscr{F}(\delta|G/N)_+)^{KN} \simeq S^0,$$

using the assumption that δ is N-closed. So, we have verified the conditions that characterize the Eilenberg-Mac Lane spectrum and conclude that

$$(H^\delta\underline{S})^N \simeq H^\delta(\theta^*\underline{S}).$$

Now consider $(H_\delta\overline{T})^N$. The computation of the homotopy groups is the same and we get that

$$\overline{\pi}_n^{G/N,\delta}((H_\delta\overline{T})^N) \cong \begin{cases} \theta^*\overline{T} & \text{if } n=0 \\ 0 & \text{if } n\neq 0. \end{cases}$$

On the other hand, we have

$$\begin{aligned} F(E\mathscr{F}(\delta|G/N)_+, (H_\delta\overline{T})^N) &\simeq F(\epsilon^* E\mathscr{F}(\delta|G/N)_+, H_\delta\overline{T})^N \\ &\simeq F(\epsilon^* E\mathscr{F}(\delta|G/N)_+ \wedge E\mathscr{F}(\delta)_+, H_\delta\overline{T})^N \\ &\simeq F(E\mathscr{F}(\delta)_+, H_\delta\overline{T})^N \\ &\simeq (H_\delta\overline{T})^N. \end{aligned}$$

From the characterization of the Eilenberg-Mac Lane spectrum we now conclude that

$$(H_\delta\overline{T})^N \simeq H_\delta(\theta^*\overline{T}).$$

□

The isomorphisms of Theorem 1.13.20 are represented as follows, for a well-based G/N-space Y and a virtual representation α of G/N:

$$\begin{aligned} [S^\alpha, H^\delta\underline{S} \wedge Y]_G &\cong [S^\alpha, (H^\delta\underline{S} \wedge Y)^N]_{G/N} \\ &\cong [S^\alpha, (H^\delta\underline{S})^N \wedge Y]_{G/N} \\ &\cong [S^\alpha, H^\delta(\theta^*\underline{S}) \wedge Y]_{G/N} \end{aligned}$$

for homology, and

$$\begin{aligned} [\Sigma^\infty Y, \Sigma^\alpha H_\delta\overline{T}]_G &\cong [\Sigma^\infty Y, (\Sigma^\alpha H_\delta\overline{T})^N]_{G/N} \\ &\cong [\Sigma^\infty Y, \Sigma^\alpha (H_\delta\overline{T})^N]_{G/N} \\ &\cong [\Sigma^\infty Y, \Sigma^\alpha H_\delta(\theta^*\overline{T})]_{G/N} \end{aligned}$$

for cohomology.

Proposition 1.13.29 *Let δ be a familial dimension function for G, let N be a normal subgroup of G, let $\overline{T}$ be a contravariant G/N-δ-Mackey functor, and let $\underline{S}$ be a covariant G/N-δ-Mackey functor. Then*

$$\epsilon^\sharp H^\delta\underline{S} \wedge \tilde{E}\mathscr{F}[N] \simeq H^\delta \operatorname{Inf}^G_{G/N}\underline{S}$$

and

$$\epsilon^\sharp H_\delta\overline{T} \wedge \tilde{E}\mathscr{F}[N] \simeq H_\delta \operatorname{Inf}^G_{G/N}\overline{T}.$$

Therefore,

$$(H^\delta \operatorname{Inf}^G_{G/N}\underline{S})^N \simeq \Phi^N H^\delta \operatorname{Inf}^G_{G/N}\underline{S} \simeq H^\delta\underline{S}$$

and

$$(H_\delta \operatorname{Inf}_{G/N}^G \overline{T})^N \simeq \Phi^N H_\delta \operatorname{Inf}_{G/N}^G \overline{T} \simeq H_\delta \overline{T}.$$

Proof First consider $\epsilon^\sharp H^\delta \underline{S} \wedge \tilde{E}\mathscr{F}[N]$. We know that

$$H^\delta \underline{S} \simeq H^\delta \underline{S} \wedge E\mathscr{F}(\delta|G/N)_+,$$

so

$$\epsilon^\sharp H^\delta \underline{S} \simeq \epsilon^\sharp H^\delta \underline{S} \wedge \epsilon^* E\mathscr{F}(\delta|G/N)_+.$$

Comparison of fixed points shows that $\epsilon^* E\mathscr{F}(\delta|G/N)_+ \wedge \tilde{E}\mathscr{F}[N] \simeq \tilde{E}\mathscr{F}[N] \wedge E\mathscr{F}(\delta)_+$, so

$$\epsilon^\sharp H^\delta \underline{S} \wedge \tilde{E}\mathscr{F}[N] \simeq \epsilon^\sharp H^\delta \underline{S} \wedge \tilde{E}\mathscr{F}[N] \wedge E\mathscr{F}(\delta)_+.$$

If K does not contain N, then $\tilde{E}\mathscr{F}[N]$ is contractible as a K-space, hence

$$\pi_n^{G,\mathscr{L}-\delta}(\epsilon^\sharp H^\delta \underline{S} \wedge \tilde{E}\mathscr{F}[N])(G/K, \mathscr{L} - \delta) = 0$$

for $K \in \mathscr{F}(\delta)$ not containing N. On the other hand, if $N \leq K$ and $K \in \mathscr{F}(\delta)$, we have

$$\begin{aligned}
&\pi_n^{G,\mathscr{L}-\delta}(\epsilon^\sharp H^\delta \underline{S} \wedge \tilde{E}\mathscr{F}[N])(G/K, \mathscr{L} - \delta) \\
&= [G_+ \wedge_K S^{-(\mathscr{L}(G/K)-\delta(G/K))+n}, \epsilon^\sharp H^\delta \underline{S} \wedge \tilde{E}\mathscr{F}[N]]_G \\
&\cong [G_+ \wedge_K S^{-(\mathscr{L}(G/K)-\delta(G/K))+n}, (\epsilon^\sharp H^\delta \underline{S} \wedge \tilde{E}\mathscr{F}[N])^N]_{G/N} \\
&\cong [G_+ \wedge_K S^{-(\mathscr{L}(G/K)-\delta(G/K))+n}, H^\delta \underline{S}]_{G/N} \\
&\cong \begin{cases} \underline{S}(G/K, \delta) & \text{if } n = 0 \\ 0 & \text{if } n \neq 0. \end{cases}
\end{aligned}$$

Therefore,

$$\pi_n^{G,\mathscr{L}-\delta}(\epsilon^\sharp H^\delta \underline{S} \wedge \tilde{E}\mathscr{F}[N]) \cong \begin{cases} \operatorname{Inf}_{G/N}^G \underline{S} & \text{if } n = 0 \\ 0 & \text{if } n \neq 0, \end{cases}$$

and we have verified the conditions that characterize the Eilenberg-MacLane spectrum, so

$$\epsilon^\sharp H^\delta \underline{S} \wedge \tilde{E}\mathscr{F}[N] \simeq H^\delta \operatorname{Inf}_{G/N}^G \underline{S}.$$

Turning to $\epsilon^\sharp H_\delta \overline{T} \wedge \tilde{E}\mathscr{F}[N]$, we calculate exactly as above that

$$\pi_n^{G,\delta}(\epsilon^\sharp H_\delta \overline{T} \wedge \tilde{E}\mathscr{F}[N]) \cong \begin{cases} \mathrm{Inf}_{G/N}^G \overline{T} & \text{if } n = 0 \\ 0 & \text{if } n \neq 0, \end{cases}$$

Further, for an arbitrary G-spectrum E we have the following isomorphisms, which follow from ones we've shown previously together with [42, II.9.2] and [42, II.9.6]:

$$\begin{aligned}
&[E \wedge E\mathscr{F}(\delta)_+, \epsilon^\sharp H_\delta \overline{T} \wedge \tilde{E}\mathscr{F}[N]]_G \\
&\quad \cong [E \wedge E\mathscr{F}(\delta)_+ \wedge \tilde{E}\mathscr{F}[N], \epsilon^\sharp H_\delta \overline{T} \wedge \tilde{E}\mathscr{F}[N]]_G \\
&\quad \cong [E \wedge \epsilon^* E\mathscr{F}(\delta|G/N)_+ \wedge \tilde{E}\mathscr{F}[N], \epsilon^\sharp H_\delta \overline{T} \wedge \tilde{E}\mathscr{F}[N]]_G \\
&\quad \cong [\Phi^N(E \wedge \epsilon^* E\mathscr{F}(\delta|G/N)_+), \Phi^N \epsilon^\sharp H_\delta \overline{T}]_{G/N} \\
&\quad \cong [\Phi^N E \wedge E\mathscr{F}(\delta|G/N)_+, H_\delta \overline{T}]_{G/N} \\
&\quad \cong [\Phi^N E, H_\delta \overline{T}]_{G/N} \\
&\quad \cong [E, \epsilon^\sharp H_\delta \overline{T} \wedge \tilde{E}\mathscr{F}[N]]_G
\end{aligned}$$

This shows that

$$\epsilon^\sharp H_\delta \overline{T} \wedge \tilde{E}\mathscr{F}[N] \simeq F(E\mathscr{F}(\delta)_+, \epsilon^\sharp H_\delta \overline{T} \wedge \tilde{E}\mathscr{F}[N]),$$

hence we've verified the conditions that characterize the Eilenberg-Mac Lane spectrum, so

$$\epsilon^\sharp H_\delta \overline{T} \wedge \tilde{E}\mathscr{F}[N] \simeq H_\delta \,\mathrm{Inf}_{G/N}^G \overline{T}.$$

The last statement of the proposition follows on taking the N-fixed points of the equivalences already shown. □

The isomorphisms of Theorem 1.13.22 are then represented as

$$\begin{aligned}
[S^\alpha, H^\delta \,\mathrm{Inf}_{G/N}^G \underline{S} \wedge X]_G &\cong [\Phi^N(S^\alpha), \Phi^N(H^\delta \,\mathrm{Inf}_{G/N}^G \underline{S} \wedge X)]_{G/N} \\
&\cong [S^{\alpha^N}, \Phi^N H^\delta \,\mathrm{Inf}_{G/N}^G \underline{S} \wedge X^N]_{G/N} \\
&\cong [S^{\alpha^N}, H^\delta \underline{S} \wedge X^N]_{G/N}
\end{aligned}$$

and

$$\begin{aligned}
[\Sigma_G^\infty X, \Sigma^\alpha H_\delta \,\mathrm{Inf}_{G/N}^G \overline{T}]_G &\cong [\Phi^N(\Sigma_G^\infty X), \Phi^N(\Sigma^\alpha H_\delta \,\mathrm{Inf}_{G/N}^G \overline{T})]_{G/N} \\
&\cong [\Sigma_{G/N}^\infty X^N, \Sigma^{\alpha^N} H_\delta \overline{T}]_{G/N}.
\end{aligned}$$

The first isomorphism in each case follows because $H_\delta \operatorname{Inf}_{G/N}^G \overline{T}$, for example, is concentrated over N in the language of [42, §II.9].

Finally, a word about taking K-fixed sets if K is not normal in G. The natural way to do this is first to restrict from G to NK (assuming $NK \in \mathscr{F}(\delta)$) and then take K fixed points, defining $a^K = (a|NK)^K$. In cohomology, for example, this gives

$$(-)^K : \tilde{H}^\alpha_{G,\delta}(X;\overline{T}) \to \tilde{H}^{\alpha^K - \delta(G/NK)^K}_{WK,\delta}(X^K;(\overline{T}|NK)^K),$$

where $WK = NK/K$. However, because $(G/K)^K = NK/K$, we have that $\mathscr{L}(G/NK)^K = 0$ and so also $\delta(G/NK)^K = 0$. Therefore, the possible shift in dimensions goes away and restriction to fixed sets defines maps

$$(-)^K : \tilde{H}^{G,\delta}_\alpha(X;\underline{S}) \to \tilde{H}^{WK,\delta}_{\alpha^K}(X^K;\underline{S}^K) \qquad \text{and}$$
$$(-)^K : \tilde{H}^\alpha_{G,\delta}(X;\overline{T}) \to \tilde{H}^{\alpha^K}_{WK,\delta}(X^K;\overline{T}^K),$$

where we write $\underline{S}^K = (\underline{S}|NK)^K$ and similarly for $\overline{T}^K$.

1.13.3 Subgroups of Quotient Groups

We now look at how induction and restriction to fixed sets interact with restriction to subgroups. Consider restriction to fixed sets, for example. Suppose that N is a normal subgroup of G, $N \le L \le G$, and δ is familial with $L \in \mathscr{F}(\delta)$. We would like to say that $(a|L)^N = a^N|(L/N)$, but, at first glance, these elements appear to live in homology groups with different coefficient systems. In cohomology, for example, if $a \in \tilde{H}^\alpha_{G,\delta}(X;\overline{T})$, then

$$(a|L)^N \in \tilde{H}^{\alpha^N - \delta(G/L)}_{L/N,\delta}(X^N;(\overline{T}|L)^N)$$

while

$$a^N|(L/N) \in \tilde{H}^{\alpha^N - \delta(G/L)}_{L/N,\delta}(X^N;\overline{T}^N|(L/N)).$$

We shall use Proposition 1.6.8 to show that $(\overline{T}|L)^N \cong \overline{T}^N|(L/N)$ and that the elements in question coincide under this identification.

The proofs for both induction and restriction to fixed sets depend on a more in-depth understanding of the stable orbit category. Recall Proposition 1.6.2, which we restate here for convenience.

Proposition 1.13.30 *Let δ be a familial dimension function for G. Then $\widehat{\mathscr{O}}_{G,\delta}((G/H,\delta),(G/K,\delta))$ is the free abelian group generated by the equivalence classes of diagrams of orbits of the form*

$$G/H \xleftarrow{p} G/J \xrightarrow{q} G/K,$$

where $J \leq H$, p is the projection, q is a (space-level) G-map with, say, $q(eJ) = gK$, and $\delta(N_H J/J) \cong \mathscr{L}(N_H J/J)$ and $\delta(N_{K^g} J/J) = 0$. Two such diagrams are equivalent if there is a diagram of the following form in which $G/J \to G/J'$ is a G-homeomorphism, the left triangle commutes, and the right triangle commutes up to G-homotopy:

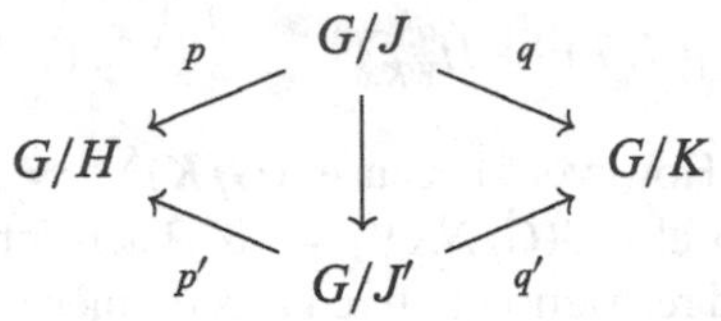

Composition is given by taking pullbacks and then replacing the resulting manifold with a disjoint union of orbits having the same Euler characteristics on all fixed sets. □

Explicitly, a pair of maps (p, q) as above corresponds to the stable map constructed as follows. Take V so large that H/J embeds in $V - \delta(G/H)$ as an H-space. We then have the following composite, in which the first map is the collapse map and the last is induced by q:

$$\begin{aligned} G_+ \wedge_H S^{V-\delta(G/H)} &\to G_+ \wedge_J S^{V-\delta(G/H)-\mathscr{L}(H/J)} \\ &\hookrightarrow G_+ \wedge_J S^{V-\delta(G/H)-\delta(H/J)} \\ &= G_+ \wedge_J S^{V-\delta(G/J)} \\ &\hookrightarrow G_+ \wedge_J S^{V-\delta(G/K^g)} \\ &\to G_+ \wedge_K S^{V-\delta(G/K)}. \end{aligned}$$

Note that the first inclusion would be null-homotopic if

$$[\mathscr{L}(H/J) - \delta(H/J)]^J = \mathscr{L}(N_H J/J) - \delta(N_H J/J) \neq 0$$

and the second would be null-homotopic if

$$\delta(K^g/J)^J = \delta(N_{K^g} J/J) \neq 0.$$

Here, we use that $\delta(H/J)^J \cong \delta(N_H J/J)$. This is clear for $\delta = \mathscr{L}$ because $(H/J)^J = N_H J/J$. In general, it follows from the fact that $\delta(H/J)$ is isomorphic to a J-subspace of $\mathscr{L}(H/J)$ and $\delta(N_H J/J)$ is isomorphic to a subspace of $\mathscr{L}(N_H J/J)$, which imply that

$$\delta(H/J)^J \cong \delta(H/N_H J)^J \oplus \delta(N_H J/J)^J = 0 \oplus \delta(N_H J/J).$$

The stable map corresponding to (p,q) is then the suspension of the space-level composite map above.

Because the corresponding stable map is 0 if G/J does not meet the dimension requirements, we shall allow ourselves to write $[G/H \leftarrow G/J \to G/K]$ for a stable map $(G/H,\delta) \to (G/K,\delta)$, with any J, with the understanding that the element may be 0.

Now, let's turn to the interaction of induction and restriction to subgroups. If $N \le L \le G$ and $L \in \mathscr{F}(\delta)$, we have the following commutative diagram (which we've used before):

$$\begin{array}{ccc} \widehat{\mathscr{O}}_{L/N,\delta} & \xrightarrow{\theta} & \widehat{\mathscr{O}}_{L,\delta} \\ {\scriptstyle i^{G/N}_{L/N}}\downarrow & & \downarrow{\scriptstyle i^G_L} \\ \widehat{\mathscr{O}}_{G/N,\delta} & \xrightarrow[\theta]{} & \widehat{\mathscr{O}}_{G,\delta} \end{array}$$

By Proposition 1.6.8, we have a natural map $\xi\colon \theta_! i^* \to i^*\theta_!$, where we write i for either i^G_L or $i^{G/N}_{L/N}$, as appropriate.

Lemma 1.13.31 *The natural transformation $\xi\colon \theta_! i^* \to i^*\theta_!$ is an isomorphism.*

Proof We give the argument for contravariant Mackey functors; the proof for covariant functors is similar or we can appeal to duality (replacing δ with $\mathscr{L} - \delta$).

By Proposition 1.6.8, the result will follow if we show that

$$\xi\colon \int^{L/J \in \widehat{\mathscr{O}}_{L/N,\delta}} \widehat{\mathscr{O}}_{G/N,\delta}(G/J, G/K) \otimes \widehat{\mathscr{O}}_{L,\delta}(L/H, L/J) \\ \to \widehat{\mathscr{O}}_{G,\delta}(G/H, G/K),$$

given by $\xi(f \otimes g) = \epsilon^\sharp f \circ (G_+ \wedge_L g)$, is an isomorphism for all L/H in $\widehat{\mathscr{O}}_{L,\delta}$ and G/K in $\widehat{\mathscr{O}}_{G/N,\delta}$.

Define a map ζ inverse to ξ as follows: Let $[G/H \xleftarrow{p} G/M \xrightarrow{q} G/K]$ be a generator of $\widehat{\mathscr{O}}_{G,\delta}(G/H, G/K)$, so $M \le H \le L$ and p is the projection. Then $p = G \times_L p'$ where $p'\colon L/M \to L/H$. By assumption, $N \le K$, so q factors as $G/M \to G/MN \to G/K$. We let

$$\zeta[G/H \leftarrow G/M \to G/K] = [G/MN \to G/K] \otimes [L/H \xleftarrow{p'} L/M \to L/MN].$$

(Notice that $MN \le L$ because L contains both M and N.) Clearly, $\xi \circ \zeta$ is the identity. On the other hand, a typical element in the coend is a sum of elements of the form

$$[G/J \xleftarrow{p} G/M \to G/K] \otimes g,$$

where $N \leq M \leq J$ and $p = G \times_L p'$. We then have

$$[G/J \xleftarrow{p} G/M \to G/K] \otimes g$$

$$\sim [G/M \to G/K] \otimes [L/J \xleftarrow{p'} L/M] \circ g$$

$$= \sum_i [G/M \to G/K] \otimes [L/H \leftarrow L/P_i \to L/M]$$

$$= \sum_i [G/M \to G/K] \otimes [L/H \leftarrow L/P_i \to L/P_iN \to L/M]$$

$$\sim \sum_i [G/P_iN \to G/M \to G/K] \otimes [L/H \leftarrow L/P_i \to L/P_iN]$$

which is in the image of ζ. So, ζ is an epimorphism, hence an isomorphism and the inverse of ξ. $\square$

Proposition 1.13.32 *Let N be a normal subgroup of G, let δ be an N-closed familial dimension function for G, let α be a virtual representation of G/N, let $\underline{S}$ be a covariant G/N-δ-Mackey functor, and let $\overline{T}$ be a contravariant G/N-δ-Mackey functor. Let $N \leq L \leq G$ with $L \in \mathscr{F}(\delta)$. If Y is a based G/N-space and $y \in \tilde{H}_\alpha^{G/N,\delta}(Y; \underline{S})$, then*

$$\epsilon^*(y|L/N) = (\epsilon^* y)|L \in \tilde{H}_\alpha^{L,\delta}(Y; \theta_!(\underline{S}|L/N)) \cong \tilde{H}_\alpha^{L,\delta}(Y; (\theta_!\underline{S})|L).$$

Similarly, if $y \in \tilde{H}_{G/N,\delta}^\alpha(Y; \overline{T})$, then

$$\epsilon^*(y|L/N) = (\epsilon^* y)|L \in \tilde{H}_{L,\delta}^\alpha(Y; \theta_!(\overline{T}|L/N)) \cong \tilde{H}_{L,\delta}^\alpha(Y; (\theta_!\overline{T})|L).$$

Proof As usual, we may assume that Y is well-based. We concentrate on cohomology; the proof for homology is similar. We need to show that the following diagram commutes:

$$
\begin{array}{ccc}
\tilde{H}_{G/N,\delta}^\alpha(Y; \overline{T}) & \xrightarrow{\epsilon^*} & \tilde{H}_{G,\delta}^\alpha(Y; \theta_!\overline{T}) \\
{\scriptstyle -|L/N} \downarrow & & \downarrow {\scriptstyle -|L} \\
\tilde{H}_{L/N,\delta}^\alpha(Y; \overline{T}|L/N) & & \tilde{H}_{L,\delta}^\alpha(Y; (\theta_!\overline{T})|L) \\
& \searrow^{\epsilon^*} & \cong \uparrow {\scriptstyle \xi_*} \\
& & \tilde{H}_{L,\delta}^\alpha(Y; \theta_!(\overline{T}|L/N))
\end{array}
$$

We can expand the top two rows of this diagram to get the following commutative diagram:

$$
\begin{array}{ccccc}
\tilde{H}^\alpha_{G/N,\delta}(Y;\overline{T}) & \longrightarrow & \tilde{H}^\alpha_{G/N,\delta}(Y;\theta^*\theta_!\overline{T}) & \overset{\cong}{\longrightarrow} & \tilde{H}^\alpha_{G,\delta}(Y;\theta_!\overline{T}) \\
\downarrow & & \downarrow & & \downarrow \\
\tilde{H}^\alpha_{G/N,\delta}(G_+\wedge_L Y;\overline{T}) & \longrightarrow & \tilde{H}^\alpha_{G/N,\delta}(G_+\wedge_L Y;\theta^*\theta_!\overline{T}) & \overset{\cong}{\longrightarrow} & \tilde{H}^\alpha_{G,\delta}(G_+\wedge_L Y;\theta_!\overline{T}) \\
\cong\downarrow & & \cong\downarrow & & \downarrow\cong \\
\tilde{H}^{L/N,\delta}_*(Y;i^*\overline{T}) & \longrightarrow & \tilde{H}^\alpha_{L/N,\delta}(Y;i^*\theta^*\theta_!\overline{T}) & \underset{\cong}{\longrightarrow} & \tilde{H}^\alpha_{L,\delta}(Y;i^*\theta_!\overline{T})
\end{array}
$$

Here, we write i^* instead of $-|L$ or $-|L/N$ for clarity. The two leftmost squares commute by naturality while the two rightmost squares commute because of the commutation $i^*\theta^* \cong \theta^*i^*$. It remains to show, then that the following diagram commutes:

$$
\begin{array}{ccc}
\tilde{H}^{L/N,\delta}_*(Y;i^*\overline{T}) & \longrightarrow \tilde{H}^\alpha_{L/N,\delta}(Y;i^*\theta^*\theta_!\overline{T}) \overset{\cong}{\longrightarrow} & \tilde{H}^\alpha_{L,\delta}(Y;i^*\theta_!\overline{T}) \\
& \searrow & \uparrow{\scriptstyle\xi_*} \\
& \tilde{H}^\alpha_{L/N,\delta}(Y;\theta^*\theta_!i^*\overline{T}) \underset{\cong}{\longrightarrow} & \tilde{H}^\alpha_{L,\delta}(Y;\theta_!i^*\overline{T})
\end{array}
$$

That this commutes follows from the following diagram of natural transformations, which commutes by the definition of ξ and properties of adjunctions:

$$
\begin{array}{ccc}
i^* & \overset{i^*\eta}{\longrightarrow} & i^*\theta^*\theta_! \\
{\scriptstyle\eta i^*}\downarrow & & \downarrow{\scriptstyle\cong} \\
\theta^*\theta_!i^* & \underset{\theta^*\xi}{\longrightarrow} & \theta^*i^*\theta_!
\end{array}
$$

$\square$

We now consider the interaction of restriction to fixed sets and restriction to subgroups. Again, let $N \leq L \leq G$ with $L \in \mathscr{F}(\delta)$. We have the following commutative diagram:

$$\begin{array}{ccc} \widehat{\mathscr{O}}_{L,\delta} & \xrightarrow{\Phi^N} & \widehat{\mathscr{O}}_{L/N,\delta} \\ {\scriptstyle i_L^G}\downarrow & & \downarrow{\scriptstyle i_{L/N}^{G/N}} \\ \widehat{\mathscr{O}}_{G,\delta} & \xrightarrow[\Phi^N]{} & \widehat{\mathscr{O}}_{G/N,\delta} \end{array}$$

By Proposition 1.6.8, we have a natural map

$$\xi\colon \Phi_!^N i^* \to i^* \Phi_!^N.$$

From Proposition 1.13.30 we get the following explicit description of Φ^N. For simplicity of notation we write G/H for the object $(G/H, \delta)$ in $\widehat{\mathscr{O}}_{G,\delta}$.

Proposition 1.13.33 *Consider the map*

$$\Phi^N\colon \widehat{\mathscr{O}}_{G,\delta}(G/H, G/K) \to \widehat{\mathscr{O}}_{G/N,\delta}((G/H)^N, (G/K)^N).$$

If either $N \not\leq H$ or $N \not\leq K$, the target is the trivial group. If $N \leq H$ and $N \leq K$, then, on generators, Φ^N is given by

$$\begin{aligned} \Phi^N[G/H \leftarrow G/J \to G/K] &= [(G/H)^N \leftarrow (G/J)^N \to (G/K)^N] \\ &= \begin{cases} [G/H \leftarrow G/J \to G/K] & \text{if } N \leq J \\ 0 & \text{otherwise.} \end{cases} \end{aligned}$$

Therefore, it is a split epimorphism with kernel generated by those diagrams

$$[G/H \leftarrow G/J \to G/K]$$

with $N \not\leq J$.

Proof That Φ^N takes the suspension of a map of spaces to the suspension of its N-fixed set map is the naturality of the equivalence $\Phi^N(\Sigma_G^\infty X) \simeq \Sigma_{G/N}^\infty X^N$. Applying this to the space-level description above of the stable map corresponding to a diagram $[G/H \leftarrow G/J \to G/K]$, we get the description of Φ^N on generators given in the statement of the proposition.

The only other comment necessary is that, if $N \leq J$, the conditions and the equivalence relations on diagrams $G/H \leftarrow G/J \to G/K$ are the same whether we consider them as diagrams of G/N-orbits or G-orbits. Therefore, the generators of $\widehat{\mathscr{O}}_{G/N,\delta}((G/H)^N, (G/K)^N)$ are in one-to-one correspondence with those generators of $\widehat{\mathscr{O}}_{G,\delta}(G/H, G/K)$ having the form $G/H \leftarrow G/J \to G/K$ with $N \leq J$. □

Now we can prove that $\xi\colon \Phi_!^N i^* \to i^* \Phi_!^N$ is an isomorphism.

Lemma 1.13.34 *If N is a normal subgroup of G, $N \le L \le G$, then $\xi\colon \Phi_!^N i^* \to i^* \Phi_!^N$ is an isomorphism on both covariant and contravariant G/N-δ-Mackey functors.*

Proof We give the argument for contravariant Mackey functors; the proof for covariant functors is similar or we can appeal to duality (replacing δ with $\mathscr{L} - \delta$).

By Proposition 1.6.8, the result will follow if we show that

$$\xi\colon \int^{L/J \in \widehat{\mathscr{O}}_{L,\delta}} \widehat{\mathscr{O}}_{G,\delta}(G/J, G/K) \otimes \widehat{\mathscr{O}}_{L/N,\delta}(L/H, (L/J)^N) \\ \to \widehat{\mathscr{O}}_{G/N,\delta}(G/H, (G/K)^N),$$

given by $\xi(f \otimes g) = \Phi^N f \circ (G_+ \wedge_L g)$, is an isomorphism for all L/H in $\widehat{\mathscr{O}}_{L/N,\delta}$ and G/K in $\widehat{\mathscr{O}}_{G,\delta}$.

If $N \not\le K$, then $(G/K)^N = *$ and the target of ξ is 0. For the source, consider a typical generator $f = [G/J \xleftarrow{p} G/M \to G/K]$ of $\widehat{\mathscr{O}}_{G,\delta}(G/J, G/K)$. Because p is a projection and $M \le J \le L$, we can write $p = G \times_L p'$ where $p'\colon L/M \to L/J$ is a projection. Because M is subconjugate to K, $N \not\le M$, hence $(L/M)^N = 0$. Therefore, for any g,

$$\begin{aligned} f \otimes g &= [G/J \xleftarrow{p} G/M \to G/K] \otimes g \\ &\sim [G/M \to G/K] \otimes [L/J \xleftarrow{p'} L/M]^N \circ g \\ &= [G/M \to G/K] \otimes [(L/J)^N \leftarrow *] \circ g \\ &= 0 \end{aligned}$$

in the coend. Therefore, the coend is 0 and ξ is an isomorphism in this case.

So, assume that $N \le K$ so $(G/K)^N = G/K$. Define a map ζ inverse to ξ as follows: If $[G/H \xleftarrow{p} G/J \to G/K]$ is a generator of $\widehat{\mathscr{O}}_{G/N,\delta}(G/H, G/K)$ with $N \le J \le H \le L$, write $p = G \times_L p'$ where $p'\colon L/J \to L/H$, and let

$$\zeta[G/H \xleftarrow{p} G/J \to G/K] = [G/J \to G/K] \otimes [L/H \xleftarrow{p'} L/J].$$

Clearly, $\xi \circ \zeta$ is the identity. On the other hand, a typical element in the coend is a sum of elements of the form

$$[G/J \xleftarrow{p} G/M \to G/K] \otimes g,$$

where we may assume $N \leq J$ because otherwise the element would live in a 0 group. Writing $p = G \times_L p'$, we then have

$$
\begin{aligned}
[G/J \overset{p}{\leftarrow} G/M \to G/K] \otimes g &\sim [G/M \to G/K] \otimes [L/J \overset{p'}{\leftarrow} L/M] \circ g \\
&= \sum_i [G/M \to G/K] \otimes [L/H \leftarrow L/P_i \to L/M] \\
&\sim \sum_i [G/P_i \to G/M \to G/K] \otimes [L/H \leftarrow L/P_i]
\end{aligned}
$$

which is in the image of ζ. So, ζ is an epimorphism, hence an isomorphism and the inverse of ξ. □

Proposition 1.13.35 *Let N be a normal subgroup of G, let δ be a familial dimension function for G, let α be a virtual representation of G, let $\underline{S}$ be a covariant G-δ-Mackey functor, and let $\overline{T}$ be a contravariant G-δ-Mackey functor. Let $N \leq L \leq G$ with $L \in \mathscr{F}(\delta)$. If X is a based G-space and $x \in \tilde{H}^{G,\delta}_\alpha(X; \underline{S})$, then*

$$(x|L)^N = x^N|L/N \in \tilde{H}^{L/N,\delta}_{\alpha^N - \delta(G/L)}(X^N; (\underline{S}|L)^N) \cong \tilde{H}^{L/N,\delta}_{\alpha^N - \delta(G/L)}(X^N; \underline{S}^N|L/N).$$

Similarly, if $x \in \tilde{H}^\alpha_{G,\delta}(X; \overline{T})$, then

$$(x|L)^N = x^N|L/N \in \tilde{H}^{\alpha^N - \delta(G/L)}_{L/N,\delta}(X^N; (\overline{T}|L)^N) \cong \tilde{H}^{\alpha^N - \delta(G/L)}_{L/N,\delta}(X^N; \overline{T}^N|L/N).$$

Proof The proof is the same as the proof of Proposition 1.13.32, but using Lemma 1.13.34. □

1.14 Products

We now turn to various pairings, including cup products, evaluation maps, and cap products. There are quite a few variations of such pairings available. All based spaces appearing in this section will be assumed to be well-based without further comment.

We begin with some geometric results about products of cell complexes.

1.14.1 Product Complexes

Consider a δ-H-CW(α) complex X and an ϵ-K-CW(β) complex Y. Their product $X \times Y$ is a $G = H \times K$ space and has a cell structure in which the cells have the form

$$
\begin{aligned}
&G \times_{J \times L} (D(V - \delta(H/J) + m) \times D(W - \epsilon(K/L) + n)) \\
&\cong G \times_{J \times L} D(V + W - \delta(H/J) - \epsilon(K/L) + m + n)
\end{aligned}
$$

with V stably equivalent to α and W stably equivalent to β. We can therefore naturally view $X \times Y$ as a $(\delta \times \epsilon)$-$(H \times K)$-CW$(\alpha + \beta)$ complex where $\delta \times \epsilon$ is the dimension function defined in Example 1.2.2(3).

The dimension function $\delta \times \epsilon$ has the defect that it is almost never complete or even familial, even if δ and ϵ are. In practice, we'll often want to use a dimension function for $H \times K$ defined on a family of subgroups. For example, suppose that X is an H-space with a δ-H-CW(α) approximation $\Gamma^\delta_\alpha X \to X$ and that Y is a K-space with an ϵ-K-CW(β) approximation $\Gamma^\epsilon_\beta Y \to Y$. Then $\Gamma^\delta_\alpha X \times \Gamma^\epsilon_\beta Y$ is a $(\delta\times\epsilon)$-$(H\times K)$-CW$(\alpha + \beta)$ complex, but what sort of approximation is $\Gamma^\delta_\alpha X \times \Gamma^\epsilon_\beta Y \to X \times Y$? In the case that all subgroups in $\mathscr{F}(\delta)$ and $\mathscr{F}(\epsilon)$ are admissible (which we can achieve in the based case after suitable suspension), $\Gamma^\delta_\alpha X \to X$ and $\Gamma^\epsilon_\beta Y \to Y$ are weak $\mathscr{F}(\delta)$- and $\mathscr{F}(\epsilon)$-equivalences, respectively, so their product is a weak $\mathscr{F}(\delta) \times \mathscr{F}(\epsilon)$-equivalence. Thus, we get the following.

Proposition 1.14.1 *Let δ be a familial dimension function for H and let ϵ be a familial dimension function for K. Suppose that α is a virtual representation of H such that every subgroup in $\mathscr{F}(\delta)$ is δ-α-admissible and suppose that β is a virtual representation of K such that every subgroup in $\mathscr{F}(\epsilon)$ is ϵ-β-admissible. If X is an H-space with a δ-H-CW(α) approximation $\Gamma^\delta_\alpha X \to X$ and Y is a K-space with an ϵ-K-CW(β) approximation $\Gamma^\epsilon_\beta Y \to Y$, then $\Gamma^\delta_\alpha X\times\Gamma^\epsilon_\beta Y$ is a $(\delta\times\epsilon)$-$(H\times K)$-CW$(\alpha+\beta)$ complex and $\Gamma^\delta_\alpha X \times \Gamma^\epsilon_\beta Y \to X \times Y$ is a weak $\mathscr{F}(\delta) \times \mathscr{F}(\epsilon)$-equivalence.* □

Corollary 1.14.2 *Let X be an H-space, Y a K-space, Z an $(H \times K)$-space, and $f: Z \to X \times Y$ an $(H \times K)$-map. Let δ be a familial dimension function for H, ϵ a familial dimension function for K, and ζ a familial dimension function for $H \times K$ with $\zeta \succcurlyeq \delta \times \epsilon$. Let α be a virtual representation of H such that every subgroup in $\mathscr{F}(\delta)$ is δ-α-admissible and let β be a virtual representation of K such that every subgroup in $\mathscr{F}(\epsilon)$ is ϵ-β-admissible. Let $\Gamma^\delta_\alpha X \to X$ be a δ-H-CW(α) approximation, $\Gamma^\epsilon_\beta Y \to Y$ an ϵ-K-CW(β) approximation, and $\Gamma^\zeta_{\alpha+\beta} Z \to Z$ a ζ-$(H\times K)$-CW$(\alpha+\beta)$ approximation. Then we may find a cellular map Γf, unique up to cellular $(H\times K)$-homotopy, making the following diagram homotopy commute:*

$$\begin{array}{ccc} \Gamma^\zeta_{\alpha+\beta} Z & \overset{\Gamma f}{\dashrightarrow} & \Gamma^\delta_\alpha X \times \Gamma^\epsilon_\beta Y \\ \downarrow & & \downarrow \\ Z & \longrightarrow & X \times Y \end{array}$$

Proof The preceding proposition and the Whitehead theorem show that there exists a unique homotopy class of maps making the diagram commute. Cellular approximation of maps (in the form of Theorem 1.4.12) shows that this homotopy class contains a cellular map and that any two cellular maps in the class are cellularly homotopic. □

Of course, there are relative and based versions of these statements as well. Corollary 1.14.2 follows from the universal case $Z = X \times Y$, but it's useful to think of the result in the more general form.

So, for particular δ and ϵ we look for a $\zeta \succcurlyeq \delta \times \epsilon$. For the approximation in the preceding corollary to be nontrivial, we should look for as small a ζ as possible. The most interesting case is that of $G \times G$. Here, we're particularly interested in the diagonal G-map $X \to X \times X$, which we view in adjunct form as the $(G \times G)$-map $(G \times G) \times_\Delta X \to X \times X$ where $\Delta \leq G \times G$ is the diagonal copy of G.

Definition 1.14.3 Let $\mathscr{F}_\Delta$ denote the family of subgroups of $G \times G$ given by

$$\mathscr{F}_\Delta = \{H \leq G \times G \mid H \text{ is conjugate to a subgroup of } \Delta\}.$$

If δ is a complete dimension function for G, let δ_Δ be the familial dimension function for $G \times G$, with underlying family $\mathscr{F}_\Delta$, given by

$$\delta_\Delta((G \times G)/H) = g^{-1} \cdot \delta(\Delta/H^g)$$

where $H^g \leq \Delta$.

Then $\delta_\Delta \succcurlyeq \delta \times 0$ and also $\delta_\Delta \succcurlyeq 0 \times \delta$. To see this, let $H \leq G$ and consider the diagonal copy of H, which we'll call $\Delta_H \leq \Delta$ (the case of a conjugate of Δ_H is similar). The smallest subgroup in $\mathscr{F}(\delta \times 0)$ containing Δ_H is $H \times H$. By definition,

$$(\delta \times 0)((G \times G)/(H \times H)) \cong \delta(G/H)$$

and

$$\delta_\Delta((G \times G)/\Delta_H) \cong \delta(G/H)$$

as well. Hence, $\delta_\Delta \succcurlyeq \delta \times 0$ as claimed, and similarly $\delta_\Delta \succcurlyeq 0 \times \delta$.

Note also that $\delta_\Delta|\Delta = \delta$ as a dimension function on G.

Now, let X be a G-space and consider a 0-G-CW(α) approximation $\Gamma^0_\alpha X$, a δ-G-CW(β) approximation $\Gamma^\delta_\beta X$, and a δ-G-CW($\alpha + \beta$) approximation $\Gamma^\delta_{\alpha+\beta} X$. Then, using Proposition 1.4.6 and the fact that $(\delta_\Delta)((G \times G)/\Delta) = 0$, we have that $(G \times G) \times_\Delta \Gamma^\delta_{\alpha+\beta} X$ is a δ_Δ-$(G \times G)$-CW($\alpha + \beta$) complex. Assuming all subgroups of H and K admissible, Corollary 1.14.2 then gives us a cellular map, unique up to cellular homotopy, making the following diagram commute and thus approximating the diagonal:

$$\begin{array}{ccc} (G \times G) \times_\Delta \Gamma^\delta_{\alpha+\beta} X & \dashrightarrow & \Gamma^0_\alpha X \times \Gamma^\delta_\beta X \\ \downarrow & & \downarrow \\ (G \times G) \times_\Delta X & \longrightarrow & X \times X \end{array}$$

Here is a similar case that we'll need. Suppose again that δ is a complete dimension function for G, with dual $\mathscr{L}-\delta$. Let $\mathscr{L}_\Delta$ be the result of applying Definition 1.14.3 to $\mathscr{L}$. Then $\mathscr{L}_\Delta|\Delta = \mathscr{L}$ as a dimension function for G and $\mathscr{L}_\Delta \succcurlyeq \delta \times (\mathscr{L}-\delta)$. To see the latter relationship, suppose that $H \leq G$ and $\Delta_H \leq \Delta$ is the corresponding subgroup of the diagonal. Again, the smallest subgroup of $\mathscr{F}(\delta\times(\mathscr{L}-\delta))$ containing Δ_H is $H \times H$ and we have

$$(\delta \times (\mathscr{L}-\delta))((G\times G)/(H\times H)) = \delta(G/H) \oplus (\mathscr{L}-\delta)(G/H) \cong \mathscr{L}(G/H)$$

and

$$\mathscr{L}_\Delta((G\times G)/\Delta_H) \cong \mathscr{L}(G/H).$$

Let X be a G-space and consider a δ-G-CW(α) approximation $\Gamma^\delta_\alpha X$, an $(\mathscr{L}-\delta)$-G-CW(β) approximation $\Gamma^{\mathscr{L}-\delta}_\beta X$, and an $\mathscr{L}$-G-CW($\alpha+\beta$) approximation $\Gamma^{\mathscr{L}}_{\alpha+\beta} X$. Then, using Proposition 1.4.6 and the fact that $\mathscr{L}_\Delta((G\times G)/\Delta) = 0$, we have that $(G\times G)\times_\Delta \Gamma^{\mathscr{L}}_{\alpha+\beta}X$ is an $\mathscr{L}_\Delta$-$(G\times G)$-CW($\alpha+\beta$) complex. Assuming all subgroups of H and K admissible, Corollary 1.14.2 then gives us a cellular map, unique up to cellular homotopy, making the following diagram commute, so again approximating the diagonal:

$$\begin{array}{ccc} (G\times G)\times_\Delta \Gamma^{\mathscr{L}}_{\alpha+\beta}X & \dashrightarrow & \Gamma^\delta_\alpha X \times \Gamma^{\mathscr{L}-\delta}_\beta X \\ \downarrow & & \downarrow \\ (G\times G)\times_\Delta X & \longrightarrow & X\times X \end{array}$$

1.14.2 Cup Products

The product of two orbits of G is not generally an orbit of G; if G is infinite the product is not generally even a disjoint union of orbits. Thus, if X and Y are G-CW complexes, their product does not have a canonical G-CW structure. However, it does have a canonical $(G\times G)$-CW structure. Following where the geometry leads us, we therefore start with an *external* cup product, pairing a based G-space X and a based K-space Y to get a based $(G\times K)$-space $X\wedge Y$. In the case of two G-spaces, we can then internalize by restricting to the diagonal $G \leq G\times G$ using the restriction to subgroups discussed in detail in the preceding section.

We begin by defining the appropriate tensor product of Mackey functors (see also [22] and [36]).

Definition 1.14.4 Let G and K be two compact Lie groups. Let δ be a dimension function for G, let ϵ be a dimension function for K, and let $\delta\times\epsilon$ denote their product as in Example 1.2.2(3). Let ζ be a dimension function for $G \times K$ with $\zeta \succcurlyeq \delta \times \epsilon$, as in Definition 1.2.8.

(1) Let $\widehat{\mathscr{O}}_{G,\delta} \otimes \widehat{\mathscr{O}}_{K,\epsilon}$ denote the preadditive category whose objects are pairs of objects (a, b), as in the product category, with

$$(\widehat{\mathscr{O}}_{G,\delta} \otimes \widehat{\mathscr{O}}_{K,\epsilon})((a,b),(c,d)) = \widehat{\mathscr{O}}_{G,\delta}(a,c) \otimes \widehat{\mathscr{O}}_{K,\epsilon}(b,d).$$

Let

$$p\colon \widehat{\mathscr{O}}_{G,\delta} \otimes \widehat{\mathscr{O}}_{K,\epsilon} \to \mathrm{Ho}(G \times K)\mathscr{P}$$

be the restriction of the external smash product, i.e., the additive functor taking

$$\begin{aligned}((G/H,\delta),(K/L,\epsilon)) &\mapsto (G_+ \wedge_H S^{-\delta(G/H)}) \wedge (K_+ \wedge_L S^{-\epsilon(K/L)}) \\ &\cong (G \times K)_+ \wedge_{H\times L} S^{-(\delta\times\epsilon)((G\times K)/(H\times L))}.\end{aligned}$$

Similarly, pairs of maps are taken to their smash products.

Let $i\colon \widehat{\mathscr{O}}_{G\times K,\zeta} \to \mathrm{Ho}(G \times K)\mathscr{P}$ denote the inclusion of the subcategory.

(2) If $\overline{T}$ is a contravariant δ-G-Mackey functor and $\overline{U}$ is a contravariant ϵ-K-Mackey functor, let $\overline{T} \otimes \overline{U}$ denote the $(\widehat{\mathscr{O}}_{G,\delta} \otimes \widehat{\mathscr{O}}_{K,\epsilon})$-module defined by

$$(\overline{T} \otimes \overline{U})(a,b) = \overline{T}(a) \otimes \overline{U}(b).$$

Define $\underline{S} \otimes \underline{V}$ similarly for covariant Mackey functors.

(3) Let $\overline{T} \boxtimes \overline{U} = \overline{T} \boxtimes_\zeta \overline{U}$ (the *external box product*) be the contravariant ζ-$(G\times K)$-Mackey functor defined by

$$\overline{T} \boxtimes \overline{U} = \overline{T} \boxtimes_\zeta \overline{U} = i^* p_!(\overline{T} \otimes \overline{U}),$$

where i^* and $p_!$ are defined in Definition 1.6.5. Define the external box product of covariant Mackey functors similarly.

(4) Let $\Delta \leq G \times G$ denote the diagonal subgroup. If $K = G$, so $\overline{T}$ and $\overline{U}$ are both G-Mackey functors, and $\Delta \in \mathscr{F}(\zeta)$, define the *(internal) box product* to be

$$\overline{T} \,\square\, \overline{U} = \overline{T} \,\square_\zeta\, \overline{U} = (\overline{T} \boxtimes \overline{U})|\Delta.$$

Define the box product of covariant Mackey functors similarly.

Remark 1.14.5 In order to define the internal box product we required $\Delta \in \mathscr{F}(\zeta)$, hence $\Delta \in \mathscr{F}(\delta) \times \mathscr{F}(\epsilon)$. However, the smallest product subgroup containing Δ is all of $G \times G$, so the internal box product can be defined only if δ and ϵ are both complete. Hence, this is the case of most interest to us. We introduced more general dimension functions largely so that we could properly handle $\delta \times \epsilon$.

Proposition 1.14.6 *Let δ be a dimension function for G, let ϵ be a dimension function for K, and let $\zeta \succcurlyeq \delta \times \epsilon$. Then*

$$\overline{A}_{G/H,\delta} \boxtimes_\zeta \overline{A}_{K/L,\epsilon} \cong \mathrm{Ho}(G \times K)\mathscr{P}(-, (G \times K)_+ \wedge_{H\times L} S^{-(\delta\times\epsilon)((G\times K)/(H\times L))})$$

and

$$\underline{A}^{G/H,\delta} \boxtimes_\zeta \underline{A}^{K/L,\epsilon} \cong \mathrm{Ho}(G \times K)\mathscr{P}((G \times K)_+ \wedge_{H\times L} S^{-(\delta\times\epsilon)((G\times K)/(H\times L))}, -).$$

Proof It's clear from the definitions that

$$\overline{A}_{G/H,\delta} \otimes \overline{A}_{K/L,\epsilon} \cong (\widehat{\mathscr{O}}_{G,\delta} \otimes \widehat{\mathscr{O}}_{K,\epsilon})(-, ((G/H, \delta), (K/L, \epsilon))).$$

Applying $p_!$ and using Proposition 1.6.6 gives the result. The proof for covariant functors is similar. □

The following proposition shows that $\overline{A}_{G/G} = \overline{A}_{G/G,0}$ acts as a unit for the internal box product. If δ is a complete dimension function for G, recall from Definition 1.14.3 that we have the familial dimension function δ_Δ on $G\times G$ with the properties that $\delta_\Delta \succcurlyeq 0 \times \delta$, $\delta_\Delta(G \times G/\Delta) = 0$, and $\delta_\Delta|\Delta = \delta$.

Proposition 1.14.7 *Let δ be a complete dimension function for G and let $\overline{T}$ be a contravariant δ-G-Mackey functor. Then*

$$\overline{A}_{G/G} \,\square_{\delta_\Delta}\, \overline{T} \cong \overline{T}.$$

Proof By writing $\overline{T} \cong \int^{(G/H,\delta)} \overline{T}(G/H, \delta) \otimes \overline{A}_{G/H,\delta}$, we see that it suffices to prove the result for $\overline{T} = \overline{A}_{G/H,\delta}$. From the preceding proposition we know that

$$\overline{A}_{G/G} \boxtimes_{\delta_\Delta} \overline{A}_{G/H,\delta} \cong \mathrm{Ho}(G \times G)\mathscr{P}(-, (G \times G)_+ \wedge_{G\times H} S^{-\delta(G/H)}),$$

with $G \times H$ acting on $\delta(G/H)$ via the projection $G \times H \to H$. Let $i\colon \Delta \to G \times G$ be the inclusion of the diagonal. Then, using that the restriction of the spectrum $(G \times G)_+ \wedge_{G\times H} S^{-\delta(G/H)}$ to the diagonal is $G_+ \wedge_H S^{-\delta(G/H)}$, we have

$$\begin{aligned}
&(\overline{A}_{G/G} \,\square_{\delta_\Delta}\, \overline{A}_{G/H,\delta})(G/K, \delta)\\
&\quad = [(\mathrm{Ho}(G \times G)\mathscr{P}(-, (G \times G)_+ \wedge_{G\times H} S^{-\delta(G/H)}))|\Delta](G/K, \delta)\\
&\quad = \mathrm{Ho}(G \times G)\mathscr{P}((G \times G)_+ \wedge_\Delta (G_+ \wedge_K S^{-\delta(G/K)}), (G \times G)_+ \wedge_{G\times H} S^{-\delta(G/H)})\\
&\quad \cong \widehat{\mathscr{O}}_{G,\delta}(G_+ \wedge_K S^{-\delta(G/K)}, G_+ \wedge_H S^{-\delta(G/H)})\\
&\quad = \overline{A}_{G/H,\delta}(G/K, \delta).
\end{aligned}$$

Thus $\overline{A}_{G/G} \,\square_{\delta_\Delta}\, \overline{A}_{G/H,\delta} \cong \overline{A}_{G/H,\delta}$, which implies the result in general. □

The following result identifies the chain complex of a product of CW complexes.

Proposition 1.14.8 *Let δ be a dimension function for G and let ϵ be a dimension function for K. If X is a based δ-G-CW(α) complex and Y is a based ϵ-K-CW(β) complex, then $X \wedge Y$ is a based $(\delta \times \epsilon)$-$(G \times K)$-CW($\alpha + \beta$) complex with*

$$\overline{C}^{G,\delta}_{\alpha+*}(X, *) \boxtimes_{\delta\times\epsilon} \overline{C}^{K,\epsilon}_{\beta+*}(Y, *) \cong \overline{C}^{G\times K,\delta\times\epsilon}_{\alpha+\beta+*}(X \wedge Y, *).$$

Moreover, this isomorphism respects suspension in each of X and Y.

Proof We give $X \wedge Y$ the filtration

$$(X \wedge Y)^{\alpha+\beta+n} = \bigcup_{i+j=n} X^{\alpha+i} \wedge Y^{\alpha+j}.$$

With this filtration, $X \wedge Y$ is a based $(\delta \times \epsilon)$-$(G \times K)$-CW($\alpha + \beta$) complex, with

$$(X \wedge Y)^{\alpha+\beta+n}/(X \wedge Y)^{\alpha+\beta+n-1} = \bigvee_{i+j=n} X^{\alpha+i}/X^{\alpha+i-1} \wedge Y^{\beta+j}/Y^{\beta+j-1}.$$

We now define

$$\kappa : \overline{C}^{G,\delta}_{\alpha+*}(X, *) \boxtimes_{\delta\times\epsilon} \overline{C}^{K,\epsilon}_{\beta+*}(Y, *) \to \overline{C}^{G\times K,\delta\times\epsilon}_{\alpha+\beta+*}(X \wedge Y, *)$$

by $\kappa(x \otimes y) = (-1)^{pq} x \wedge y$, where

$$x \in \overline{C}^{G,\delta}_{\alpha+p}(X, *)(G/H) = [G_+ \wedge_H S^{-\delta(G/H)+\alpha+p}, \Sigma^\infty_G X^{\alpha+p}/X^{\alpha+p-1}]_G,$$

$$y \in \overline{C}^{K,\epsilon}_{\beta+q}(Y, *)(K/L) = [K_+ \wedge_L S^{-\epsilon(K/L)+\beta+q}, \Sigma^\infty_K Y^{\beta+q}/Y^{\beta+q-1}]_K,$$

and $x \wedge y \in \overline{C}^{G\times K,\delta\times\epsilon}_{\alpha+\beta+*}(X \wedge Y, *)(G\times K/H\times L)$ is their smash product. With the sign convention $d(x\otimes y) = dx\otimes y+(-1)^p x\otimes dy$, we see that κ is an isomorphism of chain complexes, using Proposition 1.14.6 and the calculation $d(x\wedge y) = (-1)^q dx\wedge y+x\wedge dy$ as in [48, §13.4]. That the isomorphism respects suspension is straightforward. □

Because of the limitations of $\delta\times\epsilon$, this result by itself is not as useful as we would like, in particular for discussing general spaces or internal products. Therefore, we need the following results. First, we note that level-wise smash product induces a map

$$\wedge : G\mathscr{P}\mathscr{V} \otimes K\mathscr{P}\mathscr{W} \to (G\times K)\mathscr{P}(\mathscr{V} \oplus \mathscr{W})$$

where $\mathscr{V} \oplus \mathscr{W} = \{V_i \oplus W_i\}$. This pairing extends to semistable maps as well.

Proposition 1.14.9 *Let δ be a familial dimension function for G and let ϵ be a familial dimension function for K. Let $\Gamma^\delta\Sigma_G^\infty X \to \Sigma_G^\infty X$ and $\Gamma^\epsilon\Sigma_K^\infty Y \to \Sigma_K^\infty Y$ be, respectively, δ-G-CW(α) and ϵ-K-CW(β) approximations. Then, there exists a nonnegative integer N such that*

$$(\Gamma^\delta\Sigma_G^\infty X \wedge \Gamma^\epsilon\Sigma_K^\infty Y)(V_i \oplus W_i) \to \Sigma_G^{V_i} X \wedge \Sigma_K^{W_i} Y$$

is an $\overline{\mathscr{F}(\delta) \times \mathscr{F}(\epsilon)}$-equivalence for all $i \geq N$.

Proof This is an immediate corollary of Proposition 1.14.1 and the fact that, for large enough i, $(\Gamma^\delta\Sigma_G^\infty X)(V_i) \to \Sigma_G^{V_i} X$ is an $\mathscr{F}(\delta)$-approximation, and similarly for Y. □

Corollary 1.14.10 *Let δ be a familial dimension function for G, let ϵ be a familial dimension function for K, and let ζ be a familial dimension function for $G \times K$ with $\zeta \succcurlyeq \delta \times \epsilon$. Let $\Gamma^\delta\Sigma_G^\infty X \to \Sigma_G^\infty X$ and $\Gamma^\epsilon\Sigma_K^\infty Y \to \Sigma_K^\infty Y$ be, respectively, δ-G-CW(α) and ϵ-K-CW(β) approximations, and let $\Gamma^\zeta\Sigma_{G\times K}^\infty(X \wedge Y) \to \Sigma_{G\times K}^\infty(X \wedge Y)$ be a ζ-($G \times K$)-CW($\alpha + \beta$) approximation. Then there exists a semistable cellular map*

$$\mu\colon \Gamma^\zeta\Sigma_{G\times K}^\infty(X \wedge Y) \to \Gamma^\delta\Sigma_G^\infty X \wedge \Gamma^\epsilon\Sigma_K^\infty Y$$

over $\Sigma_{G\times K}^\infty(X \wedge Y)$, unique up to semistable cellular homotopy.

Proof This follows from the preceding proposition and Corollary 1.14.2. □

Definition 1.14.11 Let δ be a familial dimension function for G, let ϵ be a familial dimension function for K, and let ζ be a familial dimension function for $G \times K$ with $\zeta \succcurlyeq \delta \times \epsilon$. If X is a based G-space and Y is a based K-space, let

$$\mu_*\colon \overline{C}_{\alpha+\beta+*}^{G\times K,\zeta}(X \wedge Y) \to \overline{C}_{\alpha+*}^{G,\delta}(X) \boxtimes_\zeta \overline{C}_{\beta+*}^{K,\epsilon}(Y)$$

be the chain map induced by the map μ from the preceding corollary (using Proposition 1.14.8 to identify the chain complex on the right). It is well-defined up to chain homotopy.

We can now define pairings in cohomology. Let δ, ϵ, and ζ be familial with $\zeta \succcurlyeq \delta \times \epsilon$, as above, let X be a based G-space, let Y be a based K-space, let $\overline{T}$ be a δ-G-Mackey functor, and let $\overline{U}$ be an ϵ-K-Mackey functor. The external box product $\boxtimes = \boxtimes_\zeta$ and the map μ_* induce a natural chain map

$$\begin{aligned}
&\mathrm{Hom}_{\widehat{\mathscr{O}}_{G,\delta}}(\overline{C}_{\alpha+*}^{G,\delta}(X), \overline{T}) \otimes \mathrm{Hom}_{\widehat{\mathscr{O}}_{K,\epsilon}}(\overline{C}_{\beta+*}^{K,\epsilon}(Y), \overline{U}) \\
&\to \mathrm{Hom}_{\widehat{\mathscr{O}}_{G\times K,\zeta}}(\overline{C}_{\alpha+*}^{G,\delta}(X) \boxtimes \overline{C}_{\beta+*}^{K,\epsilon}(Y), \overline{T} \boxtimes \overline{U}) \\
&\to \mathrm{Hom}_{\widehat{\mathscr{O}}_{G\times K,\zeta}}(\overline{C}_{\alpha+\beta+*}^{G\times K,\zeta}(X \wedge Y), \overline{T} \boxtimes \overline{U}).
\end{aligned}$$

This induces the (external) cup product

$$- \cup -: \tilde{H}^{\alpha}_{G,\delta}(X; \overline{T}) \otimes \tilde{H}^{\beta}_{K,\epsilon}(Y; \overline{U}) \to \tilde{H}^{\alpha+\beta}_{G\times K,\zeta}(X \wedge Y; \overline{T} \boxtimes \overline{U}).$$

When $G = K$ and $\mathscr{F}(\zeta)$ contains the diagonal subgroup $\Delta \leq G \times G$, we can follow the external cup product with the restriction to Δ. This gives the internal cup product

$$- \cup -: \tilde{H}^{\alpha}_{G,\delta}(X; \overline{T}) \otimes \tilde{H}^{\beta}_{G,\epsilon}(Y; \overline{U}) \to \tilde{H}^{\alpha+\beta-\zeta(G\times G/\Delta)}_{G,\zeta|\Delta}(X \wedge Y; \overline{T} \,\Box\, \overline{U}).$$

Of course, when $X = Y$ we can apply restriction along the diagonal $X \to X \wedge X$ to get

$$- \cup -: \tilde{H}^{\alpha}_{G,\delta}(X; \overline{T}) \otimes \tilde{H}^{\beta}_{G,\epsilon}(X; \overline{U}) \to \tilde{H}^{\alpha+\beta-\zeta(G\times G/\Delta)}_{G,\zeta|\Delta}(X; \overline{T} \,\Box\, \overline{U}).$$

A useful special case is $\delta_\Delta \succcurlyeq 0 \times \delta$ with δ complete, which gives us pairings

$$- \cup -: \tilde{H}^{\alpha}_{G}(X; \overline{T}) \otimes \tilde{H}^{\beta}_{G,\delta}(Y; \overline{U}) \to \tilde{H}^{\alpha+\beta}_{G,\delta}(X \wedge Y; \overline{T} \,\Box\, \overline{U})$$

and

$$- \cup -: \tilde{H}^{\alpha}_{G}(X; \overline{T}) \otimes \tilde{H}^{\beta}_{G,\delta}(X; \overline{U}) \to \tilde{H}^{\alpha+\beta}_{G,\delta}(X; \overline{T} \,\Box\, \overline{U}).$$

Recall that a contravariant G-Mackey functor (i.e., 0-G-Mackey functor) $\overline{T}$ is a *ring* (also called a *Green functor*) if there is an associative multiplication $\overline{T} \,\Box\, \overline{T} \to \overline{T}$. This then makes $\tilde{H}^*_G(X; \overline{T})$ a ring. For example, consider $\overline{A}_{G/G} = \overline{A}_{G/G,0}$, which is a ring by Proposition 1.14.7. By that same proposition, $\overline{A}_{G/G} \,\Box_{\delta_\Delta}\, \overline{T} \cong \overline{T}$ for any contravariant δ-G-Mackey functor $\overline{T}$. This makes $\overline{T}$ a *module* over $\overline{A}_{G/G}$, in the sense that there is an associative pairing $\overline{A}_{G/G} \Box_{\delta_\Delta} \overline{T} \to \overline{T}$ (namely, the isomorphism). This makes every cellular cohomology theory with complete δ a module over ordinary cohomology with coefficients in $\overline{A}_{G/G}$, using the cup product

$$- \cup -: \tilde{H}^{\alpha}_{G}(X; \overline{A}_{G/G}) \otimes \tilde{H}^{\beta}_{G,\delta}(X; \overline{T}) \to \tilde{H}^{\alpha+\beta}_{G,\delta}(X; \overline{T}).$$

Obviously, there are other variations possible. The following theorem records the main properties of the external cup product, from which similar properties of the other products follow by naturality.

Theorem 1.14.12 *Let δ be a familial dimension function for G, let ϵ be a familial dimension function for K, and let ζ be a familial dimension function for $G \times K$ with $\zeta \succcurlyeq \delta \times \epsilon$. The external cup product*

$$- \cup -: \tilde{H}^{\alpha}_{G,\delta}(X; \overline{T}) \otimes \tilde{H}^{\beta}_{K,\epsilon}(Y; \overline{U}) \to \tilde{H}^{\alpha+\beta}_{G\times K,\zeta}(X \wedge Y; \overline{T} \boxtimes \overline{U})$$

generalizes the nonequivariant cup product and satisfies the following.

(1) *It is natural:* $f^*(x) \cup g^*(y) = (f \wedge g)^*(x \cup y)$.

(2) *It respects suspension: For any representation V of G, the following diagram commutes:*

$$\begin{array}{ccc}
\tilde{H}^{\alpha}_{G,\delta}(X;\overline{T}) \otimes \tilde{H}^{\beta}_{K,\epsilon}(Y;\overline{U}) & \xrightarrow{\ \cup\ } & \tilde{H}^{\alpha+\beta}_{G\times K,\zeta}(X \wedge Y;\overline{T} \boxtimes \overline{U}) \\
\Big\downarrow{\scriptstyle \sigma^V \otimes \mathrm{id}} & & \\
\tilde{H}^{\alpha+V}_{G,\delta}(\Sigma^V X;\overline{T}) \otimes \tilde{H}^{\beta}_{K,\epsilon}(Y;\overline{U}) & & \Big\downarrow{\scriptstyle \sigma^V} \\
\Big\downarrow{\scriptstyle \cup} & & \\
\tilde{H}^{\alpha+V+\beta}_{G\times K,\zeta}(\Sigma^V X \wedge Y;\overline{T} \boxtimes \overline{U}) & \xrightarrow[\cong]{} & \tilde{H}^{\alpha+\beta+V}_{G\times K,\zeta}(\Sigma^V (X \wedge Y);\overline{T} \boxtimes \overline{U})
\end{array}$$

The horizontal isomorphism at the bottom of the diagram comes from the identification $\alpha + V + \beta \cong \alpha + \beta + V$. The similar diagram for suspension of Y also commutes.

(3) *It is associative:* $(x \cup y) \cup z = x \cup (y \cup z)$ *when we identify gradings using the obvious identification* $(\alpha + \beta) + \gamma \cong \alpha + (\beta + \gamma)$.

(4) *It is commutative: If* $x \in \tilde{H}^{\alpha}_{G,\delta}(X;\overline{T})$ *and* $y \in \tilde{H}^{\beta}_{K,\epsilon}(Y;\overline{U})$ *then* $x \cup y = \iota(y \cup x)$ *where ι is the evident isomorphism between* $\tilde{H}^{\alpha+\beta}_{G\times K,\zeta}(X \wedge Y;\overline{T} \boxtimes \overline{U})$ *and* $\tilde{H}^{\beta+\alpha}_{K\times G,\tilde{\zeta}}(Y \wedge X;\overline{U} \boxtimes \overline{T})$*; $\tilde{\zeta}$ is the dimension function on $K \times G$ induced by ζ and ι uses the isomorphism of $\alpha + \beta$ and $\beta + \alpha$ that switches the direct summands.*

(5) *It is unital: The map*

$$\tilde{H}^0_G(S^0;\overline{A}_{G/G}) \otimes \tilde{H}^{\alpha}_{G,\delta}(X;\overline{T}) \to \tilde{H}^{\alpha}_{G,\delta}(X;\overline{T})$$

takes $1 \otimes x \mapsto x$, *where* $1 \in \tilde{H}^0_G(S^0;\overline{A}_{G/G}) \cong A(G)$ *is the unit.*

(6) *It respects the Wirthmüller isomorphism: Suppose that* $J \leq G$ *and* $L \leq K$, *that* $J \in \mathscr{F}(\delta)$, $L \in \mathscr{F}(\epsilon)$, *and* $J \times L \in \mathscr{F}(\zeta)$, *and that* $\zeta(G \times K/J \times L) = \delta(G/K) \oplus \epsilon(K/L)$. *Then the following diagram commutes:*

$$\begin{array}{ccc}
\tilde{H}^{\alpha}_{G,\delta}(G_+ \wedge_J X;\overline{T}) \otimes \tilde{H}^{\beta}_{K,\epsilon}(K_+ \wedge_L Y;\overline{U}) & \xrightarrow{\ \cong\ } & \tilde{H}^{\alpha-\delta(G/J)}_{J,\delta}(X;\overline{T}|J) \otimes \tilde{H}^{\beta-\epsilon(K/L)}_{L,\epsilon}(Y;\overline{U}|L) \\
& & \Big\downarrow{\scriptstyle -\cup-} \\
\Big\downarrow{\scriptstyle -\cup-} & & \tilde{H}^{\alpha+\beta-\zeta(G\times K/J\times L)}_{J\times L,\zeta}(X \wedge Y;(\overline{T}|J) \boxtimes (\overline{U}|L)) \\
& & \Big\downarrow \\
\tilde{H}^{\alpha+\beta}_{G\times K,\zeta}((G\times K)_+ \wedge_{J\times L} (X \wedge Y);\overline{T} \boxtimes \overline{U}) & \xrightarrow[\cong]{} & \tilde{H}^{\alpha+\beta-\zeta(G\times K/J\times L)}_{J\times L,\zeta}(X \wedge Y;(\overline{T} \boxtimes \overline{U})|(J\times L))
\end{array}$$

(7) *It respects restriction to subgroups: Under the same assumptions as the previous point,*

$$(x|J) \cup (y|L) = (x \cup y)|(J \times L).$$

(But see Remark 1.14.13.)

(8) *It respects restriction to fixed sets:*

$$x^J \cup y^L = (x \cup y)^{J \times L}.$$

(But, again, see Remark 1.14.13.)

The proofs are all standard except for the last three points.

Proof (Proof of Parts (6) and (7) of Theorem 1.14.12) (Refer to Sect. 1.13 for the algebra involved in the construction of the restriction to subgroups.) Note that the assumptions given are necessary for the statement to even make sense; see Proposition 1.14.15 for a related result. From the commutativity of the diagram

$$\begin{array}{ccccc}
\widehat{\mathscr{O}}_{J,\delta} \otimes \widehat{\mathscr{O}}_{L,\epsilon} & \xrightarrow{p} & \mathrm{Ho}(J \times L)\mathscr{P} & \xleftarrow{j} & \widehat{\mathscr{O}}_{J\times L,\zeta} \\
\Big\downarrow{\scriptstyle i_J^G \otimes i_L^K} & & \Big\downarrow{\scriptstyle i_{J\times L}^{G\times K}} & & \Big\downarrow{\scriptstyle i_{J\times L}^{G\times K}} \\
\widehat{\mathscr{O}}_{G,\delta} \otimes \widehat{\mathscr{O}}_{K,\epsilon} & \xrightarrow[p]{} & \mathrm{Ho}(G \times K)\mathscr{P} & \xleftarrow[j]{} & \widehat{\mathscr{O}}_{G\times K,\zeta}
\end{array}$$

(where j denotes the inclusions) it follows that there is a natural isomorphism

$$(G \times_J \overline{C}) \boxtimes (K \times_L \overline{D}) \cong (G \times K) \times_{J \times L} (\overline{C} \boxtimes \overline{D}).$$

The diagram and the $p_!$-p^* adjunction lead to a natural homomorphism

$$(\overline{T}|J) \boxtimes (\overline{U}|L) \to (\overline{T} \boxtimes \overline{U})|(J \times L)$$

(not necessarily an isomorphism). Together these give the following diagram.

$$\begin{array}{ccc}
\mathrm{Hom}_{\widehat{\mathscr{O}}_{G,\delta}}(G\times_J\overline{C},\overline{T}) \otimes \mathrm{Hom}_{\widehat{\mathscr{O}}_{K,\epsilon}}(K\times_L\overline{D},\overline{U}) & \xrightarrow{\cong} & \mathrm{Hom}_{\widehat{\mathscr{O}}_{J,\delta}}(\overline{C},\overline{T}|J) \otimes \mathrm{Hom}_{\widehat{\mathscr{O}}_{L,\epsilon}}(\overline{D},\overline{U}|L) \\
\Big\downarrow & & \Big\downarrow \\
\mathrm{Hom}_{\widehat{\mathscr{O}}_{G\times K,\zeta}}((G\times_J\overline{C})\boxtimes(K\times_L\overline{D}),\overline{T}\boxtimes\overline{U}) & & \mathrm{Hom}_{\widehat{\mathscr{O}}_{J\times L,\zeta}}(\overline{C}\boxtimes\overline{D},(\overline{T}|J)\boxtimes(\overline{U}|L)) \\
\cong\Big\downarrow & & \Big\downarrow \\
\mathrm{Hom}_{\widehat{\mathscr{O}}_{G\times K,\zeta}}((G\times K)\times_{J\times L}(\overline{C}\boxtimes\overline{D}),\overline{T}\boxtimes\overline{U}) & \xrightarrow{\cong} & \mathrm{Hom}_{\widehat{\mathscr{O}}_{J\times L,\zeta}}(\overline{C}\boxtimes\overline{D},(\overline{T}\boxtimes\overline{U})|(J\times L))
\end{array}$$

It is an elementary exercise to show that this diagram commutes. Letting $\overline{C} = \overline{C}_*^{J,\delta}(X)$ and $\overline{D} = \overline{C}_*^{L,\epsilon}(Y)$ shows that the cup product respects the Wirthmüller isomorphism on the chain level, so does so also in cohomology.

Since restriction from G to J is defined as the composite

$$\tilde{H}^{\alpha}_{G,\delta}(X;\overline{T}) \to \tilde{H}^{\alpha}_{G,\delta}(G_+ \wedge_J X;\overline{T}) \cong \tilde{H}^{\alpha-\delta(G/J)}_{J,\delta}(X;\overline{T}|J)$$

and the cup product is natural, that the cup product respects restriction to subgroups follows from the fact that it respects the Wirthmüller isomorphism. □

Proof of Part (8) of Theorem 1.14.12 (Refer to Sect. 1.13 for the algebra involved in the construction of restriction to fixed points.) By part (7) we know that the cup product respects the restriction to the normalizers NJ and NL, so we may as well assume that J is normal in G and L is normal in K. The naturality of the cup product means that it suffices to show that the cup product preserves the isomorphism of Theorem 1.13.22.

From the commutativity of the diagram

$$\begin{array}{ccccc}
\widehat{\mathscr{O}}_{G,\delta} \otimes \widehat{\mathscr{O}}_{K,\epsilon} & \xrightarrow{p} & \mathrm{Ho}(G\times K)\mathscr{P} & \xleftarrow{j} & \widehat{\mathscr{O}}_{G\times K,\zeta} \\
\downarrow{\scriptstyle \Phi^J\otimes\Phi^L} & & \downarrow{\scriptstyle \Phi^{J\times L}} & & \downarrow{\scriptstyle \Phi^{J\times L}} \\
\widehat{\mathscr{O}}_{G/J,\delta} \otimes \widehat{\mathscr{O}}_{K/L,\epsilon} & \xrightarrow[p]{} & \mathrm{Ho}(G\times K/J\times L)\mathscr{P} & \xleftarrow[j]{} & \widehat{\mathscr{O}}_{G\times K/J\times L,\zeta}
\end{array}$$

(where j denotes the inclusions) it follows that there is a natural isomorphism

$$\overline{C}^J \boxtimes \overline{D}^L \cong (\overline{C} \boxtimes \overline{D})^{J\times L}$$

and a natural homomorphism

$$\mathrm{Inf}^G_{G/J}\,\overline{T} \boxtimes \mathrm{Inf}^K_{K/L}\,\overline{U} \to \mathrm{Inf}^{G\times K}_{G\times K/J\times L}(\overline{T} \boxtimes \overline{U}).$$

The result now follows from the following commutative diagram:

$$\begin{array}{ccc}
\mathrm{Hom}_{\widehat{\mathscr{O}}_{G,\delta}}(\overline{C},\mathrm{Inf}^G_{G/J}\overline{T})\otimes\mathrm{Hom}_{\widehat{\mathscr{O}}_{K,\epsilon}}(\overline{D},\mathrm{Inf}^K_{K/L}\overline{U}) & \xrightarrow{\cong} & \mathrm{Hom}_{\widehat{\mathscr{O}}_{G/J,\delta}}(\overline{C}^J,\overline{T})\otimes\mathrm{Hom}_{\widehat{\mathscr{O}}_{K/L,\epsilon}}(\overline{D}^L,\overline{U}) \\
\downarrow & & \downarrow \\
\mathrm{Hom}_{\widehat{\mathscr{O}}_{G\times K,\zeta}}(\overline{C}\boxtimes\overline{D},\mathrm{Inf}^G_{G/J}\overline{T}\boxtimes\mathrm{Inf}^K_{K/L}\overline{U}) & & \mathrm{Hom}_{\widehat{\mathscr{O}}_{G\times K/J\times L,\zeta}}(\overline{C}^J\boxtimes\overline{D}^L,\overline{T}\boxtimes\overline{U}) \\
\downarrow & & \downarrow{\scriptstyle\cong} \\
\mathrm{Hom}_{\widehat{\mathscr{O}}_{G\times K,\zeta}}(\overline{C}\boxtimes\overline{D},\mathrm{Inf}^{G\times K}_{G\times K/J\times L}(\overline{T}\boxtimes\overline{U})) & \xrightarrow{\cong} & \mathrm{Hom}_{\widehat{\mathscr{O}}_{G\times K/J\times L,\zeta}}((\overline{C}\boxtimes\overline{D})^{J\times L},\overline{T}\boxtimes\overline{U})
\end{array}$$

Remark 1.14.13 Part (7) of Theorem 1.14.12 is stated a bit loosely. In fact,

$$(x|J) \cup (y|L) \in \tilde{H}^*_{J\times L,\zeta}(X \wedge Y; (\overline{T}|J) \boxtimes (\overline{U}|L))$$

while

$$(x \cup y)|(J \times L) \in \tilde{H}^*_{J\times L,\zeta}(X \wedge Y; (\overline{T} \boxtimes \overline{U})|(J \times L)).$$

The theorem should say that, when we apply the map induced by the homomorphism $(\overline{T}|J) \boxtimes (\overline{U}|L) \to (\overline{T} \boxtimes \overline{U})|(J \times L)$, the element $(x|J) \cup (y|L)$ maps to $(x \cup y)|(J \times L)$.

A similar comment applies to part (8) in general, when we have to first restrict to normalizers. However, if J is normal in G and L is normal in K, we can use the isomorphism $\overline{T}^J \boxtimes \overline{U}^L \cong (\overline{T} \boxtimes \overline{U})^{J\times L}$ to identify the two cohomology groups.

Remark 1.14.14 Part (4) of Theorem 1.14.12 states that the external product is commutative, in the sense given there. What about the internal cup product (in the case $\delta = \epsilon = 0$)

$$\tilde{H}^\alpha_G(X; \overline{T}) \otimes \tilde{H}^\beta_G(X; \overline{T}) \to \tilde{H}^{\alpha+\beta}_G(X; \overline{T})?$$

Recall that this product is defined when $\overline{T}$ is a ring, meaning that there is an associative product $\overline{T} \,\Box\, \overline{T} \to \overline{T}$. Let us further assume that the product is *commutative*, meaning that the following diagram commutes, where γ is the interchange map:

$$\begin{array}{ccc} \overline{T}\,\Box\,\overline{T} & \xrightarrow{\ \gamma\ } & \overline{T}\,\Box\,\overline{T} \\ & \searrow \qquad \swarrow & \\ & \overline{T}. & \end{array}$$

After restricting to diagonals, Theorem 1.14.12 tells us that, if $x \in \tilde{H}^\alpha_G(X; \overline{T})$ and $y \in \tilde{H}^\beta_G(X; \overline{T})$, then $x \cup y = \iota(y \cup x)$, where ι is the self map of $H^{\alpha+\beta}_G(X; \overline{T})$ induced by reversing the summands in $\alpha + \beta$ and then identifying $\beta + \alpha$ with $\alpha + \beta$ again. To be very precise: Write

$$\alpha = \bigoplus_i V_i^{m_i} \qquad \text{and} \qquad \beta = \bigoplus_i V_i^{n_i}$$

where the V_i are the irreducible representations of G and $m_i, n_i \in \mathbb{Z}$. Then ι is given by multiplication by the unit in $A(G)$ determined by the stable self-map

$$\bigwedge_i S^{(m_i+n_i)V_i} = \bigwedge_i S^{m_iV_i} \wedge S^{n_iV_i} \xrightarrow{\gamma} \bigwedge_i S^{n_iV_i} \wedge S^{m_iV_i} = \bigwedge_i S^{(m_i+n_i)V_i}.$$

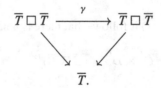

Call this element $\gamma(\alpha, \beta) \in A(G)$. Explicitly, if $\gamma_i = \gamma(V_i, V_i)$ is the unit given by the interchange map on $S^{V_i} \wedge S^{V_i}$, then

$$\gamma(\alpha, \beta) = \prod_i \gamma_i^{m_i n_i}.$$

We then we have the anti-commutativity rule

$$x \cup y = \gamma(\alpha, \beta) y \cup x$$

for the internal cup product.

We have the following result for Mackey-valued cohomology.

Proposition 1.14.15 *Let δ be a familial dimension function for G, let ϵ be a familial dimension function for K, and let ζ be a familial dimension function for $G \times K$ with $\zeta \succcurlyeq \delta \times \epsilon$. Then the external cup product gives a map of $(G \times K)$-ζ-Mackey functors*

$$- \cup - : \overline{H}^{\alpha}_{G,\delta}(X; \overline{T}) \boxtimes_{\zeta} \overline{H}^{\beta}_{K,\epsilon}(Y; \overline{U}) \to \overline{H}^{\alpha+\beta}_{G\times K,\zeta}(X \wedge Y; \overline{T} \boxtimes_{\zeta} \overline{U}).$$

When $G = K$ and $\mathscr{F}(\zeta)$ contains the diagonal subgroup Δ, the cup product restricts to give a map of G-ζ-Mackey functors

$$- \cup - : \overline{H}^{\alpha}_{G,\delta}(X; \overline{T}) \,\square_{\zeta}\, \overline{H}^{\beta}_{G,\epsilon}(Y; \overline{U}) \to \overline{H}^{\alpha+\beta-\zeta(G\times G/\Delta)}_{G,\zeta}(X \wedge Y; \overline{T} \,\square_{\zeta}\, \overline{U}).$$

Proof For $J \in \mathscr{F}(\delta)$, $L \in \mathscr{F}(\epsilon)$, and $M \in \mathscr{F}(\zeta)$, to make the following readable we introduce the abbreviations

$$A = G_+ \wedge_J S^{-\delta(G/J)},$$
$$B = K_+ \wedge_L S^{-\delta(K/L)},$$

and

$$C = (G \times K)_+ \wedge_M S^{-\zeta(G\times K/M)}.$$

We then have the composite

$$\begin{aligned}
&\tilde{H}^{\alpha}_{G,\delta}(X \wedge A; \overline{T}) \otimes \tilde{H}^{\beta}_{K,\epsilon}(Y \wedge B; \overline{U}) \otimes (G \times K)\mathscr{P}(C, A \wedge B) \\
&\xrightarrow{-\cup-} \tilde{H}^{\alpha+\beta}_{G\times K,\zeta}(X \wedge Y \wedge A \wedge B; \overline{T} \boxtimes \overline{U}) \otimes (G \times K)\mathscr{P}(C, A \wedge B) \\
&\to \tilde{H}^{\alpha+\beta}_{G\times K,\zeta}(X \wedge Y \wedge C; \overline{T} \boxtimes \overline{U})
\end{aligned}$$

Naturality of the cup product implies that this passes to the coend over A and B to give a map of Mackey functors

$$- \cup -: \overline{H}^{\alpha}_{G,\delta}(X;\overline{T}) \boxtimes \overline{H}^{\beta}_{K,\epsilon}(Y;\overline{U}) \to \overline{H}^{\alpha+\beta}_{G\times K,\zeta}(X \wedge Y; \overline{T} \boxtimes \overline{U}).$$

Restriction to the diagonal subgroup gives the internal version. □

The pairings in the preceding proposition were defined in terms of the pairing of G- and K-cohomology. Note, however, that Theorem 1.14.12(6) says that, if $J \times L \in \mathscr{F}(\zeta)$ and $\zeta(G \times K/J \times L) = \delta(G/J) \oplus \epsilon(K/L)$, then the component

$$\tilde{H}^{\alpha}_{J,\delta}(X;\overline{T}|J) \otimes \tilde{H}^{\beta}_{L,\epsilon}(Y;\overline{U}|L) \to H^{\alpha+\beta}_{J\times L,\zeta}(X \wedge Y; (\overline{T} \boxtimes \overline{U})|J \times L)$$

of the Mackey functor pairing, corresponding to the identity map on $(G \times K)_+ \wedge_{J\times L} S^{-\zeta(G\times K/J\times L)}$, agrees with the cup product pairing J- and L-cohomology.

Now we look at how the cup product is represented. Given the G-spectrum $H_\delta\overline{T}$ and K-spectrum $H_\epsilon\overline{U}$, we can form the $(G \times K)$-spectrum $H_\delta\overline{T} \wedge H_\epsilon\overline{U}$. The external cup product should then be represented by a $(G\times K)$-map $H_\delta\overline{T} \wedge H_\epsilon\overline{U} \to H_\zeta(\overline{T} \boxtimes \overline{U})$ that is an isomorphism in $\overline{\pi}_0^{G\times K,\zeta}$. An explicit construction is based on the following calculation.

Proposition 1.14.16 *Let δ be a familial dimension function for G, let ϵ be a familial dimension function for K, and let $\zeta \succcurlyeq \delta \times \epsilon$. Let $\overline{T}$ be a δ-G-Mackey functor and let $\overline{U}$ be an ϵ-K-Mackey functor. Then we have*

$$\overline{\pi}_n^{G\times K,\zeta}(H_\delta\overline{T} \wedge H_\epsilon\overline{U}) = \begin{cases} 0 & \text{if } n < 0 \\ \overline{T} \boxtimes_\zeta \overline{U} & \text{if } n = 0. \end{cases}$$

Proof There should be a proof based on a Künneth spectral sequence as developed in [26] and [40]. However, the former is the nonequivariant case and the latter deals only with finite groups. Developing the spectral sequence in the compact Lie case would take us too far afield and may not be available in the parametrized case we discuss later, anyway. So, we give a more elementary argument here.

We can assume that our Eilenberg-MacLane spectra are constructed as in Construction 1.12.5 as, for example $H_\delta\overline{T} = F(E\mathscr{F}(\delta)_+, P_\delta\overline{T})$. Consider the cofibrations

$$R_\delta\overline{T} \to F_\delta\overline{T} \to C_\delta\overline{T}$$

and

$$R_\epsilon\overline{U} \to F_\epsilon\overline{U} \to C_\epsilon\overline{U}$$

used in Construction 1.12.5, which, on applying $\overline{\pi}_0$, give the exact sequences

$$\overline{\pi}_0^{G,\delta} R_\delta \overline{T} \to \overline{\pi}_0^{G,\delta} F_\delta \overline{T} \to \overline{T} \to 0$$

and

$$\overline{\pi}_0^{K,\epsilon} R_\epsilon \overline{U} \to \overline{\pi}_0^{K,\epsilon} F_\epsilon \overline{U} \to \overline{U} \to 0.$$

If C is the cofiber of

$$(R_\delta \overline{T} \wedge F_\epsilon \overline{U}) \vee (F_\delta \overline{T} \wedge R_\epsilon \overline{U}) \to F_\delta \overline{T} \wedge F_\epsilon \overline{U},$$

we have an exact sequence

$$\begin{aligned}&\overline{\pi}_0^{G\times K,\zeta}(R_\delta \overline{T} \wedge F_\epsilon \overline{U}) \oplus \overline{\pi}_0^{G\times K,\zeta}(F_\delta \overline{T} \wedge R_\epsilon \overline{U}) \\ &\quad \to \overline{\pi}_0^{G\times K,\zeta}(F_\delta \overline{T} \wedge F_\epsilon \overline{U}) \to \overline{\pi}_0^{G\times K,\zeta}(C) \to 0.\end{aligned}$$

Because F_δ, R_δ, F_ϵ, and R_ϵ are wedges of spheres, using the calculation

$$\operatorname{Ho} G\mathscr{P}(-, (G/J, \delta) \wedge (K/L, \epsilon)) \cong \overline{A}_{G/J,\delta} \boxtimes_\zeta \overline{A}_{K/L,\delta}$$

on $\widehat{\mathscr{O}}_{G\times K,\zeta}$, the exact sequence above is isomorphic to

$$\begin{aligned}&(\overline{\pi}_0^{G,\delta}(R_\delta \overline{T}) \boxtimes \overline{\pi}_0^{K,\epsilon}(F_\epsilon \overline{U})) \oplus (\overline{\pi}_0^{G,\delta}(F_\delta \overline{T}) \boxtimes \overline{\pi}_0^{K,\epsilon}(R_\epsilon \overline{U})) \\ &\quad \to \overline{\pi}_0^{G,\delta}(F_\delta \overline{T}) \boxtimes \overline{\pi}_0^{K,\epsilon}(F_\epsilon \overline{U}) \to \overline{\pi}_0^{G\times K,\zeta}(C) \to 0,\end{aligned}$$

from which it follows that $\overline{\pi}_0^{G\times K,\zeta}(C) \cong \overline{T} \boxtimes \overline{U}$. Further, looking at the connectivities, we have

$$\overline{\pi}_0^{G\times K,\zeta}(C_\delta \overline{T} \wedge C_\epsilon \overline{U}) \cong \overline{\pi}_0^{G\times K,\zeta}(C) \cong \overline{T} \boxtimes \overline{U}.$$

Passing to P and then H, we conclude that

$$\overline{\pi}_0^{G\times K,\zeta}(H_\delta \overline{T} \wedge H_\epsilon \overline{U}) \cong \overline{T} \boxtimes \overline{U}$$

as claimed. The vanishing of the homotopy groups for negative n follows from the construction. □

As usual, for example, by killing higher homotopy groups, it follows that there is a map of $(G \times K)$-spectra

$$H_\delta \overline{T} \wedge H_\epsilon \overline{U} \to H_\zeta(\overline{T} \boxtimes_\zeta \overline{U})$$

that is an isomorphism in $\overline{\pi}_0^{G\times K,\zeta}$. That this represents the cup product in cohomology that we constructed on the chain level follows by considering its effect on the quotients $X^{\alpha+m}/X^{\alpha+m-1}$ and $Y^{\beta+n}/Y^{\beta+n-1}$ when X and Y are CW complexes.

When $G = K$ we can restrict to the diagonal $\Delta \leq G \times G$. Doing so gives us a map of G-spectra

$$H_\delta \overline{T} \wedge H_\epsilon \overline{U} \to \Sigma^{-\zeta(G\times G/\Delta)} H_\zeta(\overline{T} \square_\zeta \overline{U})$$

that is an isomorphism in $\overline{\pi}_0^{G,\zeta}$.

Finally, let's point out the specializations to several interesting choices of δ and ϵ, using the cup products internal in G.

Remark 1.14.17 The general cup product gives us the following special cases.

(1) Taking $\delta = \epsilon = 0$ on G and $\zeta = 0$ on $G \times G$, we have the cup product

$$\tilde{H}_G^\alpha(X; \overline{T}) \otimes \tilde{H}_G^\beta(Y; \overline{U}) \to \tilde{H}_G^{\alpha+\beta}(X \wedge Y; \overline{T} \square \overline{U}).$$

The Mackey functor-valued version is a pairing

$$\overline{H}_G^\alpha(X; \overline{T}) \square \overline{H}_G^\beta(Y; \overline{U}) \to \overline{H}_G^{\alpha+\beta}(X \wedge Y; \overline{T} \square \overline{U}).$$

This product is represented by a G-map

$$H\overline{T} \wedge H\overline{U} \to H(\overline{T} \square \overline{U}).$$

(2) The completely internal version of the preceding example is the following, where $\overline{T}$ is a ring:

$$\tilde{H}_G^\alpha(X; \overline{T}) \otimes \tilde{H}_G^\beta(X; \overline{T}) \to \tilde{H}_G^{\alpha+\beta}(X; \overline{T}).$$

This was discussed in more detail in Remark 1.14.14.

(3) Taking $\delta = \epsilon = \mathscr{L}$ on G and $\zeta = \mathscr{L}$ on $G \times G$, we have the cup product

$$\tilde{\mathscr{H}}_G^\alpha(X; \underline{R}) \otimes \tilde{\mathscr{H}}_G^\beta(Y; \underline{S}) \to \tilde{\mathscr{H}}_G^{\alpha+\beta-\mathscr{L}(G)}(X \wedge Y; \underline{R} \square_{\mathscr{L}} \underline{S}).$$

The Mackey functor-valued version is a pairing

$$\underline{\mathscr{H}}_G^\alpha(X; \underline{R}) \square \underline{\mathscr{H}}_G^\beta(Y; \underline{S}) \to \underline{\mathscr{H}}_G^{\alpha+\beta-\mathscr{L}(G)}(X \wedge Y; \underline{R} \square_{\mathscr{L}} \underline{S}).$$

These Mackey functors can all be thought of as either covariant G-Mackey functors or contravariant G-$\mathscr{L}$-Mackey functors. Here and in the following cases the notation indicates their variance as G-Mackey functors. This product is represented by a G-map

$$H_{\mathscr{L}}\underline{R} \wedge H_{\mathscr{L}}\underline{S} \to \Sigma^{-\mathscr{L}(G)} H_{\mathscr{L}}(\underline{R} \,\Box_{\mathscr{L}}\, \underline{S}).$$

(4) The completely internal version of the preceding example is the following, where $\underline{R}$ is a ring:

$$\tilde{\mathscr{H}}_G^{\alpha}(X;\underline{R}) \otimes \tilde{\mathscr{H}}_G^{\beta}(X;\underline{R}) \to \tilde{\mathscr{H}}_G^{\alpha+\beta-\mathscr{L}(G)}(X;\underline{R}).$$

(5) Taking $\delta = 0$, $\epsilon = \mathscr{L}$, and $\zeta = \mathscr{L}_\Delta$, we have the cup product

$$\tilde{H}_G^{\alpha}(X;\overline{T}) \otimes \tilde{\mathscr{H}}_G^{\beta}(Y;\underline{S}) \to \tilde{\mathscr{H}}_G^{\alpha+\beta}(X \wedge Y;\overline{T} \,\Box_{\mathscr{L}_\Delta}\, \underline{S}).$$

In terms of Mackey functor-valued theories we get

$$\overline{\tilde{H}_G^{\alpha}}(X;\overline{T}) \,\Box_{\mathscr{L}_\Delta}\, \underline{\tilde{\mathscr{H}}_G^{\beta}}(Y;\underline{S}) \to \underline{\tilde{\mathscr{H}}_G^{\alpha+\beta}}(X \wedge Y;\overline{T} \,\Box_{\mathscr{L}_\Delta}\, \underline{S}).$$

This product is represented by a G-map

$$H\overline{T} \wedge H_{\mathscr{L}}\underline{S} \to H_{\mathscr{L}}(\overline{T} \,\Box_{\mathscr{L}_\Delta}\, \underline{S}).$$

Note that $\mathscr{L}_\Delta$ restricts to $\mathscr{L}$ on the diagonal, so $\overline{T} \,\Box_{\mathscr{L}_\Delta}\, \underline{S}$ is a contravariant $\mathscr{L}$-Mackey functor.

(6) Taking $\delta = 0$, $\epsilon = \mathscr{L}$, and $\zeta = \mathscr{L} - \mathscr{L}_\Delta$, we have the cup product

$$\tilde{H}_G^{\alpha}(X;\overline{T}) \otimes \tilde{\mathscr{H}}_G^{\beta}(Y;\underline{S}) \to \tilde{H}_G^{\alpha+\beta-\mathscr{L}(G)}(X \wedge Y;\overline{T} \,\Box_{\mathscr{L}-\mathscr{L}_\Delta}\, \underline{S}).$$

In terms of Mackey functor-valued theories we get

$$\overline{\tilde{H}_G^{\alpha}}(X;\overline{T}) \,\Box_{\mathscr{L}-\mathscr{L}_\Delta}\, \underline{\tilde{\mathscr{H}}_G^{\beta}}(Y;\underline{S}) \to \overline{\tilde{H}_G^{\alpha+\beta-\mathscr{L}(G)}}(X \wedge Y;\overline{T} \,\Box_{\mathscr{L}-\mathscr{L}_\Delta}\, \underline{S}).$$

This product is represented by a G-map

$$H\overline{T} \wedge H_{\mathscr{L}}\underline{S} \to \Sigma^{-\mathscr{L}(G)} H(\overline{T} \,\Box_{\mathscr{L}-\mathscr{L}_\Delta}\, \underline{S}).$$

Note that $\overline{T} \,\Box_{\mathscr{L}-\mathscr{L}_\Delta}\, \underline{S}$ is an ordinary contravariant Mackey functor because $\mathscr{L} - \mathscr{L}_\Delta$ restricts to 0 on the diagonal.

1.14.3 Slant Products, Evaluations, and Cap Products

We now construct evaluation maps and cap products, but we start with the slant product that underlies both. As May points out in [50], the literature does not agree on how the slant product should be formulated. Unfortunately, if we were to follow

May's suggestion of using the earliest definition that appeared, the resulting cap product would not make homology be a module over cohomology. Given that, we use what we think is the most useful formulation for us.

A little bit of algebra first.

Definition 1.14.18 Let δ be a dimension function for G, let ϵ be a dimension function for K, and let ζ be a dimension function for $G\times K$ with $\zeta \succcurlyeq \delta\times\epsilon$. Suppose that $\overline{T}$ is a contravariant ϵ-K-Mackey functor and $\underline{U}$ is a covariant ζ-$(G\times K)$-Mackey functor. Then we define a covariant δ-G-Mackey functor $\overline{T} \lhd \underline{U}$ by

$$(\overline{T} \lhd \underline{U})(G/J,\delta) = \overline{T} \otimes_{\widehat{\mathscr{O}}_{K,\epsilon}} (p^* i_! \underline{U})((G/J,\delta)\otimes -)$$

where $p\colon \widehat{\mathscr{O}}_{G,\delta} \otimes \widehat{\mathscr{O}}_{K,\epsilon} \to \mathrm{Ho}(G\times K)\mathscr{P}$ and $i\colon \widehat{\mathscr{O}}_{G\times K,\zeta} \to \mathrm{Ho}(G\times K)\mathscr{P}$ are the functors given in Definition 1.14.4.

Example 1.14.19 As an example and as a calculation we'll need later, we show that

$$\overline{A}_{K/L,\epsilon} \lhd \underline{A}^{G\times K/M,\zeta} \cong \mathrm{Ho}(G\times K)\mathscr{P}((G\times K/M,\zeta), - \wedge (K/L,\epsilon)).$$

For, if $(G/J,\delta)$ is an object in $\widehat{\mathscr{O}}_{G,\delta}$, we have

$$\begin{aligned}
&(\overline{A}_{K/L,\epsilon} \lhd \underline{A}^{G\times K/M,\zeta})(G/J,\delta) \\
&\quad = \overline{A}_{K/L,\epsilon} \otimes_{\widehat{\mathscr{O}}_{K,\epsilon}} (p^* i_! \underline{A}^{G\times K/M,\zeta})((G/J,\delta)\otimes -) \\
&\quad = \overline{A}_{K/L,\epsilon} \otimes_{\widehat{\mathscr{O}}_{K,\epsilon}} \mathrm{Ho}(G\times K)\mathscr{P}((G\times K/M,\zeta),(G/J,\delta)\wedge -) \\
&\quad \cong \mathrm{Ho}(G\times K)\mathscr{P}((G\times K/M,\zeta),(G/J,\delta)\wedge (K/L,\epsilon)).
\end{aligned}$$

For X a based G-space and Y a based K-space, the external slant product will be a map

$$-\setminus -\colon \tilde{H}^{\beta}_{K,\epsilon}(Y;\overline{T}) \otimes \tilde{H}^{G\times K,\zeta}_{\alpha+\beta}(X\wedge Y;\underline{U}) \to \tilde{H}^{G,\delta}_{\alpha}(X;\overline{T}\lhd\underline{U}).$$

On the chain level, we take the following map:

$$\begin{aligned}
&\mathrm{Hom}_{\widehat{\mathscr{O}}_{K,\epsilon}}(\overline{C}^{K,\epsilon}_{\beta+*}(Y),\overline{T}) \otimes (\overline{C}^{G\times K,\zeta}_{\alpha+\beta+*}(X\wedge Y) \otimes_{\widehat{\mathscr{O}}_{G\times K,\zeta}} \underline{U}) \\
&\to \mathrm{Hom}_{\widehat{\mathscr{O}}_{K,\epsilon}}(\overline{C}^{K,\epsilon}_{\beta+*}(Y),\overline{T}) \otimes ((\overline{C}^{G,\delta}_{\alpha+*}(X) \boxtimes_\zeta \overline{C}^{K,\epsilon}_{\beta+*}(Y)) \otimes_{\widehat{\mathscr{O}}_{G\times K,\zeta}} \underline{U}) \\
&\xrightarrow{\nu} (\overline{C}^{G,\delta}_{\alpha+*}(X) \boxtimes_\zeta \overline{T}) \otimes_{\widehat{\mathscr{O}}_{G\times K,\zeta}} \underline{U} \\
&\cong (\overline{C}^{G,\delta}_{\alpha+*}(X) \otimes \overline{T}) \otimes_{\widehat{\mathscr{O}}_{G,\delta}\otimes\widehat{\mathscr{O}}_{K,\epsilon}} p^* i_! \underline{U} \\
&\cong \overline{C}^{G,\delta}_{\alpha+*}(X) \otimes_{\widehat{\mathscr{O}}_{G,\delta}} (\overline{T} \otimes_{\widehat{\mathscr{O}}_{K,\epsilon}} p^* i_! \underline{U}) \\
&= \overline{C}^{G,\delta}_{\alpha+*}(X) \otimes_{\widehat{\mathscr{O}}_{G,\delta}} (\overline{T} \lhd \underline{U}).
\end{aligned}$$

The first arrow is induced by the map μ_* from Definition 1.14.11. The map ν is evaluation, with the sign

$$\nu(y \otimes a \otimes b \otimes u) = (-1)^{pq} a \otimes y(b) \otimes u$$

if $y \in \mathrm{Hom}_{\widehat{\mathcal{O}}_{K,\epsilon}}(\overline{C}^{K,\epsilon}_{\beta+p}(Y), \overline{T})$ and $a \in \overline{C}^{G,\delta}_{\alpha+q}(X)$. Our various sign conventions imply that the composite above is a chain map. Taking homology defines our slant product. The following properties follow easily from the definition, except for the last, which follows from examining chains and arguing much as in the proof of Part (6) of Theorem 1.14.12.

Theorem 1.14.20 *Let δ be a familial dimension function for G, let ϵ be a familial dimension function for K, and let ζ be a familial dimension function for $G \times K$ with $\zeta \succcurlyeq \delta \times \epsilon$. Let α be a virtual representation of G and let β be a virtual representation of K. The slant product*

$$- \setminus -: \tilde{H}^{\beta}_{K,\epsilon}(Y; \overline{T}) \otimes \tilde{H}^{G\times K,\zeta}_{\alpha+\beta}(X \wedge Y; \underline{U}) \to \tilde{H}^{G,\delta}_{\alpha}(X; \overline{T} \lhd \underline{U}).$$

has the following properties.

(1) *It is natural in the following sense: Given a G-map $f: X \to X'$, a K-map $g: Y \to Y'$, and elements $y' \in \tilde{H}^{\beta}_{K,\epsilon}(Y'; \overline{T})$ and $z \in \tilde{H}^{G\times K,\zeta}_{\alpha+\beta}(X \wedge Y; \underline{U})$, we have*

$$y' \setminus (f \wedge g)_*(z) = f_*(g^*(y') \setminus z).$$

Put another way, the slant product is a natural transformation in its adjoint form

$$\tilde{H}^{G\times K,\zeta}_{\alpha+\beta}(X \wedge Y; \underline{U}) \to \mathrm{Hom}(\tilde{H}^{\beta}_{K,\epsilon}(Y; \overline{T}), \tilde{H}^{G,\delta}_{\alpha}(X; \overline{T} \lhd \underline{U})).$$

(2) *It respects suspension in the sense that*

$$(\sigma^W y) \setminus (\sigma^{V+W} z) = \sigma^V (y \setminus z).$$

(3) *It is associative in the following sense. Suppose given three groups, G, K, and L, with respective familial dimension functions δ, ϵ, and ζ. Suppose that $\eta \succcurlyeq \delta \times \epsilon$, $\theta \succcurlyeq \epsilon \times \zeta$, and $\kappa \succcurlyeq \eta \times \zeta$, $\kappa \succcurlyeq \delta \times \theta$ are also familial. Given $y \in \tilde{H}^{\beta}_{K,\epsilon}(Y; \overline{R})$, $z \in \tilde{H}^{\gamma}_{L,\zeta}(Z; \overline{S})$, and $w \in \tilde{H}^{G\times K\times L,\theta}_{\alpha+\beta+\gamma}(X \wedge Y \wedge Z; \underline{U})$, we have*

$$(y \cup z) \setminus w = y \setminus (z \setminus w).$$

That is, the following diagram commutes:

$$\begin{array}{ccc}
\tilde{H}^{\beta}_{K,\epsilon}(Y;\overline{R})\otimes\tilde{H}^{\gamma}_{L,\zeta}(Z;\overline{S})\otimes\tilde{H}^{G\times K\times L,\kappa}_{\alpha+\beta+\gamma}(X\wedge Y\wedge Z;\underline{U}) & \longrightarrow & \tilde{H}^{\beta}_{K,\epsilon}(Y;\overline{R})\otimes\tilde{H}^{G\times K,\eta}_{\alpha+\beta}(X\wedge Y;\overline{S}\lhd\underline{U}) \\
\downarrow & & \downarrow \\
\tilde{H}^{\beta+\gamma}_{K\times L,\theta}(Y\wedge Z;\overline{R}\boxtimes\overline{S})\otimes\tilde{H}^{G\times K\times L,\kappa}_{\alpha+\beta+\gamma}(X\wedge Y\wedge Z;\underline{U}) & \longrightarrow & \tilde{H}^{G,\delta}_{\alpha}(X;\overline{R}\lhd(\overline{S}\lhd\underline{U}))
\end{array}$$

Note that we're using the algebraic identity $(\overline{R}\boxtimes\overline{S})\lhd\underline{U}\cong\overline{R}\lhd(\overline{S}\lhd\underline{U})$.

(4) *It respects the Wirthmüller isomorphism: Suppose that* $J\leq G$ *and* $L\leq K$, *that* $J\in\mathscr{F}(\delta)$, $K\in\mathscr{F}(\epsilon)$, *and* $J\times L\in\mathscr{F}(\zeta)$, *and that* $\zeta(G\times K/J\times L)=\delta(G/K)\oplus\epsilon(K/L)$. *Then the following diagram commutes:*

$$\begin{array}{ccc}
\tilde{H}^{\beta}_{K,\epsilon}(K_+\wedge_L Y;\overline{T})\otimes\tilde{H}^{G\times K,\zeta}_{\alpha+\beta}(G_+\wedge_J X\wedge K_+\wedge_L Y;\underline{U}) & \longrightarrow & \tilde{H}^{G,\delta}_{\alpha}(G_+\wedge_J X;\overline{T}\lhd\underline{U}) \\
\cong\downarrow & & \downarrow\cong \\
\tilde{H}^{\beta-\epsilon(K/L)}_{L,\epsilon}(Y;\overline{T}|L)\otimes\tilde{H}^{J\times L,\zeta}_{\alpha+\beta-\zeta(G\times K/J\times L)}(X\wedge Y;\underline{U}|J\times L) & \longrightarrow & \tilde{H}^{J,\delta}_{\alpha-\delta(G/J)}(X;(\overline{T}\lhd\underline{U})|J)
\end{array}$$

In the bottom row we are implicitly using a map $(\overline{T}|L)\lhd(\underline{U}|J\times L)\to(\overline{T}\lhd\underline{U})|J$ *that need not be an isomorphism.* □

Naturality implies that we have the following version of the slant product for the Mackey functor-valued theories:

$$-\backslash-:\overline{H}^{\beta}_{K,\epsilon}(Y;\overline{T})\lhd\underline{H}^{G\times K,\zeta}_{\alpha+\beta}(X\wedge Y;\underline{U})\to\underline{H}^{G,\delta}_{\alpha}(X;\overline{T}\lhd\underline{U}).$$

We're most interested in the internalization of the slant product to the diagonal $\Delta\leq G\times G$. So, we also call the following a slant product.

Definition 1.14.21 Let δ and ϵ be familial dimension functions for G and let $\zeta\succcurlyeq\delta\times\epsilon$ be a familial dimension function for $G\times G$; assume that $\Delta\in\mathscr{F}(\zeta)$ and write ζ again for $\zeta|\Delta$. (Recall that the only way we can have $\Delta\in\mathscr{F}(\zeta)$ is for δ and ϵ to be complete.) If $\overline{T}$ is a contravariant ϵ-G-Mackey functor and $\underline{U}$ is a covariant ζ-G-Mackey functor, write

$$\overline{T}\lhd\underline{U}=\overline{T}\lhd[(G\times G)\times_{\Delta}\underline{U}],$$

a covariant δ-G-Mackey functor defined using the external version of $\lhd$ on the right. If δ and ϵ are complete, we define the *internal slant product*

$$-\backslash-:\tilde{H}^{\beta}_{G,\epsilon}(Y;\overline{T})\otimes\tilde{H}^{G,\zeta}_{\alpha+\beta-\zeta(G\times G/\Delta)}(X\wedge Y;\underline{U})\to\tilde{H}^{G,\delta}_{\alpha}(X;\overline{T}\lhd\underline{U})$$

to be the composite

$$\begin{aligned}
&\tilde{H}^{\beta}_{G,\epsilon}(Y;\overline{T}) \otimes \tilde{H}^{G,\zeta}_{\alpha+\beta-\zeta(G\times G/\Delta)}(X \wedge Y;\underline{U}) \\
&\to \tilde{H}^{\beta}_{G,\epsilon}(Y;\overline{T}) \otimes \tilde{H}^{G,\zeta}_{\alpha+\beta-\zeta(G\times G/\Delta)}(X \wedge Y;((G\times G)\times_\Delta \underline{U})|\Delta) \\
&\cong \tilde{H}^{\beta}_{G,\epsilon}(Y;\overline{T}) \otimes \tilde{H}^{G\times G,\zeta}_{\alpha+\beta}((G\times G)_+ \wedge_\Delta (X\wedge Y);(G\times G)\times_\Delta \underline{U}) \\
&\to \tilde{H}^{\beta}_{G,\epsilon}(Y;\overline{T}) \otimes \tilde{H}^{G\times G,\zeta}_{\alpha+\beta}(X\wedge Y;(G\times G)\times_\Delta \underline{U}) \\
&\to \tilde{H}^{G,\delta}_{\alpha}(X;\overline{T} \lhd \underline{U}).
\end{aligned}$$

The first map is the unit $\underline{U} \to ((G\times G)\times_\Delta \underline{U})|\Delta$, the second is the Wirthmüller isomorphism, the third is induced by the $(G\times G)$-map $(G\times G)_+ \wedge_\Delta (X\wedge Y) \to X\wedge Y$, and the last map is the external slant product.

Of course, this gives an internal slant product of Mackey functor-valued theories as well:

$$-\backslash -: \overline{H}^{\beta}_{G,\epsilon}(Y;\overline{T}) \lhd \underline{H}^{G,\zeta}_{\alpha+\beta-\zeta(G\times G/\Delta)}(X\wedge Y;\underline{U}) \to \underline{H}^{G,\delta}_{\alpha}(X;\overline{T}\lhd \underline{U})$$

We can now use the internal slant product to define evaluation and the cap product.

Definition 1.14.22 Let δ be a complete dimension function for G. The *evaluation map*

$$\langle -,-\rangle: \tilde{H}^{\beta}_{G,\delta}(X;\overline{T}) \otimes \tilde{H}^{G,\delta}_{\alpha+\beta}(X;\underline{U}) \to \tilde{H}^{G,0}_{\alpha}(S^0;\overline{T}\lhd \underline{U})$$

is the slant product

$$-\backslash -: \tilde{H}^{\beta}_{G,\delta}(X;\overline{T}) \otimes \tilde{H}^{G,\delta_\Delta}_{\alpha+\beta}(S^0 \wedge X;\underline{U}) \to \tilde{H}^{G,0}_{\alpha}(S^0;\overline{T}\lhd \underline{U}).$$

Here we are using $\delta_\Delta \succcurlyeq 0\times\delta$ and the fact that $\delta_\Delta(G\times G/\Delta) = 0$ and $\delta_\Delta|\Delta = \delta$.

Note that we can express the naturality of evaluation by saying that the adjoint map

$$\tilde{H}^{\beta}_{G,\delta}(X;\overline{T}) \to \mathrm{Hom}(\tilde{H}^{G,\delta}_{\alpha+\beta}(X;\underline{U}), \tilde{H}^{G,0}_{\alpha}(S^0;\overline{T}\lhd \underline{U}))$$

is natural in X. When $\alpha = n \in \mathbb{Z}$, $\tilde{H}^{G,0}_{n}(S^0;\overline{T}\lhd \underline{U})$ is nonzero only when $n = 0$ and is then $(\overline{T}\lhd \underline{U})(G/G) \cong \overline{T} \otimes_{\widehat{\mathscr{O}}_G} \underline{U}$, giving the evaluation

$$\tilde{H}^{\beta}_{G,\delta}(X;\overline{T}) \to \mathrm{Hom}(\tilde{H}^{G,\delta}_{\beta}(X;\underline{U}), \overline{T} \otimes_{\widehat{\mathscr{O}}_G} \underline{U}).$$

There are many other interesting variations available, which we leave to the imagination of the reader.

Definition 1.14.23 Let δ, ϵ, and ζ be as in Definition 1.14.21. The *cap product*

$$-\cap -\colon \tilde{H}^{\beta}_{G,\epsilon}(X;\overline{T}) \otimes \tilde{H}^{G,\zeta}_{\alpha+\beta-\zeta(G\times G/\Delta)}(X;\underline{U}) \to \tilde{H}^{G,\delta}_{\alpha}(X;\overline{T} \lhd \underline{U})$$

is the composite

$$\begin{aligned}
&\tilde{H}^{\beta}_{G,\epsilon}(X;\overline{T}) \otimes \tilde{H}^{G,\zeta}_{\alpha+\beta-\zeta(G\times G/\Delta)}(X;\underline{U}) \\
&\quad \to \tilde{H}^{\beta}_{G,\epsilon}(X;\overline{T}) \otimes \tilde{H}^{G,\zeta}_{\alpha+\beta-\zeta(G\times G/\Delta)}(X\wedge X;\underline{U}) \\
&\quad \to \tilde{H}^{G,\delta}_{\alpha}(X;\overline{T} \lhd \underline{U})
\end{aligned}$$

where the first map is induced by the diagonal $X \to X \wedge X$.

We get interesting special cases by considering particular choices of δ, ϵ, and ζ. One we'll use later is the case $\delta = \mathscr{L} - \epsilon$ and $\zeta = \mathscr{L}_\Delta$. Using the facts that $\mathscr{L}_\Delta(G \times G/\Delta) = 0$ and $\mathscr{L}_\Delta|\Delta = \mathscr{L}$, the cap product in this case takes the form

$$-\cap -\colon \tilde{H}^{\beta}_{G,\epsilon}(X;\overline{T}) \otimes \tilde{\mathscr{H}}^{G}_{\alpha+\beta}(X;\underline{U}) \to \tilde{H}^{G,\mathscr{L}-\epsilon}_{\alpha}(X;\overline{T} \lhd \underline{U}).$$

Specializing further, we can use $\underline{U} = \overline{A}_{G/G}$ (considered as a covariant $\mathscr{L}$-G-Mackey functor) and the following calculation.

Proposition 1.14.24 *If $\overline{T}$ is an ϵ-G-Mackey functor, then $\overline{T} \lhd \overline{A}_{G/G} \cong \overline{T}$ as covariant $(\mathscr{L} - \epsilon)$-G-Mackey functors.*

Proof Write $\delta = \mathscr{L} - \epsilon$. We first compute

$$\begin{aligned}
&(p^* i_!(G \times G) \times_\Delta \overline{A}_{G/G})((G/H, \delta) \otimes (G/K, \epsilon)) \\
&\quad \cong \mathrm{Ho}(G \times G)\mathscr{P}(\Sigma^\infty (G \times G)/\Delta_+, (G_+ \wedge_H S^{-\delta(G/H)}) \wedge (G_+ \wedge_K S^{-\epsilon(G/K)})) \\
&\quad \cong \mathrm{Ho}\, G\mathscr{P}(S, (G_+ \wedge_H S^{-\delta(G/H)}) \wedge (G_+ \wedge_K S^{-\epsilon(G/K)})) \\
&\quad \cong \mathrm{Ho}\, G\mathscr{P}(G_+ \wedge_H S^{-\epsilon(G/H)}, G_+ \wedge_K S^{-\epsilon(G/K)}) \\
&\quad = \underline{A}^{(G/H,\epsilon)}(G/K, \epsilon),
\end{aligned}$$

where we use duality to get the third isomorphism. Therefore, we have

$$\begin{aligned}
(\overline{T} \lhd \overline{A}_{G/G})(G/H, \delta) &= \overline{T} \otimes_{\widehat{\mathscr{O}}_{G,\epsilon}} (p^* i_!(G \times G) \times_\Delta \overline{A}_{G/G})((G/H, \delta) \otimes -) \\
&\cong \overline{T} \otimes_{\widehat{\mathscr{O}}_{G,\epsilon}} \underline{A}^{(G/H,\epsilon)} \\
&\cong \overline{T}(G/H, \epsilon),
\end{aligned}$$

as claimed. □

Thus, we have a cap product

$$-\cap -: \tilde{H}^{\beta}_{G,\epsilon}(X;\overline{T}) \otimes \tilde{\mathscr{H}}^{G}_{\alpha+\beta}(X;\overline{A}_{G/G}) \to \tilde{H}^{G,\mathscr{L}-\epsilon}_{\alpha}(X;\overline{T}).$$

Both the evaluation map and the cap product inherit properties from Theorem 1.14.20. In particular, the associativity property gives us the following:

$$(x \cup y) \cap z = x \cap (y \cap z)$$

and

$$\langle x \cup y, z \rangle = \langle x, y \cap z \rangle$$

when x, y, and z lie in appropriate groups.

Let us now look at these pairings on the spectrum level. The main result is the following calculation.

Proposition 1.14.25 *Let δ be a familial dimension function for G, let ϵ be a familial dimension function for K, and let ζ be a familial dimension function for $G \times K$ with $\zeta \succcurlyeq \delta \times \epsilon$. Let $\overline{T}$ be a contravariant K-ϵ-Mackey functor and let $\underline{U}$ be a covariant $(G \times K)$-ζ-Mackey functor. Then*

$$\pi^{G,\mathscr{L}-\delta}_{n}((H_\epsilon \overline{T} \wedge H^\zeta \underline{U})^K) \cong \begin{cases} 0 & \text{if } n < 0 \\ \overline{T} \lhd \underline{U} & \text{if } n = 0. \end{cases}$$

Proof The proof is, in outline, the same as that of Proposition 1.14.16. The calculational input this time is that

$$\begin{aligned} &\mathrm{Ho}(G \times K)\mathscr{P}((G/J, \mathscr{L} - \delta), (K/L, \epsilon) \wedge (G \times K/M, \mathscr{L} - \zeta)) \\ &\quad \cong \mathrm{Ho}(G \times K)\mathscr{P}((G \times K/M, \zeta), (G/J, \delta) \wedge (K/L, \epsilon)) \end{aligned}$$

so that

$$\mathrm{Ho}(G \times K)\mathscr{P}(-, (K/L, \epsilon) \wedge (G \times K/M, \mathscr{L} - \zeta)) \cong \overline{A}_{K/L,\epsilon} \lhd \underline{A}^{G\times K/M,\zeta}$$

as modules over $\widehat{\mathscr{O}}_{G,\delta}$, using Example 1.14.19. The result then follows by analyzing the structure provided by Construction 1.12.5, as in the proof of Proposition 1.14.16. □

It follows that there is a map

$$(H_\epsilon \overline{T} \wedge H^\zeta \underline{U})^K \to P^\delta(\overline{T} \lhd \underline{U})$$

of G-spectra that is an isomorphism on $\overline{\pi}_0^{G,\mathscr{L}-\delta}$, where $P^\delta(\overline{T} \lhd \underline{U})$ is a spectrum with $\overline{\pi}_n^{G,\mathscr{L}-\delta}$ homotopy concentrated in dimension 0. We then use that $E\mathscr{F}(\zeta) \times p_1^* E\mathscr{F}(\delta) \simeq E\mathscr{F}(\zeta)$ because $\mathscr{F}(\zeta) \subset \mathscr{F}(\delta) \times \mathscr{A}(K)$, where $\mathscr{A}(K)$ is the collection of all subgroups of K, so that $H^\zeta \underline{U} \wedge p_1^* E\mathscr{F}(\delta)_+ \simeq H^\zeta \underline{U}$, to get a map

$$
\begin{aligned}
(H_\epsilon \overline{T} \wedge H^\zeta \underline{U})^K &\simeq (H_\epsilon \overline{T} \wedge H^\zeta \underline{U} \wedge p_1^* E\mathscr{F}(\delta)_+)^K \\
&\simeq (H_\epsilon \overline{T} \wedge H^\zeta \underline{U})^K \wedge E\mathscr{F}(\delta)_+ \\
&\to P^\delta(\overline{T} \lhd \underline{U}) \wedge E\mathscr{F}(\delta)_+ \\
&\simeq H^\delta(\overline{T} \lhd \underline{U}).
\end{aligned}
$$

The slant product

$$
-\backslash -: \tilde{H}^\beta_{K,\epsilon}(Y;\overline{T}) \otimes \tilde{H}^{G\times K,\zeta}_{\alpha+\beta}(X \wedge Y; \underline{U}) \to \tilde{H}^{G,\delta}_\alpha(X; \overline{T} \lhd \underline{U}).
$$

is then represented as follows:

$$
\begin{aligned}
&[Y, H_\epsilon \overline{T} \wedge S^\beta]_K \otimes [S^{\alpha+\beta}, H^\zeta \underline{U} \wedge X \wedge Y]_{G\times K} \\
&\to [S^{\alpha+\beta}, H^\zeta \underline{U} \wedge X \wedge H_\epsilon \overline{T} \wedge S^\beta]_{G\times K} \\
&\cong [S^\alpha, H_\epsilon \overline{T} \wedge H^\zeta \underline{U} \wedge X]_{G\times K} \\
&\cong [S^\alpha, (H_\epsilon \overline{T} \wedge H^\zeta \underline{U} \wedge X)^K]_G \\
&\cong [S^\alpha, (H_\epsilon \overline{T} \wedge H^\zeta \underline{U})^K \wedge X]_G \\
&\to [S^\alpha, H^\delta(\overline{T} \lhd \underline{U}) \wedge X]_G.
\end{aligned}
$$

1.15 The Thom Isomorphism and Poincaré Duality

Two of the most useful calculational results in nonequivariant homology are the Thom isomorphism and Poincaré duality. We give versions here for the $RO(G)$-graded theories. These results apply only under very restrictive conditions, so we present them here not so much for their own sake but as prelude to the more general results we shall show for the $RO(\Pi B)$-graded theories we shall discuss later.

1.15.1 The Thom Isomorphism

Definition 1.15.1 Let $p: E \to B$ be a V-bundle as in Example 1.1.5(5). Let $D(p)$ and $S(p)$ denote the disk and sphere bundles of p, respectively, and let $T(p) = D(p)/S(p)$, the *Thom space* of p, be the quotient space. A *Thom class* for p is an element $t \in \tilde{H}_G^V(T(p); \overline{A}_{G/G})$ such that, for each G-map $b: G/K \to B$,

$$\begin{aligned} b^*(t) &\in \tilde{H}_G^V(T(b^*p); \overline{A}_{G/G}) \\ &\cong \tilde{H}_G^V(\Sigma^V G/K_+; \overline{A}_{G/G}) \\ &\cong A(K) \end{aligned}$$

is a generator of $A(K)$ as an $A(K)$-module, i.e., a unit.

Remarks 1.15.2 A Thom class must live in ordinary cohomology $\tilde{H}^*_{G,\delta}$ with $\delta = 0$. The isomorphism $\tilde{H}_G^V(\Sigma^V G/K_+; \overline{A}_{G/G}) \cong A(K)$ would not hold for all orbits with any other δ. We could generalize the definition to use the dimension function δ given by the restriction of the 0 dimension function to a family $\mathscr{F}$, if we require the base space B to be an $\mathscr{F}$-space (that is, have all its isotropy in $\mathscr{F}$), but there seems to be no advantage to doing so. In particular, the ordinary cohomology and the δ-cohomology of the Thom space would be isomorphic under these conditions.

The requirement that p be a V-bundle is highly restrictive. One of our motivations for discussing $RO(\Pi B)$-graded cohomology later is to remove this restriction.

A Thom class for p is related to Thom classes for the fixed-point bundles $p^K: E^K \to B^K$ as follows.

Proposition 1.15.3 *The following are equivalent for a cohomology class* $t \in \tilde{H}_G^V(T(p); \overline{A}_{G/G})$.

(1) *t is a Thom class for p.*
(2) *For every subgroup $K \leq G$, $t|K \in \tilde{H}_K^V(T(p); \overline{A}_{K/K})$ is a Thom class for p as a K-bundle.*
(3) *For every subgroup $K \leq G$, $t^K \in \tilde{H}_{WK}^{V^K}(T(p^K); \overline{A}_{WK/WK})$ is a Thom class for p^K as a WK-bundle.*
(4) *For every subgroup $K \leq G$, $t^K|e \in \tilde{H}^{|V^K|}(T(p^K); \mathbb{Z})$ is a Thom class for p^K as a nonequivariant bundle.*

Proof To show that (1) $\Rightarrow$ (2), let $b: K/J \to B$ be a K-map, let $\tilde{b}: G/J \to B$ be the extension of b to a G-map, and consider the following diagram.

$$\begin{array}{ccc} \tilde{H}_G^V(T(p); \overline{A}_{G/G}) & \longrightarrow & \tilde{H}_K^V(T(p); \overline{A}_{K/K}) \\ \Big\downarrow {\scriptstyle \tilde{b}^*} & & \Big\downarrow {\scriptstyle b^*} \\ \tilde{H}_G^V(T(\tilde{b}^*p); \overline{A}_{G/G}) & \xrightarrow[\cong]{} & \tilde{H}_K^V(T(b^*p); \overline{A}_{K/K}) \end{array}$$

We need to show that the image of t, under the composite across the top and right, is a generator of the copy of $A(J)$ on the bottom right. However, by definition, the image of t is a generator in the bottom left, and the bottom map is an example of the Wirthmüller isomorphism.

To show that (2) $\Rightarrow$ (3), it suffices to assume that K is normal in G (otherwise, begin by restricting to NK). Let $b\colon G/L \to B^K$ where $K \le L \le G$, and consider the following diagram:

$$\begin{array}{ccc}
\tilde{H}_G^V(T(p);\overline{A}_{G/G}) & \longrightarrow & \tilde{H}_{G/K}^{V^K}(T(p^K);\overline{A}_{(G/K)/(G/K)}) \\
\downarrow{\scriptstyle b^*} & & \downarrow{\scriptstyle b^*} \\
\tilde{H}_G^V(T(b^*p);\overline{A}_{G/G}) & \longrightarrow & \tilde{H}_{G/K}^{V^K}(T(b^*p^K);\overline{A}_{(G/K)/(G/K)}) \\
\downarrow{\scriptstyle \cong} & & \downarrow{\scriptstyle \cong} \\
\tilde{H}_G^V(\Sigma^V G/L_+;\overline{A}_{G/G}) & \xrightarrow[\cong]{} & \tilde{H}_{G/K}^{V^K}(\Sigma^{V^K} G/L_+;\overline{A}_{(G/K)/(G/K)})
\end{array}$$

Again, we need to show that the image of t across the top and down the right is a generator. By definition, the image down the left is a generator, and the bottom map is an isomorphism because restriction to fixed sets respects suspension, as shown in Sect. 1.13.2.

That (3) $\Rightarrow$ (4) follows from the implication (1) $\Rightarrow$ (2). That (4) $\Rightarrow$ (1) follows from the fact that, for every K, an element of $A(K) \cong \operatorname{colim}_W[S^W, S^W]_K$ is a generator if and only if its restriction to $\operatorname{colim}_W[S^{W^J}, S^{W^J}]$ is a unit (that is, $\pm 1 \in \mathbb{Z}$) for each subgroup J of K. (This follows from, for example, [42, V.1.9].)$\square$

In order to state the Thom isomorphism, we need the correct definition of orientation. The appropriate notion for $RO(G)$-graded cohomology is the following, which was called a "naive" orientation in [20, 11.2].

Definition 1.15.4 Let $p\colon E \to B$ be a V-bundle. If $b\colon G/H \to B$ is a G-map, let $E_b = p^{-1}(b(eH))$ be the corresponding fiber and define an orientation of E_b to be a choice of a homotopy class $\varphi(b)$ of stable spherical H-maps $E_b \to V$, i.e., stable homotopy equivalences $\Sigma_H^\infty S^{E_b} \to \Sigma_H^\infty S^V$. A *strict orientation* of p is a compatible collection $\{\varphi(b)\}$ of orientations of each E_b, where compatibility means that the following two conditions are satisfied:

(1) Let $b\colon G/H \to B$ and $\alpha\colon G/K \to G/H$, with $\alpha(eK) = gH$. Then

$$\varphi(b \circ \alpha) = g\varphi(b)g^{-1}.$$

(2) Let $\omega\colon G/H \times I \to B$ be a homotopy from ω_0 to ω_1 and let $\bar{\omega}\colon E_{\omega_0} \to E_{\omega_1}$ be the induced homotopy class of H-linear isometries. Then

$$\varphi(\omega_0) = \varphi(\omega_1) \circ \bar{\omega}.$$

We say that p is *strictly orientable* if there exists a strict orientation of p.

Theorem 1.15.5 (Thom Isomorphism) *If $p: E \to B$ is a V-bundle, then p is strictly orientable if and only if it has a Thom class $t \in \tilde{H}^V_G(T(p); \overline{A}_{G/G})$. Moreover, there is a one-to-one correspondence between strict orientations of p and Thom classes for p. For any Thom class t, the map*

$$t \cup -: H^{\alpha}_{G,\delta}(B; \overline{U}) \to \tilde{H}^{\alpha+V}_{G,\delta}(T(p); \overline{U})$$

is an isomorphism for any familial δ.

In the statement, the cup product uses the special case of $p_2^*\delta \succcurlyeq 0 \times \delta$ (see the discussion preceding Theorem 1.14.12) and is internalized along the diagonal map $T(p) \to T(p) \wedge B_+$.

Proof The theorem is clear when p is trivial; the correspondence between Thom classes and orientations comes from the correspondence between units in $A(K)$ and stable homotopy classes of self-equivalences of spheres. The general case follows, as it does nonequivariantly, by a Mayer-Vietoris patching argument (see [52] or [14]). □

1.15.2 Poincaré Duality

We now outline a proof of Poincaré duality for compact oriented V-manifolds. The noncompact case can be handled in the usual way using cohomology with compact supports, as in [52] or [14].

Definition 1.15.6 Let M be a closed V-manifold as in Example 1.1.5(6). A *fundamental class* of M is a class $[M] \in \mathscr{H}^G_V(M; \overline{A}_{G/G})$ such that, for each point $m \in M$ with G-orbit $Gm \subset M$ and tangent plane $\tau_m \cong V$, the image of $[M]$ in

$$\begin{aligned}\mathscr{H}^G_V(M, M - Gm; \overline{A}_{G/G}) &\cong \tilde{\mathscr{H}}^G_V(G_+ \wedge_{G_m} S^{V-\mathscr{L}(G/G_m)}; \overline{A}_{G/G}) \\ &\cong \tilde{\mathscr{H}}^{G_m}_V(S^V; \overline{A}_{G_m/G_m}) \\ &\cong A(G_m)\end{aligned}$$

is a generator.

Recall that $\mathscr{H}^G_* = H^{G,\mathscr{L}}_*$. The first isomorphism above comes from excision, using a tubular neighborhood of the orbit Gm. The second isomorphism is the Wirth-müller isomorphism, which would not be true in this form if we were to use any δ other than $\mathscr{L}$, because there would then be a shift in dimension. A fundamental class must live in dual homology. (As we mentioned in Remarks 1.15.2, we could instead use the restriction of $\mathscr{L}$ to a family $\mathscr{F}$, as long as we require that M be an $\mathscr{F}$-manifold, but there is no obvious advantage to making this generalization.)

The fundamental class $[M]$ is related to fundamental classes of the fixed submanifolds M^K as follows. The proof is similar to that of Proposition 1.15.3.

Proposition 1.15.7 *Let M be a closed V-manifold. The following are equivalent for a homology class $\mu \in \mathscr{H}_V^G(M;\overline{A}_{G/G})$:*

(1) *μ is a fundamental class for M as a G-manifold.*
(2) *For every subgroup $K \leq G$, $\mu|K \in \mathscr{H}_V^K(M;\overline{A}_{K/K})$ is a fundamental class for M as a K-manifold.*
(3) *For every subgroup $K \leq G$, $\mu^K \in \mathscr{H}_{V^K}^{WK}(M^K;\overline{A}_{WK/WK})$ is a fundamental class for M^K as a WK-manifold.*
(4) *For every subgroup $K \leq G$, $\mu^K|e \in H_{|V^K|}(M^K;\mathbb{Z})$ is a fundamental class for M^K as a nonequivariant manifold.* □

We say that a V-manifold is strictly orientable if its tangent bundle is strictly orientable in the sense of the preceding section.

Theorem 1.15.8 (Poincaré Duality) *A closed V-dimensional G-manifold M is strictly orientable if and only if it has a fundamental class $[M] \in \mathscr{H}_V^G(M;\overline{A}_{G/G})$. If M does have a fundamental class $[M]$, then*

$$- \cap [M] \colon H^\alpha_{G,\delta}(M;\overline{T}) \to H^{G,\mathscr{L}-\delta}_{V-\alpha}(M;\overline{T})$$

is an isomorphism for every familial δ such that M is an $\mathscr{F}(\delta)$-manifold (i.e., $\mathscr{F}(\delta)$ contains every isotropy subgroup of M).

Here, we are using the special case of the general cap product given after Proposition 1.14.24. Notice the following particular cases: When $\delta = 0$ we get the isomorphism

$$- \cap [M] \colon H^\alpha_G(M;\overline{T}) \to \mathscr{H}^G_{V-\alpha}(M;\overline{T})$$

and when $\delta = \mathscr{L}$ we get the isomorphism

$$- \cap [M] \colon \mathscr{H}^\alpha_G(M;\overline{T}) \to H^G_{V-\alpha}(M;\overline{T}).$$

Proof We can adapt the standard geometric proof, as given in [52] or [14]. Recall that the argument starts with the local case of a tubular neighborhood of an orbit G/G_m in M. This neighborhood will have the form $G \times_{G_m} \bar{D}(V - \mathscr{L}(G/G_m))$. Letting $\mu \in \mathscr{H}_V^G(G \times_{G_m} \bar{D}(V - \mathscr{L}(G/G_m));\overline{A}_{G/G})$ be the restriction of $[M]$, we then have the following diagram (recall that we write $\bar{D}$ as shorthand for the pair (D, S)):

$$\begin{array}{ccc}
H^{\alpha}_{G,\delta}(G\times_{G_m}\bar{D}(V-\mathscr{L}(G/G_m));\overline{T}) & \xrightarrow{-\cap\mu} & H^{G,\mathscr{L}-\delta}_{V-\alpha}(G\times_{G_m}D(V-\mathscr{L}(G/G_m));\overline{T}) \\
\cong\big\downarrow & & \big\downarrow\cong \\
H^{\alpha-\delta(G/G_m)}_{G_m,\delta}(\bar{D}(V-\mathscr{L}(G/G_m));\overline{T}|G_m) & \xrightarrow{-\cap\mu|G_m} & H^{G_m,\mathscr{L}-\delta}_{V-\alpha-\mathscr{L}(G/G_m)+\delta(G/G_m)}(D(V-\mathscr{L}(G/G_m));\overline{T}|G_m) \\
\cong\big\downarrow & & \big\downarrow\cong \\
H^{\alpha-V+\mathscr{L}(G/G_m)-\delta(G/G_m)}_{G_m,\delta}(*;\overline{T}|G_m) & \underset{\cong}{\xrightarrow{-\cap u}} & H^{G_m,\mathscr{L}-\delta}_{V-\alpha-\mathscr{L}(G/G_m)+\delta(G/G_m)}(*;\overline{T}|G_m)
\end{array}$$

The top square commutes because the cap product respects the Wirthmüller isomorphism; the bottom square commutes because it respects suspension. The element $u \in \mathscr{H}_0^{G_m}(*;\overline{A}_{G_m/G_m}) \cong A(G_m)$ is the image of μ and is, by assumption, a unit, making the map along the bottom an isomorphism between two copies of $\overline{T}(G/G_m)$. It follows that the other two horizontal maps are isomorphisms as well. Note, however, that this argument requires that $G_m \in \mathscr{F}(\delta)$; we cannot expect to get a local isomorphism otherwise.

The argument then proceeds via a Mayer-Vietoris patching argument as in the references above. □

As with the Thom isomorphism theorem, we will lift both the orientability assumption and the assumption that M is V-dimensional in Sect. 3.11.

We can also prove Poincaré duality from the Thom isomorphism by using the representing spectra. For this, we use the fact that, if M is embedded in a G-representation W with normal bundle ν, then the stable dual of M_+ is $\Sigma^{-W}T\nu$ [42]. Thus, we have the following chain of isomorphisms, where the first is the Thom isomorphism.

$$\begin{aligned}
H^{\alpha}_{G,\delta}(M;\overline{T}) &\cong \tilde{H}^{\alpha+W-V}_{G,\delta}(T\nu;\overline{T}) \\
&\cong [T\nu, \Sigma^{\alpha+W-V}H_\delta\overline{T}]_G \\
&\cong [\Sigma^W D(M_+), \Sigma^{\alpha+W-V}H_\delta\overline{T}]_G \\
&\cong [S^W, \Sigma^{\alpha+W-V}H_\delta\overline{T}\wedge M_+]_G \\
&\cong H^{G,\mathscr{L}-\delta}_{V-\alpha}(M;\overline{T}).
\end{aligned}$$

For the last isomorphism, we use the fact that, because $\mathscr{F}(\delta)$ contains every isotropy subgroup of M, we have $M \simeq E\mathscr{F}(\delta)\times M$, hence

$$H_\delta\overline{T}\wedge M_+ \simeq H_\delta\overline{T}\wedge E\mathscr{F}(\delta)_+\wedge M_+ \simeq H^{\mathscr{L}-\delta}\overline{T}\wedge M_+.$$

As usual, we can take as a fundamental class $[M] \in \mathscr{H}_V^G(M;\overline{A}_{G/G})$ the image of a Thom class in $\tilde{H}_G^{W-V}(T\nu;\overline{A}_{G/G})$. The maps displayed above can then be shown to be given by capping with this fundamental class.

If M is a compact orientable V-manifold with boundary, then we can prove relative, or Lefschetz, duality. We state the results.

Definition 1.15.9 Let M be a compact V-manifold with boundary. A *fundamental class* of M is a class $[M,\partial M] \in \mathscr{H}_V^G(M,\partial M;\overline{A}_{G/G})$ such that, for each point $m \in M - \partial M$ with G-orbit $Gm \subset M$ and tangent plane $\tau_m \cong V$, the image of $[M,\partial M]$ in

$$\begin{aligned}\mathscr{H}_V^G(M,M-Gm;\overline{A}_{G/G}) &\cong \tilde{\mathscr{H}}_V^G(G_+\wedge_{G_m} S^{V-\mathscr{L}(G/G_m)};\overline{A}_{G/G})\\ &\cong \tilde{\mathscr{H}}_V^{G_m}(S^V;\overline{A}_{G_m/G_m})\\ &\cong A(G_m)\end{aligned}$$

is a generator.

There is an obvious relative version of Proposition 1.15.7. Finally, we have the following relative duality.

Theorem 1.15.10 (Lefschetz Duality) *A compact V-dimensional G-manifold M is strictly orientable if and only if it has a fundamental class $[M,\partial M]$. If M does have a fundamental class, then the following are isomorphisms for every familial dimension function δ such that M is an $\mathscr{F}(\delta)$-manifold:*

$$-\cap[M,\partial M]\colon H_{G,\delta}^{\alpha}(M;\overline{T}) \to H_{V-\alpha}^{G,\mathscr{L}-\delta}(M,\partial M;\overline{T})$$

and

$$-\cap[M,\partial M]\colon H_{G,\delta}^{\alpha}(M,\partial M;\overline{T}) \to H_{V-\alpha}^{G,\mathscr{L}-\delta}(M;\overline{T}).$$

□

Note that these versions of the cap product are obtained by internalizing the slant product along the diagonal map $M/\partial M \to M/\partial M \wedge M_+$ in the first case and $M/\partial M \to M_+ \wedge M/\partial M$ in the second.

Again, we have the following special cases when we take $\delta = 0$ or $\delta = \mathscr{L}$:

$$\begin{aligned}&-\cap[M,\partial M]\colon H_G^{\alpha}(M;\overline{T}) \to \mathscr{H}_{V-\alpha}^G(M,\partial M;\overline{T}),\\ &-\cap[M,\partial M]\colon H_G^{\alpha}(M,\partial M;\overline{T}) \to \mathscr{H}_{V-\alpha}^G(M;\overline{T}),\\ &-\cap[M,\partial M]\colon \mathscr{H}_G^{\alpha}(M;\overline{T}) \to H_{V-\alpha}^G(M,\partial M;\overline{T}), \qquad \text{and}\\ &-\cap[M,\partial M]\colon \mathscr{H}_G^{\alpha}(M,\partial M;\overline{T}) \to H_{V-\alpha}^G(M;\overline{T}).\end{aligned}$$

1.16 An Example: The Rotating Sphere

We illustrate Poincaré duality by considering the action of $G = S^1$ on the sphere S^2 by rotation, as considered in Examples 1.1.3(3), 1.1.5(2), and 1.1.8(3).

Let V be the representation of G given by rotation on $\mathbb{R}^2$, so the G-space we're considering is S^V, a V-manifold. Observe that S^V is strictly orientable, hence satisfies Poincaré duality: If τ is the tangent bundle and ν is the normal bundle to the evident embedding in $V \oplus \mathbb{R}$, then ν is trivial, so $\tau + \mathbb{R}$ is trivial, giving a strict orientation of τ as in Definition 1.15.4. Notice that, if we did not allow stable maps in that definition, but required unstable ones, then τ would not be strictly orientable, as the north and south poles must be oppositely oriented with respect to their identifications with V.

Noting that the fundamental class of S^V lives in $\mathscr{H}_V^G(S^V; \overline{A}_{G/G})$, we first calculate $\mathscr{H}_{V+i}^G(S^V; \overline{A}_{G/G})$.

We need to know the structure of $\overline{A}_{G/G}$. We recall that the Burnside ring $A(H)$ of an abelian group is the free abelian group on the orbits H/J.

Proposition 1.16.1 *If $G = S^1$, then*

$$\overline{A}_{G/G}(G/H) \cong A(H).$$

If $G/K \to G/H$ is a G-map, the induced induction map

$$\tau: \overline{A}_{G/G}(G/K) \to \overline{A}_{G/G}(G/H)$$

is given by $\tau(K/J) = H \times_K (K/J) = H/J$ for $J \leq K$. The induced restriction map

$$\rho: \overline{A}_{G/G}(G/H) \to \overline{A}_{G/G}(G/K)$$

is described as follows: If $H = G$, then

$$\rho(G/J) = \begin{cases} K/K & \text{if } J = G \\ 0 & \text{if } J \neq G. \end{cases}$$

If $H \neq G$, then

$$\rho(H/J) = \frac{|H| \cdot |K \cap J|}{|K| \cdot |J|} K/(K \cap J).$$

Proof For the calculation that $\overline{A}_{G/G}(G/H) \cong A(H)$, see [42, V.9.4] or our Proposition 1.6.2. Those results describe $\overline{A}_{G/G}(G/H)$ as the free abelian group on diagrams of the form $G/H \leftarrow G/J \to G/G$, which we identify with the generator H/J of $A(H)$.

The descriptions of induction and restriction follow from this description, with induction the easy case. For the restriction, if $H = G$, $\rho(G/J)$ is obtained by taking the pullback P in the following diagram:

$$\begin{array}{ccccccc} & & P & & & & \\ & \swarrow & & \searrow & & & \\ G/K & & & & G/J & & \\ & \searrow & & \swarrow & & \searrow & \\ & & G/G & & & & G/G \end{array}$$

We need to find a finite union of orbits G/L, with $L \leq K$, having the same Euler characteristics as P on all fixed sets. But, if $J \neq G$, then P is a torus, and all of its fixed sets will be either tori or empty, hence have 0 Euler characteristic. Hence, $\rho(G/J) = 0$ if $J \neq G$. If $J = G$, then clearly $P = G/K$ and $\rho(G/K) = K/K$ as an element of $A(K)$.

On the other hand, if $H \neq G$, so H and K are finite, then the description of $\rho(H/J)$ given in the statement of the proposition is just the decomposition of H/J into orbits as a K-set. Alternatively, viewed as a stable map from G/K to G/G, $\rho(H/J)$ is obtained by taking the pullback P in the following diagram:

$$\begin{array}{ccccccc} & & P & & & & \\ & \swarrow & & \searrow & & & \\ G/K & & & & G/J & & \\ & \searrow & & \swarrow & & \searrow & \\ & & G/H & & & & G/G \end{array}$$

In this case, $P = G \times_K (H/J)$, so is the disjoint union of $(|H| \cdot |K \cap J|)/(|K| \cdot |J|)$ copies of $G/(K \cap J)$, from which the statement in the proposition follows. □

Now we describe the dual V-chain complex of S^V. We use the dual cell structure described in Example 1.1.8(3), in which there is one dual $(V-1)$-cell $S^1 \times D(V-2)$, the equator, and two dual V-cells $S \times_{S^1} D(V)$, the northern and southern hemispheres. This gives us the chain complex

$$\begin{array}{ccccccc} \overline{C}^{G,\mathscr{L}}_{V-2}(S^V) & \longleftarrow & \overline{C}^{G,\mathscr{L}}_{V-1}(S^V) & \xleftarrow{\ \partial\ } & \overline{C}^{G,\mathscr{L}}_{V}(S^V) & \longleftarrow & \overline{C}^{G,\mathscr{L}}_{V+1}(S^V) \\ \| & & \| & & \| & & \| \\ 0 & & \underline{A}^{G/e} & & \underline{A}^{G/G} \oplus \underline{A}^{G/G} & & 0 \end{array}$$

where we need to describe the map ∂. Note that we are thinking of the chains, which are contravariant Mackey functors on $\widehat{\mathscr{O}}_{G,\mathscr{L}}$, as covariant Mackey functors on $\widehat{\mathscr{O}}_G$. The generators of $\overline{C}_V^{G,\mathscr{L}}(S^V)$ are the northern and southern hemispheres. Writing $(S^V)^V$ for the V-skeleton, the connecting map $(S^V)^V/(S^V)^{V-1} \to \Sigma(S^V)^{V-1}/(S^V)^{V-2}$ is, when restricted to the northern hemisphere, the dual of the projection $G/e \to G/G$. Similarly, when restricted to the southern hemisphere, it is the negative of that dual, taking the orientations of the cells that match the strict orientation given earlier. It follows that

$$\partial(a,b) = \rho(a) - \rho(b)$$

where $\rho\colon \underline{A}^{G/G} \to \underline{A}^{G/e}$ is the map induced by $G/e \to G/G$. Now, if we tensor with $\overline{A}_{G/G}$, we get the complex

$$0 \longleftarrow \mathbb{Z} \xleftarrow{\;\partial\;} A(G) \oplus A(G) \longleftarrow 0$$

where $\partial(a,b) = \rho(a) - \rho(b)$ for ρ being the restriction map described in the proposition above. This map is clearly onto, and we can describe its kernel as

$$\ker \partial \cong A(G) \oplus I(G),$$

where $I(G) = \ker \rho$ is the *augmentation ideal*, and the $A(G)$ is the diagonal in $A(G) \oplus A(G)$. In other words, we get

$$\mathscr{H}^G_{V-n}(S^V;\overline{A}_{G/G}) \cong \begin{cases} A(G) \oplus I(G) & \text{if } n = 0 \\ 0 & \text{otherwise.} \end{cases}$$

In this notation, a fundamental class for S^V is $1 \in A(G)$, which we can think of as the sum of the two hemispheres.

These groups are dual to the ordinary cohomology groups in integer grading, which are calculated as follows. We use the G-CW structure from Example 1.1.3(3), which has two fixed 0-cells, the poles, and one free 1-cell connecting them. The resulting chain complex is

$$\begin{array}{ccccc} \overline{C}_0^G(S^V) & \xleftarrow{\;\partial\;} & \overline{C}_1^G(S^V) & \longleftarrow & \overline{C}_2^G(S^V) \\ \| & & \| & & \| \\ \overline{A}_{G/G} \oplus \overline{A}_{G/G} & & \overline{A}_{G/e} & & 0 \end{array}$$

Taking maps from this chain complex to $\overline{A}_{G/G}$ gives

$$A(G) \oplus A(G) \xrightarrow{\;\delta\;} \mathbb{Z} \longrightarrow 0$$

where δ is, in fact, the same map we called ∂ above. Hence, the calculation is exactly the same as the one we just did, and we get

$$H_G^n(S^V; \overline{A}_{G/G}) \cong \begin{cases} A(G) \oplus I(G) & \text{if } n = 0 \\ 0 & \text{otherwise.} \end{cases}$$

Of course, these groups are isomorphic to the corresponding dual homology groups, as they must be by Poincaré duality. The isomorphism is given by capping with a fundamental class.

Another interesting calculation, which we leave to the reader, is to take the G-CW(V) structure given in Example 1.1.5(2) and calculate the ordinary homology or cohomology in gradings $V - n$. Then, write down the corresponding dual G-CW(0) structure and calculate the dual cohomology or homology in integer gradings, and verify that Poincaré duality holds for these groups. The reader can also see then that there is no duality between the integer-graded ordinary groups and the $(V - n)$-graded ordinary groups.

1.17 A Survey of Calculations

Calculations in the literature of $RO(G)$-graded homology or cohomology are sparse, and those that are published are only for the finite groups $G = \mathbb{Z}/p$ with p prime. The first was the calculation of the $RO(G)$-graded cohomology of a point for $G = \mathbb{Z}/p$ by Stong in 1980, circulated but never published by him. This calculation was published in 1988 by Lewis as an appendix in [37]. To state the results, we first need to describe the various Mackey functors that occur. For $G = \mathbb{Z}/p$, we can picture a Mackey functor $\overline{T}$ using a diagram of the following form:

$$\begin{array}{c} \overline{T}(G/G) \\ \rho \downarrow \;\; \uparrow \tau \\ \overline{T}(G/e) \\ \circlearrowleft \\ t^* \end{array}$$

Here, ρ is the map induced by the projection $G/e \to G/G$ while τ is the map induced by its dual. $\overline{T}(G/G)$ is a module over the Burnside ring, with the action of $g = [G/e] \in A(G)$ given by $\tau \circ \rho$. The map t^* is induced by the map $t \colon G/e \to G/e$ given by multiplication by a generator of G. We must also have $\rho \circ \tau = 1 + t^*$. Here are the Mackey functors that show up in the calculation:

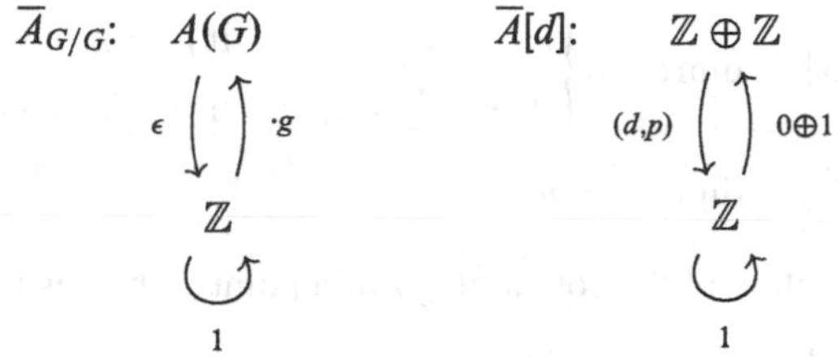

Here, $d \in \mathbb{Z}$. We have $\overline{A}[1] \cong \overline{A}_{G/G}$ and $\overline{A}[d_1] \cong \overline{A}[d_2]$ if $d_1 \equiv \pm d_2 \mod p$.

$$\langle \mathbb{Z} \rangle: \quad \mathbb{Z}$$
$$\begin{pmatrix} \uparrow \\ \downarrow \end{pmatrix}$$
$$0$$
$$\circlearrowright$$

$$\langle \mathbb{Z}/p \rangle: \quad \mathbb{Z}/p$$
$$\begin{pmatrix} \uparrow \\ \downarrow \end{pmatrix}$$
$$0$$
$$\circlearrowright$$

$$\overline{L}: \quad \mathbb{Z}$$
$$\cdot p \begin{pmatrix} \uparrow \\ \downarrow \end{pmatrix} 1$$
$$\mathbb{Z}$$
$$\circlearrowright$$
$$1$$

$$\overline{R}: \quad \mathbb{Z}$$
$$1 \begin{pmatrix} \uparrow \\ \downarrow \end{pmatrix} \cdot p$$
$$\mathbb{Z}$$
$$\circlearrowright$$
$$1$$

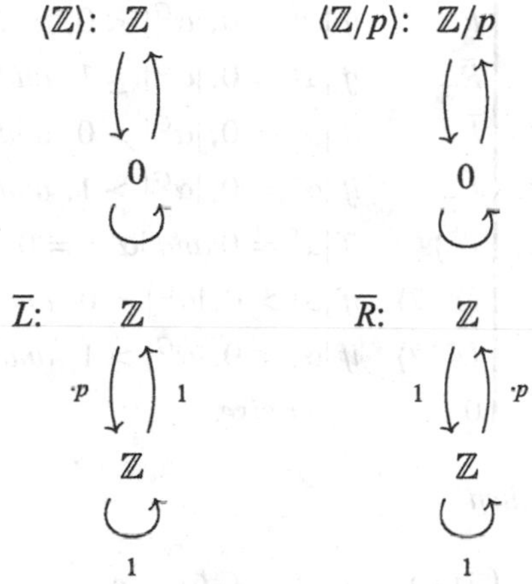

The following two occur only for $G = \mathbb{Z}/2$:

$$\overline{L}_-: \quad \mathbb{Z}/2$$
$$0 \begin{pmatrix} \uparrow \\ \downarrow \end{pmatrix} \pi$$
$$\mathbb{Z}$$
$$\circlearrowright$$
$$-1$$

$$\overline{R}_-: \quad 0$$
$$\begin{pmatrix} \uparrow \\ \downarrow \end{pmatrix}$$
$$\mathbb{Z}$$
$$\circlearrowright$$
$$-1$$

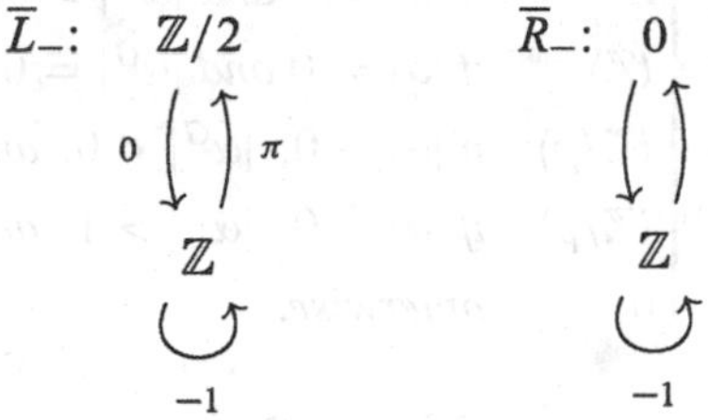

For $p = 2$, $RO(\mathbb{Z}/2)$ is generated by $\mathbb{R}$ and Λ, where Λ is the nontrivial one-dimensional representation. For p odd, $RO(\mathbb{Z}/p)$ is generated by $\mathbb{R}$ and $(p-1)/2$ irreducible two-dimensional representations Ω_k, where a chosen generator of $\mathbb{Z}/p$

acts on Ω_k as $e^{2\pi ik/p}$. If $\alpha = n_0\mathbb{R} + n_1\Lambda$ or $\alpha = n_0\mathbb{R} + \sum_{k\neq 0} n_k\Omega_k$, then

$$|\alpha| = \dim\alpha = \begin{cases} n_0 + n_1 & \text{if } p = 2 \\ n_0 + 2\sum_{k\neq 0} n_k & \text{if } p > 2, \text{ and} \end{cases}$$
$$|\alpha^G| = \dim\alpha^G = n_0.$$

The additive structure of the cohomology of a point is then as follows.

Theorem 1.17.1 *[37, 2.1 & 2.3] If $G = \mathbb{Z}/2$, then*

$$\overline{H}_G^\alpha(S^0;\overline{A}_{G/G}) = \begin{cases} \overline{A}_{G/G} & \textit{if } |\alpha| = |\alpha^G| = 0, \\ \overline{R} & \textit{if } |\alpha| = 0, |\alpha^G| < 0, \textit{ and } |\alpha^G| \textit{ is even}, \\ \overline{R}_- & \textit{if } |\alpha| = 0, |\alpha^G| \le 1, \textit{ and } |\alpha^G| \textit{ is odd}, \\ \overline{L} & \textit{if } |\alpha| = 0, |\alpha^G| > 0, \textit{ and } |\alpha^G| \textit{ is even}, \\ \overline{L}_- & \textit{if } |\alpha| = 0, |\alpha^G| > 1, \textit{ and } |\alpha^G| \textit{ is odd}, \\ \langle\mathbb{Z}\rangle & \textit{if } |\alpha| \neq 0 \textit{ and } |\alpha^G| = 0, \\ \langle\mathbb{Z}/2\rangle & \textit{if } |\alpha| > 0, |\alpha^G| < 0, \textit{ and } |\alpha^G| \textit{ is even}, \\ \langle\mathbb{Z}/2\rangle & \textit{if } |\alpha| < 0, |\alpha^G| > 1, \textit{ and } |\alpha^G| \textit{ is odd}, \\ 0 & \textit{otherwise}. \end{cases}$$

If $G = \mathbb{Z}/p$ with $p > 2$, then

$$\overline{H}_G^\alpha(S^0;\overline{A}_{G/G}) = \begin{cases} \overline{A}[d_\alpha] & \textit{if } |\alpha| = |\alpha^G| = 0, \\ \overline{R} & \textit{if } |\alpha| = 0 \textit{ and } |\alpha^G| < 0, \\ \overline{L} & \textit{if } |\alpha| = 0 \textit{ and } |\alpha^G| > 0, \\ \langle\mathbb{Z}\rangle & \textit{if } |\alpha| \neq 0 \textit{ and } |\alpha^G| = 0, \\ \langle\mathbb{Z}/p\rangle & \textit{if } |\alpha| > 0, |\alpha^G| < 0, \textit{ and } |\alpha^G| \textit{ is even}, \\ \langle\mathbb{Z}/p\rangle & \textit{if } |\alpha| < 0, |\alpha^G| > 1, \textit{ and } |\alpha^G| \textit{ is odd}, \\ 0 & \textit{otherwise}. \end{cases}$$

Here, when $|\alpha| = |\alpha^G| = 0$, so $\alpha = \sum_{k\neq 0} n_k\Omega_k$, $d_\alpha \equiv \pm\prod_{k\neq 0} k^{n_k} \mod p$. □

These results are best understood by arranging the Mackey functors in a lattice of $|\alpha^G|$ versus $|\alpha|$; see [37] for the pictures. Stong also computed the multiplicative structure for $G = \mathbb{Z}/2$ and $\mathbb{Z}/3$, and Lewis gives the multiplicative structure for all $\mathbb{Z}/p$ in [37, 4.3 & 4.9]. It's complicated, but for use in stating some other calculations, we give the simplest part of it. In the case $G = \mathbb{Z}/2$, there are elements $\epsilon \in \tilde{H}_G^\Lambda(S^0)$ and $\xi \in \tilde{H}_G^{-2+2\Lambda}(S^0)$ that together multiplicatively generate the subring consisting of the groups $\tilde{H}_G^\alpha(S^0)$ with $|\alpha| \ge 0$ and $|\alpha^G| \le 0$. The relations between these generators are completely determined by $2\epsilon\xi = 0$.

There are further, partial calculations of the cohomology of a point for $G = \mathbb{Z}/p$, for other coefficient systems, in [10] and [6]. Dugger gives a complete calculation for $G = \mathbb{Z}/2$ and constant $\mathbb{Z}$ coefficients in the appendix of [23].

Shortly before Lewis's paper appeared, the second author, in [68], published some results on and calculations of the cohomology of classifying spaces $E\mathscr{F}$ for certain families of subgroups $\mathscr{F}$. The main result of that paper is the following: Let G be a finite group and V a representation of G with no trivial summand. Let $\mathscr{F}(V)$ be the family of subgroups H such that $V^H \neq 0$. Let $e_V \in \tilde{H}^V_G(S^0; \overline{A}_{G/G})$ be the Euler class of V, the restriction of the generator of $\tilde{H}^V_G(S^V; \overline{A}_{G/G})$ along the inclusion $S^0 \to S^V$. Then, for any $\overline{T}$, multiplication by e_V induces a map

$$H_G^{nV+m}(E\mathscr{F}(V); \overline{T}) \to H_G^{(n+1)V+m}(E\mathscr{F}(V); \overline{T})$$

that is an isomorphism when $n \geq 0$, $m \geq 0$, and at least one is positive, and an epimorphism when $n = m = 0$. When G acts freely on the unit sphere in V, so $E\mathscr{F}(V) = EG$, this result generalizes known results about periodicity in the cohomology of spherical space form groups (groups that act freely on spheres). It also leads to explicit calculations of the groups $H_G^{nV+m}(E\mathscr{F}(V); \overline{T})$ for $n \geq 0$ and $m \geq 0$.

The purpose of Lewis's paper was actually another calculation, the $RO(G)$-graded cohomology of complex projective spaces, for $G = \mathbb{Z}/p$ and $\overline{A}_{G/G}$ coefficients. The easiest of his results to state is the case $G = \mathbb{Z}/2$ and the infinite complex projective space. We write $B_GU(1)$ for the space of one-dimensional complex subspaces of $\mathbb{C}^\infty \oplus \Lambda_c^\infty$, where $\mathbb{C}$ has trivial G-action and Λ_c is $\mathbb{C}$ with G acting nontrivially.

Theorem 1.17.2 *[37, 5.1] Let $G = \mathbb{Z}/2$ and take all cohomology with coefficients in $\overline{A}_{G/G}$. As an $RO(G)$-graded algebra over $\overline{H}_G^*(S^0)$, $\overline{H}_G^*(B_GU(1)_+)$ is generated by two elements, $\gamma \in \tilde{H}_G^{2\Lambda}(B_GU(1)_+)$ and $\Gamma \in \tilde{H}_G^{2+2\Lambda}(B_GU(1)_+)$, with one relation*

$$\gamma^2 = \epsilon^2\gamma + \xi\Gamma,$$

where ϵ and ξ are the elements in the cohomology of S^0 mentioned above. □

Lewis also gives calculations for truncated projective spaces and for the case $G = \mathbb{Z}/p$ with p odd. They're complicated.

Kronholm, in [35], does similar calculations for real projective spaces, for $G = \mathbb{Z}/2$ and constant $\mathbb{Z}/2$ coefficients. His result for the infinite real projective space, $B_GO(1)$, can be stated as follows.

Theorem 1.17.3 *[35, 5.5] Let $G = \mathbb{Z}/2$ and take all cohomology with constant $\mathbb{Z}/2$ coefficients. As an $RO(G)$-graded algebra over $\overline{H}_G^*(S^0)$, $\overline{H}_G^*(B_GO(1)_+)$ is generated by two elements, $\omega \in \tilde{H}_G^{\Lambda}(B_GO(1)_+)$ and $\Omega \in \tilde{H}_G^{1+\Lambda}(B_GO(1)_+)$, with one relation*

$$\omega^2 = \epsilon\omega + \xi\Omega$$

where ϵ is as above and $\xi \in \tilde{H}_G^{-1+\Lambda}(S^0)$. □

Recently, Dugger, in [24], used Kronholm's result to calculate the $RO(G)$-graded cohomology of $B_G O(n)$, again for $G = \mathbb{Z}/2$ and constant $\mathbb{Z}/2$ coefficients. His result can be stated as follows.

Theorem 1.17.4 *[24, 1.2] Let $G = \mathbb{Z}/2$ and take all cohomology with constant $\mathbb{Z}/2$ coefficients. The map*

$$\overline{H}_G^*(B_G O(n)_+) \to [\overline{H}_G^*(B_G O(1)_+)^{\otimes n}]^{\Sigma_n}$$

is an isomorphism in $RO(G)$ grading. □

These are the only published calculations to date that we are aware of. In particular, no calculations have been published for infinite groups, so we invite the reader to pursue such.

1.18 Relationship to Borel Homology

Classically, Borel cohomology is an integer-graded equivariant cohomology theory defined as follows.

Definition 1.18.1 Let M be a $\mathbb{Z}[\pi_0 G]$-module and let X be a G-space. The *Borel cohomology* of X is

$$BH_G^n(X; M) = H^n(EG \times_G X; M),$$

where we consider M as defining a local coefficient system on $BG = EG/G$. We extend this definition to pairs and to a reduced theory in the usual way.

This theory has suspension isomorphisms of the following form for certain representations V.

Proposition 1.18.2 *If V is a representation of G on which the action of G preserves orientation, then*

$$\widetilde{BH}_G^n(\Sigma^V X; M) \cong \widetilde{BH}_G^{n-|V|}(X; M).$$

Proof This follows from the nonequivariant Thom isomorphism applied to the orientable bundle $(EG \times X) \times_G V \to (EG \times X)/G$. □

When G does not preserve the orientation of V we need not have such an isomorphism, which prevents Borel cohomology from being an $RO(G)$-graded theory in the same sense that, for example, tom Dieck's homotopical bordism is [9, 2.1]. However, we have the following definition and result.

Definition 1.18.3 If M is a $\mathbb{Z}[\pi_0 G]$-module, let $\overline{R}M$ be the contravariant Mackey functor defined by

$$\overline{R}M(G/H) = \mathrm{Hom}_{\mathbb{Z}[\pi_0 G]}(\mathbb{Z}[\pi_0(G/H)], M) \cong \mathrm{Hom}_{\widehat{\mathscr{O}}_G(G/e,G/e)}(\widehat{\mathscr{O}}_G(G/e, G/H), M).$$

Here, we use the calculation

$$\widehat{\mathscr{O}}_G(G/e, G/H) \cong \mathbb{Z}[\pi_0(G/H)],$$

which comes from Proposition 1.6.2. Note that $\overline{R}M(G/e) = M$.

Proposition 1.18.4 *Let δ be the dimension function 0 restricted to the family $\mathscr{F}(\delta) = \{G/e\}$. Then*

$$BH^n_G(X; M) \cong H^n_G(EG \times X; \overline{R}M) \cong H^n_{G,\delta}(X; M).$$

Proof The isomorphism of $H^*_G(EG \times X; \overline{R}M)$ with $H^*_{G,\delta}(X; M)$ follows from Corollary 1.11.3, because, considering both as cohomology theories in X, both take $\mathscr{F}(\delta)$-equivalences to isomorphisms and they obey the same dimension axiom for the one orbit G/e in that family.

Now, if Y is any free G-space, we have the isomorphism

$$H^n_G(Y; \overline{R}M) \cong H^n(Y/G; M),$$

where M is considered as a local coefficient system on Y/G via the map $Y/G \to BG$ induced from the universal G-map $Y \to EG$. The isomorphism follows from the definitions of the two sides as cohomology of cellular chain complexes, because the two can be expressed in terms of the same complex of $\mathbb{Z}[\pi_0 G]$-modules. (Alternatively, we can view both sides as defining cohomology theories on spaces over BG, observe that they satisfy the same dimension axiom for points, and apply a nonequivariant uniqueness result.) Applying this isomorphism to $Y = EG \times X$, we get the first isomorphism of the proposition. □

Thus, Borel cohomology is the restriction to integer grading of the $RO(G)$-graded theory $H^*_G(EG \times X; \overline{R}M) \cong H^*_{G,\delta}(X; M)$. It follows then, from Theorem 1.15.8, that Borel cohomology exhibits Poincaré duality for free G-manifolds, with the dual theory being shifted Borel homology, i.e., for an m-dimensional free G-manifold M,

$$BH^k_G(M; A) \cong BH^G_{m-|G|-k}(M; A).$$

The Borel homology and cohomologies of M are really the nonequivariant homology and cohomology of M/G, which is an $(m - |G|)$-dimensional manifold, so the duality above is exactly what we should expect—it is just nonequivariant Poincaré duality applied to M/G. We should not expect Poincaré duality to hold in Borel cohomology for nonfree manifolds, as it fails for orbits G/H when H is nontrivial.

In another direction, the second author showed, in the unpublished [66], that, if G is finite and V is a unitary representation of G, there exist "Chern classes" $c_{V-2i} \in \tilde{H}_G^{V-2i}(S^0;\overline{A})$, $0 \leq i \leq |V|/2$, where $|V| = \dim_{\mathbb{R}} V$. Further, if V contains a free orbit of G, then, on inverting the class $c_{V-|V|}$, we have an isomorphism

$$BH_G^{|V|+n}(X;M) \cong c_{V-|V|}^{-1} H_G^{V+n}(X;\overline{R}M)$$

for $n \in \mathbb{Z}$. This can be seen explicitly for $G = \mathbb{Z}/p$ from the calculation of the cohomology of a point outlined in the preceding section. For example, when $G = \mathbb{Z}/2$, the class $c_{2\Lambda-2}$ is the element ξ mentioned earlier, and, from the known multiplicative structure, we can compute that inverting ξ gives copies of the cohomology of BG along the lines $2n\Lambda + *$. This and uniqueness imply that the resulting cohomology theory is Borel cohomology. The authors intend to extend this and other results from that preprint in a future publication.

See Sect. 3.12 for more about these Chern classes.

1.19 Miscellaneous Remarks

1.19.1 Ordinary Homology of G-Spectra

In the nonparametrized context, there is no significant difficulty to developing a theory of G-CW spectra and resulting cellular homology and cohomology theories. The reason we didn't do this is that there are significant problems in the parametrized case, as discussed in [51, Ch. 24]. We sketch here how the nonparametrized case would go.

Given a dimension function δ for G and a virtual representation $\alpha = V \ominus W$, we define a *δ-α-cell* to be a spectrum of the form

$$G_+ \wedge_H S^{\alpha-\delta(G/H)+q} = G_+ \wedge_H \Sigma_H^{V+q} \Sigma_{W+\delta(G/H)}^{\infty} S^0,$$

where $\Sigma_{W+\delta(G/H)}^{\infty}$ is the shift desuspension functor of [42, I.4.1] or [44, II.4.7] and $G_+ \wedge_H -$ is the functor defined in [42, II.4.1] or [44, V.2.3]. If E is a G-spectrum, we say that the cofiber of a map $G_+ \wedge_H S^{\alpha-\delta(G/H)+q-1} \to E$ is the result of *attaching a cell of dimension* $\alpha + q$ to E.

We can then generalize the definitions of [42, I.5]. A *δ-α-cell spectrum* is a G-spectrum E together with a sequence of subspectra E_n, $n \geq 0$, such that E is the union of the E_n, with $E_0 = *$ and E_{n+1} obtained from E_n as the cofiber of a map $J_n \to E_n$ where J_n is a wedge of δ-α-cells. A *δ-G-CW(α) spectrum* is a G-cell spectrum E such that each attaching map

$$G_+ \wedge_H S^{\alpha-\delta(G/H)+q-1} \to E_n$$

factors through a cell subspectrum containing only cells of dimension $< \alpha + q$. We define the $(\alpha + n)$*-skeleton* $E^{\alpha+n}$ to be the union of the cells of dimension at most $\alpha + n$.

If E is any G-spectrum, let

$$\pi^{H,\delta}_{\alpha+n}(E) = [G_+ \wedge_H S^{\alpha-\delta(G/H)+n}, E]_G.$$

We say that a map $f: E \to F$ is a δ*-weak*$_\alpha$ *equivalence* if, for each subgroup $H \leq G$ and integer n, $\pi^{H,\delta}_{\alpha+n}(E) \to \pi^{H,\delta}_{\alpha+n}(F)$ is an isomorphism. Using Corollary 1.4.14 and Lemma 1.4.15 we can show that $f: E \to F$ is a δ-weak$_\alpha$ equivalence if and only if it is an $\mathscr{F}(\delta)$-equivalence.

With these definitions we get a HELP lemma, approximation of spectra by δ-G-CW(α) spectra (up to δ-weak$_\alpha$ equivalence), and cellular approximation of maps between δ-G-CW(α) spectra.

To define the chains of an arbitrary G-spectrum E we first take a δ-G-CW(α) approximation $\Gamma E \to E$ and let

$$\overline{C}^{G,\delta}_{\alpha+n}(E)(G/H, \delta) = \pi^{H,\delta}_{\alpha+n}((\Gamma E)^{\alpha+n}/(\Gamma E)^{\alpha+n-1}).$$

This gives a chain complex, independent of the choice of approximation ΓE up to chain homotopy equivalence. We then define homology and cohomology groups of E using this chain complex.

Homology and cohomology theories defined on spectra are representable. In particular, for each contravariant δ-Mackey functor $\overline{T}$ we get a spectrum $H_\delta \overline{T}$ with the same characterization as in Sect. 1.12. Of importance to the theories defined on spaces, we can now conclude that $H_\delta \overline{T}$ is unique *up to unique stable equivalence*. On the other hand, if $\underline{S}$ is a covariant δ-Mackey functor, then Theorem 1.12.2 tells us that $H^\delta \underline{S} \simeq H_{\mathscr{L}-\delta} \underline{S} \wedge E\mathscr{F}(\delta)_+$ and we get a similar strong uniqueness result for this spectrum.

1.19.2 Model Categories

Although we've deliberately kept the discussion as elementary as possible, there were various points at which we could have phrased things using the language of model categories. Indeed, for a given dimension function δ and virtual representation α, we could define a compactly generated model category structure on G-spaces in which the weak equivalences are the δ-weak$_\alpha$ equivalences and the generating cofibrations are the maps $G \times_H S(Z) \to G \times_H D(Z)$ where $G \times_H \bar{D}(Z)$ is a δ-α-cell. We leave the details to the interested reader.

Although this is an interesting point of view, it's not clear that it would add much to the exposition. Also, CW objects are at the heart of the definition of cellular homology and cohomology, but are rarely mentioned in discussions of model categories. It would be interesting to see if a general theory of CW objects in model categories would be of any use in other contexts.

Chapter 2
Parametrized Homotopy Theory and Fundamental Groupoids

In general, smooth G-manifolds are not modeled on a single representation V, so, following the geometry, we need a way of encoding the varying local representations that do appear. The machinery for doing this was developed in detail in [20], where it was used to give a theory of equivariant orientations. It assembles the various local representations into what we call a *representation of the fundamental groupoid*. The fundamental groupoid $\Pi_G X$ of a G-space X, defined by tom Dieck in [61], is a category who objects are the maps of orbits $G/H \to X$; the morphisms in this category are defined in Sect. 2.1. Representations of the fundamental groupoid can be thought of as the natural dimensions of G-vector bundles or G-manifolds, and provide the grading for the extension of ordinary homology and cohomology we will define in Chap. 3. The material from [20] that we need is recounted in Sect. 2.1.

The stable orbit category played a prominent role in Chap. 1. In our extended theory a similar role is played by the *stable fundamental groupoid* of a G-space X, which we introduce and discuss in Sect. 2.6. This is naturally defined as the full subcategory of a homotopy category of spectra parametrized by X given by the (suspension spectra of) the maps of orbits $G/H \to X$. That is, the stable fundamental groupoid has the same set of objects as the (unstable) fundamental groupoid. So, in order to define the stable fundamental groupoid, we will need to first discuss parametrized spaces and spectra.

There is another reason we need to discuss parametrized objects, as we mentioned in the Introduction: It will be convenient to fix a space B and a representation of its fundamental groupoid, and consider spaces parametrized by B. The theories we define will then be defined on a category of spaces parametrized by B and represented by spectra parametrized by B. The morphisms we use in the category of spaces parametrized by B are what we call *lax maps,* and capture the discrete nature of the representations of the fundamental groupoid of B. The idea of using parametrized spaces to model spaces with additional structure, for example, representations of their fundamental groupoids, is a tradition that goes back at least

S.R. Costenoble, S. Waner, *Equivariant Ordinary Homology and Cohomology*,
Lecture Notes in Mathematics 2178, DOI 10.1007/978-3-319-50448-3_2

as far [58] and [21]. Thus, we spend considerable time in this chapter discussing parametrized spaces and spectra, in anticipation of using them heavily in Chap. 3.

The reader interested in taking a geodesic path to Chap. 3 can read just the first two sections and Sect. 2.7, and take Theorem 2.6.4 as a definition of the stable fundamental groupoid.

2.1 The Fundamental Groupoid

The following definition was given in [20].

Definition 2.1.1 Let B be a G-space. The *equivariant fundamental groupoid* $\Pi_G B$ of B is the category whose objects are the G-maps $x\colon G/H \to B$ and whose morphisms $x \to y$, $y\colon G/K \to B$, are the pairs (ω, α), where $\alpha\colon G/H \to G/K$ is a G-map and ω is an equivalence class of paths $x \to y \circ \alpha$ in B^H. As usual, two paths are equivalent if they are homotopic rel endpoints. Composition is induced by composition of maps of orbits and the usual composition of path classes.

Recall that $\mathscr{O}_G$ is the category of G-orbits and G-maps. We have a functor $\pi\colon \Pi_G B \to \mathscr{O}_G$ defined by $\pi(x\colon G/H \to B) = G/H$ and $\pi(\omega, \alpha) = \alpha$. We topologize the mapping sets in $\Pi_G B$ in a natural way, as detailed in [20, 3.1] so, in particular, π is continuous. (When G is finite, the topology is discrete.) For each subgroup H, the subcategory $\pi^{-1}(G/H)$ of objects mapping to G/H and morphisms mapping to the identity is a copy of $\Pi(B^H)$, the nonequivariant fundamental groupoid of B^H (with discrete topology). $\Pi_G B$ itself is not a groupoid in the usual sense, but a "catégories fibrées en groupoides" [28] or a "bundle of groupoids" [20, 5.1] over $\mathscr{O}_G$ (which we usually shorten to *groupoid over* $\mathscr{O}_G$).

The *homotopy fundamental groupoid* of B is the category $h\Pi_G B$ obtained by replacing each morphism space with its set of path components. Equivalently, we identify any two morphisms that are homotopic in the following sense: Maps (ω, α), $(\xi, \beta)\colon x \to y$, where $x\colon G/H \to B$ and $y\colon G/K \to B$, are homotopic if there is a homotopy $j\colon G/H \times I \to G/K$ from α to β and a homotopy $k\colon G/H \times I \times I \to B$ from a representative of ω to a representative of ξ such that $k(a, 0, t) = x(a)$ and $k(a, 1, t) = yj(a, t)$ for all $a \in G/H$ and $t \in I$. Thus, $h\Pi_G B$ coincides with tom Dieck's "discrete fundamental groupoid" [61, 10.9]. When G is finite, $h\Pi_G B = \Pi_G B$.

Definition 2.1.2 Let $\overline{\mathscr{V}}_G$ be the category whose objects are the orthogonal G-vector bundles over orbits of G and whose morphisms are the equivalence classes of orthogonal G-vector bundle maps between them. Here, two maps are equivalent if they are G-bundle homotopic, with the homotopy inducing the constant homotopy on base spaces. Let $\pi\colon \overline{\mathscr{V}}_G \to \mathscr{O}_G$ be the functor taking the bundle $p\colon E \to G/H$ to its base space G/H, and taking a bundle map to the underlying map of base spaces. Let $\overline{\mathscr{V}}_G(n)$ be the full subcategory of $\overline{\mathscr{V}}_G$ consisting of the n-dimensional bundles.

We topologize the mapping sets in $\overline{\mathscr{V}}_G$ as in [20, 4.1]. (When G is finite, the topology is discrete.) The map

$$\pi: \overline{\mathscr{V}}_G(p,q) \to \mathscr{O}_G(\pi(p),\pi(q))$$

is then a bundle with discrete fibers.

The categories $\overline{\mathscr{V}}_G$ and $\overline{\mathscr{V}}_G(n)$ are not small, but have small skeleta which we call $\mathscr{V}_G$ and $\mathscr{V}_G(n)$, obtained by choosing one representative in each equivalence class. An explicit choice is given in [20, 2.2]. The following is [20, 2.7].

Proposition 2.1.3 *A G-vector bundle $p: E \to B$ determines by pullbacks a map*

$$p^*: \Pi_G B \to \overline{\mathscr{V}}_G$$

over $\mathscr{O}_G$. A G-vector bundle map $(\tilde{f},f): p \to q$, with $\tilde{f}: E \to E'$ the map of total spaces and $f: B \to B'$ the map of base spaces, determines a natural isomorphism $\tilde{f}_*: p^* \to q^* f_*$ over the identity. If $(\tilde{h},h): (\tilde{f}_0,f_0) \to (\tilde{f}_1,f_1)$ is a homotopy of G-vector bundle maps, then $(\tilde{f}_1)_* = (p')^* h_* \circ (\tilde{f}_0)_*$. □

The map p^* is an example of a *representation* of the fundamental groupoid $\Pi_G B$:

Definition 2.1.4

(1) An *n-dimensional orthogonal representation* of $\Pi_G B$ is a continuous functor $\gamma: \Pi_G B \to \mathscr{V}_G(n)$ over $\mathscr{O}_G$.
(2) Two representations of $\Pi_G B$ are *equivalent* if they are naturally isomorphic as functors. The set of equivalence classes of orthogonal representations of $\Pi_G B$ of all dimensions forms a ring under direct sum and tensor product, called the *representation ring* of $\Pi_G B$ and denoted $RO(\Pi_G B)$.
(3) If γ is an n-dimensional orthogonal representation of $\Pi_G B$ and γ' is an n-dimensional orthogonal representation of $\Pi_G B'$, then a *map* from γ to γ' is a pair (f,η) where $f: B \to B'$ is a G-map and $\eta: \gamma \to \gamma' \circ f_*$ is a natural isomorphism over the identity of $\mathscr{O}_G$.

Examples 2.1.5

(1) If V is an orthogonal H-representation, then we write $G \times_H \mathbb{V}$ for the orthogonal representation of $\Pi_G(G/H)$ associated to the bundle $G \times_H V \to G/H$. If B is a G-space and V is a G-representation, we also write $\mathbb{V}$ for the pullback to $\Pi_G B$ of the representation $\mathbb{V}$ of $\Pi_G(*) = \mathscr{O}_G$.
(2) If M is a G-manifold, then, associated to its tangent bundle is its *tangent representation* τ.
(3) The tangent representation of a V-manifold need not be $\mathbb{V}$. For example, consider the action of $G = S^1$ on S^2 by rotation, as in Example 1.1.3(3). This is a V-manifold, with V being $\mathbb{R}^2$ with the rotation action of G. However,

as we go from the north to the south poles via a nonequivariant path, V is carried to itself by a nonequivariant map that reverses orientation. Because V is irreducible, there is no equivariant map that reverses its orientation. So, there is no isomorphism of τ with $\mathbb{V}$. The same argument applies to S^{nV} for any n.

In addition to $\mathscr{V}_G$, there are several other small categories that are useful to use as the targets of representations of $\Pi_G B$. They fit into the following commutative diagram (see [20, §§18 & 19] for details).

$$\begin{array}{ccccccc}
\mathscr{V}_G(n) & \longrightarrow & \mathscr{PL}_G(n) & \longrightarrow & \mathscr{T}op_G(n) & \longrightarrow & \mathscr{F}_G(n) \\
\downarrow & & \downarrow & & \downarrow & & \downarrow \\
s\mathscr{V}_G(n) & \longrightarrow & s\mathscr{PL}_G(n) & \longrightarrow & s\mathscr{T}op_G(n) & \longrightarrow & s\mathscr{F}_G(n) \\
\downarrow & & \downarrow & & \downarrow & & \downarrow \\
v\mathscr{V}_G(n) & \longrightarrow & v\mathscr{PL}_G(n) & \longrightarrow & v\mathscr{T}op_G(n) & \longrightarrow & v\mathscr{F}_G(n)
\end{array}$$

To form the top row, we take the categories of G-vector bundles over orbits, with morphisms being, respectively, the fiber-homotopy classes of G-vector bundle maps, PL G-bundle maps, topological G-bundle maps, and spherical G-fibration maps. In the case of $\mathscr{F}_G(n)$, what we mean is that the objects are bundles, but a map from $G \times_H V$ to $G \times_K W$ is a based fiber-homotopy class of homotopy equivalences $G \times_H S^V \to G \times_K S^W$. The middle row is the stabilization of the top row and the bottom row gives virtual bundles. Maps into these categories give us, respectively orthogonal, PL, topological, and spherical representations, and their stable and virtual analogues.

In particular, spherical representations arise naturally from spherical fibrations. Here, by a spherical fibration we mean the following (see also [20, §23]).

Definition 2.1.6 A *spherical G-fibration* is a sectioned G-fibration $p\colon E \to B$ such that each fiber $p^{-1}(b)$ is based G_b-homotopy equivalent to S^V for some orthogonal G_b-representation V.

The various categories over $\mathscr{O}_G$ that we've discussed, including $\Pi_G B$ and the twelve categories in the diagram above, are all groupoids over $\mathscr{O}_G$. Associated to a groupoid $\mathscr{R}$ over $\mathscr{O}_G$, there is a classifying G-space $B\mathscr{R}$, defined in [20, §20]. The following classification result is proved for finite groups as [20, 24.1] and extended to compact Lie groups in [11].

Theorem 2.1.7 *The G-space $B\mathscr{R}$ classifies representations of $\Pi_G B$ in $\mathscr{R}$. That is, for G-CW complexes B, $[B, B\mathscr{R}]_G$ is in natural bijective correspondence with the set of isomorphism classes of representations $R\colon \Pi_G B \to \mathscr{R}$.*

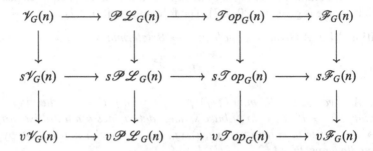

2.2 Parametrized Spaces and Lax Maps

In this section we recall the categories of parametrized spaces and introduce a relaxed notion of parametrized map. We will refer periodically to May and Sigurdsson's [51], in which they give a very careful and detailed exposition of the homotopy theory of parametrized spaces and spectra.

Definition 2.2.1 Let B be a compactly generated G-space.

(1) Let $G\mathscr{K}/B$ be the category of *G-spaces over B*: Its objects are pairs (X,p) where X is a G-k-space and $p: X \to B$ is a G-map. A map $(X,p) \to (Y,q)$ is a G-map $f: X \to Y$ such that $q \circ f = p$, i.e., a G-map over B.
(2) Let $G\mathscr{K}_B$ be the category of *ex-G-spaces over B*: Its objects are triples (X,p,σ) where (X,p) is a G-space over B and σ is a section of p, i.e., $p \circ \sigma$ is the identity. A map $(X,p,\sigma) \to (Y,q,\tau)$ is a section-preserving G-map $f: X \to Y$ over B, i.e., a G-map over and under B.

When the meaning is clear, we shall write just X for (X,p) or (X,p,σ). If (X,p) is a space over B, we write $(X,p)_+$ for the ex-G-space obtained by adjoining a disjoint section. We shall also write X_+ for $(X,p)_+$, a notation that May and Sigurdsson denigrate but should not cause confusion in context.

As May and Sigurdsson point out, in order to make these categories closed symmetric monoidal we need B to be weak Hausdorff but cannot require it of the spaces over B. Hence we assume that B is compactly generated (i.e., it is a k-space and weakly Hausdorff) but assume of the spaces over it only that they are k-spaces.

When thinking of spaces over B, the reader may be tempted to think largely about bundles or fibrations, but, in fact, many of the examples of importance to us will be neither. For this reason, it's useful to look not just at maps over B, but at what we call *lax* maps. (For another reason to consider lax maps, see the following section.) First, recall that a *Moore homotopy* from $p: X \to B$ to $q: X \to B$ is a pair (λ, l) where $\lambda: X \times [0,\infty] \to B$ and $l: X \to [0,\infty)$ with $\lambda(x,r) = \lambda(x,l(x))$ for $r \geq l(x)$, and with $\lambda(x,0) = p(x)$ and $\lambda(x,\infty) = q(x)$. We usually write λ for (λ, l) and l_λ for l, the *length* of λ.

Definition 2.2.2 A *lax map* $(X,p,\sigma) \Rightarrow (Y,q,\tau)$ is a pair (f,λ) where $f: X \to Y$ is a G-map under B and $\lambda: qf \to p$ is a Moore homotopy rel $\sigma(B)$. A lax map between unbased G-spaces over B is defined similarly, just removing the references to the sections.

In this context we shall call maps under and over B *strict* maps. We can also think of strict maps as those lax maps in which the homotopies have zero length. To avoid confusion we will use "$\Rightarrow$" to indicate lax maps, reserving "$\to$" for strict maps.

A lax map $X \Rightarrow Y$ can also be thought of as a strict map $X \to LY$, where LY is the following variant of the associated Hurewicz fibration, given in [51, §8.3]. Recall that the space of *Moore paths* in B is

$$\Lambda B = \{(\lambda, l) \in B^{[0,\infty]} \times [0, \infty) \mid \lambda(r) = \lambda(l) \text{ for } r \geq l\}.$$

Moore paths are useful largely because they have an associative composition: Suppose λ and μ are Moore paths with $\lambda(\infty) = \mu(0)$. Then we write $\lambda\mu$ for the Moore path with length $l_{\lambda\mu} = l_\lambda + l_\mu$ given by

$$(\lambda\mu)(t) = \begin{cases} \lambda(t) & \text{if } t \leq l_\lambda \\ \mu(t - l_\lambda) & \text{if } t \geq l_\lambda. \end{cases}$$

(We reverse the order of composition here compared to [51, §8.3], as it will be more useful to us in this order.) If $p: X \to B$, the *Moore path fibration* $LX = L(X, p)$ is given by

$$LX = X \times_B \Lambda B = \{(x, \lambda) \in X \times \Lambda B \mid p(x) = \lambda(0)\}.$$

We define the projection $Lp: LX \to B$ using the endpoint projection: $Lp(x, \lambda) = \lambda(\infty)$. If X is an ex-space with section σ, we define a section $L\sigma$ of LX by $L\sigma(b) = (\sigma(b), b)$, where we write b for the path of length 0 at the point b. We can now see that a lax map $(X, p, \sigma) \Rightarrow (Y, q, \tau)$ is equivalent to a strict map $(X, p, \sigma) \to (LY, Lq, L\tau)$.

There is an inclusion $\iota: X \to LX$ given by $\iota(x) = (x, p(x))$ and composition of paths defines a map $\chi: L^2X \to LX$, $\chi(x, \lambda, \mu) = (x, \lambda\mu)$. With these maps, (L, ι, χ) is a monad. The unit ι is a G-homotopy equivalence under (but not over) B, and the projection $LX \to B$ is a Hurewicz fibration.

We have an associative composition of lax maps: If $(f, \lambda): (X, p, \sigma) \Rightarrow (Y, q, \tau)$ and $(g, \mu): (Y, q, \tau) \Rightarrow (Z, r, \upsilon)$, then

$$(g, \mu) \circ (f, \lambda) = (gf, (\mu f)\lambda),$$

where $\mu f: rgf \to qf$ is the evident Moore homotopy. Thus, we can form the category $G\mathscr{K}_B^\lambda$ of ex-G-spaces and lax maps (similarly, we have the category $G\mathscr{K}^\lambda/B$ of G-spaces over B and lax maps). Thinking of lax maps as strict maps $X \to LY$, we have

$$G\mathscr{K}_B^\lambda(X, Y) = G\mathscr{K}_B(X, LY) \quad \text{and} \quad (G\mathscr{K}^\lambda/B)(X, Y) = (G\mathscr{K}/B)(X, LY)$$

and we topologize $G\mathscr{K}_B^\lambda$ and $G\mathscr{K}^\lambda/B$ using these identifications. The composite of $f: X \to LY$ and $g: Y \to LZ$ is then the composite

$$X \xrightarrow{f} LY \xrightarrow{Lg} L^2Z \xrightarrow{\chi} LZ.$$

Thus, the category of lax maps is the Kleisli category of the monad L (see [33], [43, §VI.5], and [59, §5]).

One of the points of the Kleisli category is that the inclusion $G\mathscr{K}_B \to G\mathscr{K}_B^\lambda$ has a right adjoint $\bar{L}: G\mathscr{K}_B^\lambda \to G\mathscr{K}_B$. On objects it is given by L, and, if $f: X \to LY$ represents a lax map, $\bar{L}f$ is the composite

$$LX \xrightarrow{Lf} L^2Y \xrightarrow{\chi} LY.$$

Both the inclusion $G\mathscr{K}_B \to G\mathscr{K}_B^\lambda$ and $\bar{L}: G\mathscr{K}_B^\lambda \to G\mathscr{K}_B$ are continuous.

We now give some results about the homotopy theory of lax maps.

Definition 2.2.3 Two lax maps $f, g: X \Rightarrow Y$ are *lax homotopic* if there is a lax map $h: X \wedge_B I_+ \Rightarrow Y$ such that $h(-,0) = f$ and $h(-,1) = g$. We write $\pi G\mathscr{K}_B^\lambda$ for the category obtained by identifying lax homotopic maps. We define $\pi G\mathscr{K}^\lambda/B$ similarly. We say that X and Y are *lax homotopy equivalent* if there are lax maps $f: X \Rightarrow Y$ and $g: Y \Rightarrow X$ such that $f \circ g$ and $g \circ f$ are both lax homotopic to the identity maps.

Remark 2.2.4 If G is finite, the fundamental groupoid $\Pi_G B$ coincides with the full subcategory of the homotopy category $\pi G\mathscr{K}^\lambda/B$ on the orbits over B. When G is infinite, however, the latter full subcategory coincides instead with the homotopy fundamental groupoid defined in Sect. 2.1, which is also tom Dieck's "discrete fundamental groupoid" [61, 10.9]. A lax homotopy of maps between $x: G/H \to B$ and $y: G/K \to B$ allows homotopy of the maps $G/H \to G/K$, while the definition of $\Pi_G B$ does not. However, Definition 2.1.1 seems to be the most useful definition to use.

Definition 2.2.5 A lax map $i: A \Rightarrow X$ is a *lax cofibration* if it satisfies the lax homotopy extension property, meaning that we can always fill in the dashed arrow in the following commutative diagram of lax maps:

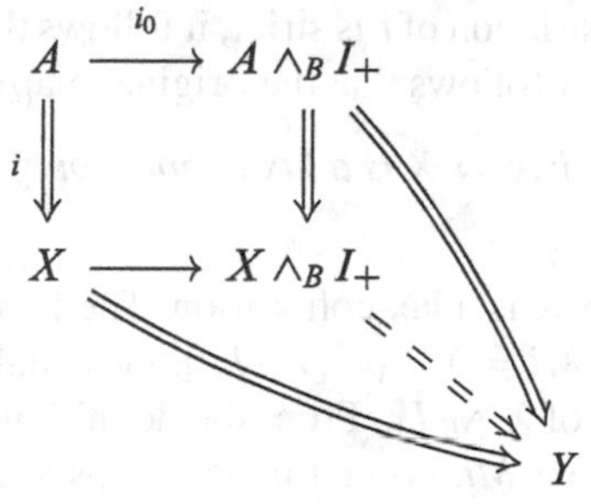

We shall show that the lax cofibrations are precisely the strict maps that are cofibrations under B.

Lemma 2.2.6 *If $(i, \lambda): A \Rightarrow X$ is a lax cofibration, then λ is a constant homotopy, so i is a map over B.*

Proof Suppose that $(i, \lambda): (A, p, \sigma) \Rightarrow (X, q, \tau)$ is a lax cofibration. Construct $\tilde{M}i$ as the following variant of the mapping cylinder: As a G-space, let $\tilde{M}i = X \cup_i (A \wedge_B I_+)$, the pushout along the inclusion $i_0: A \to A \wedge_B I_+$. Give $\tilde{M}i$ the evident section, but define $r: \tilde{M}i \to B$ as follows:

$$r(x) = q(x) \qquad \text{if } x \in X$$
$$r(a, s) = \lambda(a, l_\lambda(a) \cdot s) \qquad \text{if } (a, s) \in A \wedge_B I_+.$$

Define a lax map $(\tilde{\imath}, \tilde{\lambda}): A \wedge_B I_+ \Rightarrow \tilde{M}i$ by letting $\tilde{\imath}$ be the usual map and defining $\tilde{\lambda}$ as follows:

$$\tilde{\lambda}(a, s, t) = \lambda(a, l_\lambda(a) \cdot s + t)$$
$$l_{\tilde{\lambda}}(a, s) = l_\lambda(a) \cdot (1 - s).$$

Note that the restriction of $(\tilde{\imath}, \tilde{\lambda})$ to $A \wedge 1$ is a strict map over B. Now, from the assumption that $A \Rightarrow X$ is a lax cofibration, we can fill in a lax map f in the following diagram:

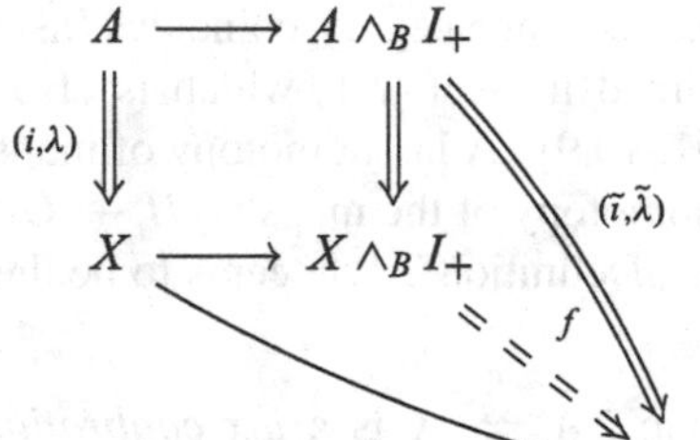

Consider the rightmost triangle restricted to $A \wedge 1$. From the definition of composition and the fact that the restriction of $\tilde{\imath}$ is strict, it follows that the map $A \wedge 1 \Rightarrow X \wedge 1$ must be strict, from which it follows that the original map $A \Rightarrow X$ must be strict. □

Theorem 2.2.7 *A lax map $i: A \Rightarrow X$ is a lax cofibration if and only i is a strict map and a cofibration under B.*

Proof Suppose that $i: A \Rightarrow X$ is a lax cofibration. The lemma above shows that i is a strict map. We now form $Mi = X \cup_A (A \wedge_B I_+)$, the usual mapping cylinder of i in $G\mathscr{K}_B$, which is a subspace of $X \wedge_B I_+$. From the definition of lax cofibration we get a lax retraction $r: X \wedge_B I_+ \Rightarrow Mi$. Forgetting the maps to B, this retraction under B shows that i is a cofibration under B.

Conversely, let $i: A \to X$ be a strict map and a cofibration under B. Let $p: X \to B$ be the projection. Let $q: Mi \to B$ be the mapping cylinder of i and let $r: X \wedge_B I_+ \to Mi$ be a retraction under B. We want to make r into a lax retraction. Define

$h: (X \wedge_B I_+) \times [0, \infty] \to B$ by

$$h(x, s, t) = \begin{cases} qr(x, s - t) & \text{if } t \leq s \\ q(x) & \text{if } t \geq s \end{cases}$$

The map h is clearly continuous, but is not quite what we need. Let $u: X \to I$ be a G-invariant map such that $u^{-1}(0) = A$, e.g., $u(x) = \sup\{t - p_2 r(x, t) \mid t \in I\}$, where p_2 is projection to I. Define $\lambda: (X \wedge_B I_+) \times [0, \infty] \to B$ by

$$\lambda(x, s, t) = \begin{cases} h(x, s, t/u(x)) & \text{if } u(x) > 0 \\ q(x) & \text{if } u(x) = 0 \end{cases}$$

and let

$$l_\lambda(x, s) = su(x).$$

Assuming that λ is continuous, the pair (r, λ) defines a lax retraction, from which it follows that i is a lax cofibration. So, we need to check that λ is continuous.

Continuity at points (x, s, t) such that $u(x) \neq 0$ is clear, so we need to check continuity at points (a, s, t) with $a \in A$. For such a point we have $\lambda(a, s, t) = p(a)$. Let V be any neighborhood of $p(a)$ in B. Because $I \times [0, \infty]$ is compact, we can find a neighborhood U of a such that $U \times I \times [0, \infty] \subset h^{-1}(V)$. It follows that $U \times I \times [0, \infty] \subset \lambda^{-1}(V)$, showing that λ is continuous. □

We are now entitled to the following result, in which C_B denotes the mapping cone over B and $/_B$ indicates the quotient over B.

Proposition 2.2.8 *Suppose that $i: A \to X$ is a strict map that is a cofibration under B. Then the natural map $c: C_B i \to X/_B A$ is a lax homotopy equivalence.*

Proof Because i is a lax cofibration, the obvious map $k: X \cup_A (A \wedge_B I_+) \to C_B i$ over B extends to a lax map $(h, \lambda): X \wedge_B I_+ \Rightarrow C_B i$. Note that $\lambda(a, t)$ is a 0-length path for every $a \in A$ and $t \in I$. If (h_1, λ_1) is the restriction to $X \wedge 1$, note that h_1 takes A to the section, hence induces a lax map $\bar{h}_1: X/_B A \Rightarrow C_B i$, which we claim is a lax homotopy inverse to c.

The composite $c \circ h: X \wedge_B I_+ \Rightarrow X/_B A$ factors through $k: X/_B A \wedge_B I_+ \Rightarrow X/_B A$, which clearly gives a lax homotopy from the identity to $c \circ \bar{h}_1$. (This uses the fact that λ is constant on A.)

On the other hand, it is easy to extend h to a lax homotopy $h': C_B i \wedge_B I_+ \to C_B i$ from the identity to $\bar{h}_1 \circ c$. □

We also have the following results on the homotopy invariance of pushouts and colimits.

Proposition 2.2.9 *Suppose given the following diagram in which the horizontal maps are strict, the vertical maps are lax homotopy equivalences, and i and i' are lax cofibrations.*

$$\begin{array}{ccccc} X & \xleftarrow{\;i\;} & A & \longrightarrow & Y \\ \Downarrow & & \Downarrow & & \Downarrow \\ X' & \xleftarrow[\;i'\;]{} & A' & \longrightarrow & Y' \end{array}$$

Then the induced map of pushouts $Y \cup_A X \Rightarrow Y' \cup_{A'} X'$ is a lax homotopy equivalence.

Proof The proof of the classical result, as in [5, 7.5.7], depends only on formal properties of the homotopy extension property and homotopies, and can be adapted to the context of $G\mathscr{K}_B^\lambda$. The precursor results to this result also appear in the more recent [48, §6.5]; May tells us that the result itself will appear in the second edition. □

Proposition 2.2.10 *Suppose given the following diagram in which the horizontal maps are lax cofibrations and the vertical maps are lax homotopy equivalences.*

$$\begin{array}{ccccccc} X_0 & \longrightarrow & X_1 & \longrightarrow & X_2 & \longrightarrow & \cdots \\ \Downarrow & & \Downarrow & & \Downarrow & & \\ Y_0 & \longrightarrow & Y_1 & \longrightarrow & Y_2 & \longrightarrow & \cdots \end{array}$$

Then the induced map $\operatorname{colim}_n X_n \Rightarrow \operatorname{colim}_n Y_n$ *is a lax homotopy equivalence.*

Proof This follows from [5, 7.4.1], specifically the special case considered in Step 1 of its proof. (It would also follow from the last proposition of [48, §6.5] using details about the forms of the homotopies used in the proof.) Again, the proof generalizes to our context. □

2.3 Lax Maps and Model Categories

We here discuss the relationship between lax maps and the model category structures on spaces over B discussed in [51].

On either $G\mathscr{K}/B$ or $G\mathscr{K}_B$ we can define a model category structure in which a map f is a weak equivalence, fibration, or cofibration if it is one as a map of G-spaces, ignoring B. Here, as usual, a weak equivalence is a weak equivalence on all components of all fixed sets, a fibration is an equivariant Serre fibration, and a cofibration is a retract of a relative G-CW complex. This is what May and Sigurdsson call the q-model structure and they call these classes of maps q-equivalences,

q-fibrations, and q-cofibrations. However, they point out serious technical difficulties with this model structure, stemming from the fact that these cofibrations do not interact well with fiberwise homotopy. To rectify this, they define another model structure, the qf-model structure, with the same weak equivalences, but with a more restrictive class of cofibrations, the qf-cofibrations, and a corresponding notion of qf-fibration. In any case, we are entitled to invert these weak equivalences, getting the homotopy categories $\operatorname{Ho} G\mathscr{K}/B$ and $\operatorname{Ho} G\mathscr{K}_B$. If X and Y are ex-G-spaces over B we write $[X, Y]_{G,B}$ for the set of maps in $\operatorname{Ho} G\mathscr{K}_B$.

In this context, inverting the weak equivalences introduces many more maps than are evident if one pays attention only to the fiber-preserving maps over B. For example, let $(X \times I, h)$ and (Y, q) be spaces over B, where $h: X \times I \to B$ does not necessarily factor through X. Let $i_0, i_1: X \to X \times I$ be the inclusions of the two endpoints and let $p_0 = hi_0$ and $p_1 = hi_1$. Then the maps $i_0: (X, p_0) \to (X \times I, h)$ and $i_1: (X, p_1) \to (X \times I, h)$ are both q-equivalences. (That we really want these to be weak equivalences is implicit already in Dold's axiom **CYL** in [21].) Hence, if $f: (X, p_0) \to (Y, q)$ is a map over B, we have in the homotopy category the map

$$[f] \circ [i_0]^{-1} \circ [i_1] \in [(X, p_1), (Y, q)]_{G,B},$$

which might not be represented by any fiberwise map $(X, p_1) \to (Y, q)$ over B. In fact, the pair (f, h) specifies a lax map $(X, p_1) \Rightarrow (Y, q)$. We now show that lax maps capture all the maps that occur after inverting weak equivalences.

We noted earlier that $\iota: Y \to LY$ is a G-weak equivalence under B, hence it is a q-equivalence. Further, $LY \to B$ is a Hurewicz fibration, hence LY is q-fibrant. If X is q-cofibrant, we then have

$$\begin{aligned} [X, Y]_{G,B} &\cong [X, LY]_{G,B} \\ &\cong \pi G\mathscr{K}_B(X, LY) \\ &\cong \pi G\mathscr{K}_B^{\lambda}(X, Y), \end{aligned}$$

where π denotes homotopy classes of maps. In other words, if X is q-cofibrant and Y is any space over B, then $[X, Y]_{G,B}$ is precisely the set of lax homotopy classes of lax maps from X to Y.

As just mentioned, $\iota: X \to LX$ is a weak equivalence; it's not hard to see that, considered as a lax map, it is a lax homotopy equivalence. Its inverse is the map $\kappa: LX \Rightarrow X$ represented by the identity $1: LX \to LX$. We record the following useful facts.

Lemma 2.3.1

(1) *If $f: X \to Y$ is a strict map that is also a lax homotopy equivalence, then it is a weak equivalence (i.e., a q-equivalence).*
(2) *If $f: X \Rightarrow Y$ is a lax homotopy equivalence, then its representing map $\hat{f}: X \to LY$ is also a lax homotopy equivalence, hence a weak equivalence.*

Proof To show the first claim, note that $\bar{L}f = Lf\colon LX \to LY$ is a fiberwise homotopy equivalence, because $\bar{L}$, being continuous, takes lax homotopy equivalences to fiberwise homotopy equivalences. It then follows, from the following diagram and the fact that ι is a weak equivalence, that f is a weak equivalence:

$$\begin{array}{ccc} X & \xrightarrow{f} & Y \\ {\scriptstyle\iota}\downarrow & & \downarrow{\scriptstyle\iota} \\ LX & \xrightarrow[Lf]{} & LY \end{array}$$

For the second claim, an easy calculation shows that $\hat{f} = \bar{L}f \circ \iota$. Again, $\bar{L}$ takes the lax homotopy equivalence f to a fiberwise homotopy equivalence, hence a lax homotopy equivalence. Combined with the fact that ι is a lax homotopy equivalence, we get that $\hat{f}$ is a lax homotopy equivalence. □

We should point out that the category $G\mathscr{K}_B^\lambda$ is wanting in several respects. It does not, in general, have limits. In particular, it has no terminal object. It also lacks colimits in general, although any diagram of strict maps has a colimit in $G\mathscr{K}_B^\lambda$, namely the image of the colimit in $G\mathscr{K}_B$ (because $G\mathscr{K}_B \to G\mathscr{K}_B^\lambda$ is a left adjoint). It is, therefore, not a candidate to be a model category. We shall use it primarily as a place to represent maps in $\operatorname{Ho} G\mathscr{K}_B$ that cannot easily be represented in $G\mathscr{K}_B$.

2.4 Parametrized Spectra

We now discuss parametrized spectra, which we will need for two purposes: We shall use them in the following section to define the stable fundamental groupoid, which will play a role in Chap. 3 similar to the role played by the stable orbit category in Chap. 1. We shall also use them as the representing objects for parametrized homology and cohomology theories.

May and Sigurdsson give a very careful treatment of equivariant parametrized spectra in [51]. They concentrate on orthogonal G-spectra over B, which give a model category with good formal properties. They also discuss LMS G-spectra over B (which they call "G-prespectra" over B), which arise more naturally as representing objects for cohomology theories and give an equivalent homotopy category. It is therefore more convenient for our purposes for us to use LMS G-spectra. As explained in [51], we need to restrict our parametrizing spaces to be compactly generated and have the homotopy types of G-CW complexes so that all of the functors and adjunctions we need pass to homotopy categories. We make this restriction from this point on.

Recall Definition 1.5.1. The following generalizes Definition 1.5.2 to the parametrized context.

Definition 2.4.1 ([51, 11.2.16 & 12.3.6])

(1) Let $\mathscr{U}$ be a complete G-universe. An *(LMS) G-spectrum E over B* consists of a collection of ex-G-spaces $E(V) \to B$, one for each finite-dimensional subrepresentation $V \subset \mathscr{U}$, and, for each inclusion $V \subset W$, a structure map

$$\sigma\colon \Sigma_B^{W-V} E(V) \to E(W)$$

over B. The σ are required to be unital and transitive in the usual way.

(2) An *Ω-G-spectrum* is a G-spectrum in which each ex-G-space $E(V)$ is qf-fibrant and the adjoint structure maps

$$\tilde{\sigma}\colon E(V) \to \Omega_B^{W-V} E(W)$$

are G-weak equivalences.

We let $G\mathscr{P}_B$ denote the category of G-spectra over B and G-maps between them, i.e., levelwise G-maps over B that respect the structure maps.

May and Sigurdsson define a stable model structure on $G\mathscr{P}_B$ [51, §12.3]. The stable equivalences (*s-equivalences* for short) are the fiberwise stable equivalences, where the fibers are those of levelwise fibrant approximations. The fibrant objects are exactly the Ω-G-spectra.

We write $[E, F]_{G,B}$ for the group of stable G-maps between two G-spectra over B, that is, $\mathrm{Ho}\, G\mathscr{P}_B(E, F)$. By [51, 12.4.5], $[E, F]_{G,B}$ is stable in the sense that, if V is any representation of G then there is an isomorphism

$$[E, F]_{G,B} \cong [\Sigma_B^V E, \Sigma_B^V F]_{G,B}.$$

It is useful to note that these groups are stable in a stronger sense: they are stable under suspension by any spherical fibration over B. The following is a direct consequence of [51, 15.1.5].

Proposition 2.4.2 *Let ξ be a spherical G-fibration over a G-CW complex B and let Σ^ξ denote the fiberwise smash product with the total space of ξ. Then, if E and F are any two G-spectra over B we have a natural isomorphism*

$$[E, F]_{G,B} \cong [\Sigma^\xi E, \Sigma^\xi F]_{G,B}.$$

□

The following functors and adjunctions are discussed in detail in [51]. Much of that book is aimed at showing that these and other relationships descend to homotopy categories. We have the functor $\Sigma_B^\infty\colon G\mathscr{K}_B \to G\mathscr{P}_B$ taking an ex-G-space X over B to the spectrum with $(\Sigma_B^\infty X)(V) = \Sigma_B^V X$. If X and Y are ex-G-spaces over B, we write

$$\{X, Y\}_{G,B} = [\Sigma_B^\infty X, \Sigma_B^\infty Y]_{G,B}$$

for the group of stable maps from X to Y. More generally, if W is a representation of G, we have the *shift desuspension* functor $\Sigma_W^\infty : G\mathscr{K}_B \to G\mathscr{P}_B$ defined by

$$\Sigma_W^\infty X(V) = \begin{cases} \Sigma_B^{V-W} X & \text{when } W \subset V \\ B & \text{when } W \not\subset V. \end{cases}$$

This functor is left adjoint to the evaluation functor Ω_W^∞ given by evaluation at W. The adjunction descends to the homotopy categories, by [51, 12.6.2]. We insert here the standard warning: Functors used on homotopy categories are the derived functors, obtained by first taking a cofibrant or fibrant approximation as appropriate. In particular, on the homotopy category, Ω_W^∞ does not return the Wth space of a spectrum but the Wth space of a stably equivalent Ω-G-spectrum.

Suppose that $\alpha : A \to B$ is a G-map. Then there are functors

$$\alpha^* : G\mathscr{K}_B \to G\mathscr{K}_A$$

and

$$\alpha^* : G\mathscr{P}_B \to G\mathscr{P}_A$$

given by taking pullbacks. These functors have left adjoints

$$\alpha_! : G\mathscr{K}_A \to G\mathscr{K}_B$$

and

$$\alpha_! : G\mathscr{P}_A \to G\mathscr{P}_B$$

given by composition with α and identification of base points. Precisely, on $G\mathscr{K}_A$, $\alpha_!(X, p, \sigma)$ is given by taking the pushout in the top square below:

$$\begin{array}{ccc} A & \xrightarrow{\alpha} & B \\ {\scriptstyle \sigma}\downarrow & & \downarrow \\ X & \dashrightarrow & \alpha_! X \\ {\scriptstyle p}\downarrow & & \downarrow \\ A & \xrightarrow[\alpha]{} & B \end{array}$$

$(\alpha_!, \alpha^*)$ is a Quillen adjoint pair. Moreover, α^* and $\alpha_!$ both commute with suspension, in the sense that

$$\Sigma_A^\infty \alpha^* X \cong \alpha^* \Sigma_B^\infty X$$

and

$$\Sigma_B^\infty \alpha_! X \cong \alpha_! \Sigma_A^\infty X.$$

Another useful property is the natural homeomorphism

$$X \wedge_B \alpha_! Y \approx \alpha_!(\alpha^* X \wedge_A Y)$$

for ex-G-spaces X over B and Y over A. (The smash product used is the usual fiberwise smash product.) This homeomorphism descends to an isomorphism in the homotopy category. Similarly, if E is a spectrum over B, then

$$E \wedge_B \alpha_! Y \cong \alpha_!(\alpha^* E \wedge_A Y).$$

We shall apply this most often to the case of $\rho\colon B \to *$, the projection to a point. (We shall use ρ generically for any projection to a point.) ρ induces functors $\rho^*\colon G\mathscr{P} \to G\mathscr{P}_B$ and $\rho_!\colon G\mathscr{P}_B \to G\mathscr{P}$. Notice that

$$\rho^* \Sigma^\infty Y = \Sigma_B^\infty \rho^* Y = \Sigma_B^\infty (B \times Y)$$

for a based G-space Y, and

$$\rho_! \Sigma_B^\infty X = \Sigma^\infty \rho_! X = \Sigma^\infty (X/\sigma(B))$$

for an ex-G-space (X, p, σ) over B.

2.5 Lax Maps of Spectra

We would like to define a category of lax maps of spectra, similar to the category $G\mathscr{K}_B^\lambda$. We will use this category in Sect. 3.4 to extend the definition of homology and cohomology from parametrized G-CW complexes to general parametrized spaces. May and Sigurdsson extended L to a functor on spectra in [51, §13.3], but there is a mistake in their definition. Moreover, they were not trying to make their L a monad, so, for example, took no care to allow for an associative composition. We begin by giving a definition that does give us a monad, so that we can again define the category of lax maps to be the Kleisli category of L.

The first difficulty with extending L to a functor on spectra is that, if K is an ex-space, there is no canonical map $LK \wedge_B S^V \to L(K \wedge_B S^V)$. So, we need to define an ad hoc map, which we call β^V. (May and Sigurdsson write β_V for their similar map.) In defining β^V, we shall think of $S^V = D(V)/S(V)$, so that a $v \in S^V$ will have $\|v\| \leq 1$. Here we differ from May and Sigurdsson; the difference is unimportant, but we find this choice easier to work with here. We first choose any homeomorphism $\varphi: [0, \infty] \to [1/2, 1]$ with $\varphi(0) = 1$ and $\varphi(\infty) = 1/2$. Now we define β^V by the following formula, in which we write l for l_λ to simplify the notation:

$$\beta^V((x,\lambda) \wedge v) = \begin{cases} (x \wedge v, \lambda) & \text{if } \|v\| \leq 1/2 \\ \left(x \wedge \varphi(\varphi^{-1}(\|v\|) - l)\frac{v}{\|v\|}, \lambda\right) & \text{if } 1/2 < \|v\| \leq \varphi(l) \\ \left(\sigma\lambda(l - \varphi^{-1}(\|v\|)), \lambda|^l_{l-\varphi^{-1}(\|v\|)}\right) & \text{if } \|v\| \geq \varphi(l) \end{cases}$$

Here, as in [51], we write $\lambda|^b_a$ for the Moore path of length $b - a$ given by

$$(\lambda|^b_a)(t) = \begin{cases} \lambda(a+t) & \text{if } t \leq b - a \\ \lambda(b) & \text{if } t \geq b - a. \end{cases}$$

Now, to extend L to spectra we need to deal with the fact that $\beta^V \circ \beta^W \neq \beta^{V \oplus W}$, so we do as in [51, §13.3] and consider spectra indexed on a fixed countable cofinal sequence $\mathscr{W} = \{0 = V_0 \subset V_1 \subset \cdots\}$. Write $W_i = V_i - V_{i-1}$ (note that our indexing of the W_i is slightly different from that of May and Sigurdsson). As explained in [51, 13.3.5], the category of spectra indexed on $\mathscr{W}$ gives us the same stable homotopy category as usual. If X is a G-spectrum, we define an indexed spectrum LX to have spaces $(LX)(V_i) = L(X(V_i))$ and structure maps $\sigma_i: LX(V_i) \wedge_B S^{W_{i+1}} \to LX(V_{i+1})$ given by the composites

$$LX(V_i) \wedge_B S^{W_{i+1}} \xrightarrow{\beta^{W_{i+1}}} L(X(V_i) \wedge_B S^{W_{i+1}}) \longrightarrow LX(V_{i+1}).$$

In general, define the structure map $\Sigma^{V_j - V_i} LX(V_i) \to LX(V_j)$ to be the composite $\sigma_{j-1} \circ \sigma_{j-2} \circ \cdots \circ \sigma_i$.

Proposition 2.5.1 *The levelwise inclusion $\iota: X \to LX$ and levelwise composition $\chi: L^2X \to LX$ make (L, ι, χ) a monad on the category of $\mathscr{W}$-indexed G-spectra over B.*

Proof The main difficulty is showing that ι and χ are actually maps of spectra, and the definition of β^V was crafted to make them so. The fact that ι is a map of spectra follows from the fact that the following diagram commutes for any ex-G-space K:

$$\begin{array}{ccc} & K \wedge_B S^V & \\ {\scriptstyle \iota\wedge 1}\swarrow & & \searrow{\scriptstyle \iota} \\ LK \wedge_B S^V & \xrightarrow[\beta^V]{} & L(K \wedge_B S^V) \end{array}$$

The fact that χ is a map of spectra comes down to the commutativity of the following diagram:

$$\begin{array}{ccc} L^2K \wedge_B S^V & \xrightarrow{(\beta^V)^2} & L^2(K \wedge_B S^V) \\ {\scriptstyle \chi\wedge 1}\downarrow & & \downarrow{\scriptstyle \chi} \\ LK \wedge_B S^V & \xrightarrow[\beta^V]{} & L(K \wedge_B S^V) \end{array}$$

The verification that these diagrams commute is by straightforward, if tedious, calculation using the definition of β^V.

That (L, ι, χ) is a monad now follows directly from the analogous fact for spaces. □

By analogy with spaces, we now make the following definition.

Definition 2.5.2 The category of *lax maps of G-spectra over B*, denoted $G\mathscr{P}_B^\lambda$, is the Kleisli category of the monad L on the category of $\mathscr{W}$-indexed G-spectra over B.

Explicitly, then, a lax map $X \Rightarrow Y$ is a strict map $X \to LY$. The composite of the maps $f: X \Rightarrow Y$ and $g: Y \Rightarrow Z$ is the following composite of strict maps:

$$X \xrightarrow{f} LY \xrightarrow{Lg} L^2Z \xrightarrow{\chi} LZ.$$

A lax map of spectra can be thought of as a collection of lax maps of ex-spaces at each level, compatible under suspension, using β^{W_i} to interpret what compatibility means. Composition of lax maps is given levelwise by composition of lax maps of ex-spaces.

Since LY is a levelwise fibrant approximation but not, in general, a qf-fibrant approximation of Y (i.e., LY is unlikely to be an Ω-G-spectrum), we don't expect to represent all stable maps by lax maps. However, lax maps give us useful examples of stable maps that are not represented by strict maps.

There are a number of cases where we will want to know that a functor on strict maps extends to one on lax maps. The general context is this: Suppose that $\mathscr{C}$ and $\mathscr{D}$ are categories, (S, η, μ) is a monad on $\mathscr{C}$, and (T, θ, ν) is a monad on $\mathscr{D}$. Write $\mathscr{C}_S$ for the Kleisli category of S and $\mathscr{D}_T$ for that of T, so $\mathscr{C}_S(X, Y) = \mathscr{C}(X, SY)$ and $\mathscr{D}_T(X, Y) = \mathscr{D}(X, TY)$. We ask what we need to extend a functor $F: \mathscr{C} \to \mathscr{D}$ to a functor $\mathscr{C}_S \to \mathscr{D}_T$. The answer is the following.

Definition 2.5.3 A *map of monads* $(F, \psi): (\mathscr{C}, S) \to (\mathscr{D}, T)$ consists of a functor $F: \mathscr{C} \to \mathscr{D}$ and a natural transformation $\psi: FS \to TF$ such that the following diagrams commute:

$$\begin{array}{ccc} & F & \\ {\scriptstyle F\eta}\swarrow & & \searrow{\scriptstyle \theta F} \\ FS & \xrightarrow[\psi]{} & TF \end{array}$$

$$\begin{array}{ccc} FS^2 & \xrightarrow{\psi^2} & T^2F \\ {\scriptstyle F\mu}\downarrow & & \downarrow{\scriptstyle \nu F} \\ FS & \xrightarrow[\psi]{} & TF \end{array}$$

(This is what Street calls a *monad opfunctor* in [59]. His monad functors have the natural transformation going in the other direction, and induce maps of Eilenberg-Moore categories.)

The general theory of [59] shows that a map of monads $(F, \psi): (\mathscr{C}, S) \to (\mathscr{D}, T)$ gives an extension of F to $\bar{F}: \mathscr{C}_S \to \mathscr{D}_T$. Explicitly, $\bar{F}(X) = F(X)$ on objects and, if $f: X \to SY$ represents a map in $\mathscr{C}_S(X, Y)$, then $\bar{F}(f)$ is the composite

$$FX \xrightarrow{Ff} FSY \xrightarrow{\psi} TFY.$$

It is straightforward to check that this defines a functor extending F. In fact, though we shall not need it, extensions of F are in one-to-one correspondence with natural transformations $\psi: FS \to TF$ such that (F, ψ) is a map of monads.

We have already seen one example of a map of monads, namely

$$(\Sigma_B^V, \beta^V): (G\mathscr{K}_B, L) \to (G\mathscr{K}_B, L).$$

The diagrams shown in the proof of Proposition 2.5.1 can be rewritten as the diagrams showing that (Σ_B^V, β^V) is a map of monads:

$$\begin{array}{ccc} & \Sigma_B^V & \\ {\scriptstyle \Sigma_B^V\iota}\swarrow & & \searrow{\scriptstyle \iota\Sigma_B^V} \\ \Sigma_B^V L & \xrightarrow[\beta^V]{} & L\Sigma_B^V \end{array}$$

$$\begin{array}{ccc} \Sigma_B^V L^2 & \xrightarrow{(\beta^V)^2} & L^2\Sigma_B^V \\ {\scriptstyle \Sigma_B^V\chi}\downarrow & & \downarrow{\scriptstyle \chi\Sigma_B^V} \\ \Sigma_B^V L & \xrightarrow[\beta^V]{} & L\Sigma_B^V \end{array}$$

Thus, we get the following.

Proposition 2.5.4 *The suspension functor Σ_B^V extends to a continuous functor we shall again call $\Sigma_B^V: G\mathscr{K}_B^\lambda \to G\mathscr{K}_B^\lambda$.* □

Explicitly, the suspension of a map $f: X \Rrightarrow Y$ is given by the composite

$$\Sigma_B^V X \xrightarrow{\Sigma_B^V f} \Sigma_B^V LY \xrightarrow{\beta^V} L\Sigma_B^V Y.$$

Note, however, that $\Sigma_B^W \Sigma_B^V \neq \Sigma_B^{V\oplus W}$ in general, because $\beta^W \circ \beta^V \neq \beta^{V\oplus W}$.

We insert here a useful fact about lax cofibrations.

Corollary 2.5.5 *Suspension preserves lax cofibrations. I.e., if $A \to X$ is a lax cofibration of ex-G-spaces, then so is $\Sigma_B^V A \to \Sigma_B^V X$.*

Proof Let $i: A \to X$ be a lax cofibration. Because Σ_B^V is a left adjoint (on $G\mathscr{K}_B$), it preserves colimits, hence $\Sigma_B^V Mi \cong M(\Sigma_B^V i)$, where Mi denotes the usual (fiberwise) mapping cylinder. Because i is a lax cofibration, Mi is a retract of $X \wedge_B I_+$ in $G\mathscr{K}_B^\lambda$, hence $M(\Sigma_B^V i) \cong \Sigma_B^V Mi$ is a retract of $\Sigma_B^V X \wedge I_+$ in $G\mathscr{K}_B^\lambda$. Therefore $\Sigma_B^V i$ is a lax cofibration. □

We would also like to extend the functors Σ_Z^∞ to lax maps.

Remark 2.5.6 Note that, because we are using indexed spectra, the functor Σ_Z^∞ is now defined by

$$\Sigma_Z^\infty X(V_i) = \begin{cases} \Sigma_B^{V_i - Z} X & \text{when } Z \subset V_i \\ B & \text{when } Z \not\subset V_i. \end{cases}$$

Let $\bar{Z}$ represent the smallest V_i such that $Z \subset V_i$. Then $\Sigma_Z^\infty K = \Sigma_{\bar{Z}}^\infty \Sigma_B^{\bar{Z}-Z} K$. From this it is clear that, in the context of indexed spectra, Σ_Z^∞ still has a right adjoint, but that right adjoint is $\Omega_B^{\bar{Z}-Z}\Omega_{\bar{Z}}^\infty$.

Definition 2.5.7 The natural transformation $\beta_Z^\infty\colon \Sigma_Z^\infty L \to L\Sigma_Z^\infty$ is defined as follows. Let k be the least integer such that $Z \subset V_k$. For $i \geq k$ we let

$$\beta_Z^\infty = \beta^{W_i} \circ \beta^{W_{i-1}} \circ \cdots \circ \beta^{W_{k+1}} \circ \beta^{V_k - Z}\colon \Sigma_B^{V_i - Z} L \to L\Sigma_B^{V_i - Z},$$

using the decomposition

$$V_i - Z = (V_k - Z) \oplus W_{k+1} \oplus \cdots \oplus W_i.$$

For $i < k$, β_Z^∞ is the trivial map between trivial ex-spaces.

Proposition 2.5.8 *The natural transformation β_Z^∞ defines a map of monads $(\Sigma_Z^\infty, \beta_Z^\infty)\colon (G\mathscr{K}_B, L) \to (G\mathscr{P}_B, L)$. Hence, Σ_Z^∞ extends to a continuous functor $\Sigma_Z^\infty\colon G\mathscr{K}_B^\lambda \to G\mathscr{P}_B^\lambda$.*

Proof We first need to verify that β_Z^∞ is a map of spectra. For this we need the following to commute:

$$\begin{array}{ccc}
\Sigma_B^{W_{i+1}}\Sigma_B^{V_i - Z} LK & \xrightarrow{\Sigma\beta_Z^\infty} & \Sigma_B^{W_{i+1}} L\Sigma_B^{V_i - Z} K \\
\Big\| & & \Big\downarrow {\scriptstyle \beta^{W_{i+1}}} \\
\Sigma_B^{V_{i+1} - Z} LK & \xrightarrow[\beta_Z^\infty]{} & L\Sigma_B^{V_{i+1} - Z} K
\end{array}$$

We defined β_Z^∞ precisely to make this diagram commute.

Now we need to know that $(\Sigma_Z^\infty, \beta_Z^\infty)$ is a map of monads. That β_Z^∞ preserves units and multiplication follows by iterating the corresponding facts about the β^Vs. □

Corollary 2.5.9 *If $f\colon X \Rightarrow Y$ is a lax homotopy equivalence of ex-G-spaces, then $\Sigma_Z^\infty f\colon \Sigma_Z^\infty X \Rightarrow \Sigma_Z^\infty Y$ is a lax homotopy equivalence of G-spectra over B.* □

The proof of the following is almost identical to that of Corollary 2.5.5.

Corollary 2.5.10 *If $A \to X$ is a lax cofibration of ex-G-spaces, then $\Sigma_Z^\infty A \to \Sigma_Z^\infty X$ is a lax cofibration of G-spectra over B.* □

We will use this result, together with the fact that pushouts of lax cofibrations are lax cofibrations, to construct lax cofibrations of G-spectra over B. The following homotopy invariance results are proved the same way as for the corresponding results for ex-G-spaces.

Proposition 2.5.11 *Suppose given the following diagram of G-spectra, in which the horizontal maps are strict, the vertical maps are lax homotopy equivalences,* *and i and i' are lax cofibrations.*

$$\begin{array}{ccccc} X & \xleftarrow{\;i\;} & A & \longrightarrow & Y \\ \Downarrow & & \Downarrow & & \Downarrow \\ X' & \xleftarrow[\;i'\;]{} & A' & \longrightarrow & Y' \end{array}$$

Then the induced map of pushouts $Y \cup_A X \Rightarrow Y' \cup_{A'} X'$ is a lax homotopy equivalence. □

Proposition 2.5.12 *Suppose given the following diagram of G-spectra over B, in which the horizontal maps are strict and are lax cofibrations, and the vertical maps are lax homotopy equivalences.*

$$\begin{array}{ccccccc} X_0 & \longrightarrow & X_1 & \longrightarrow & X_2 & \longrightarrow & \cdots \\ \Downarrow & & \Downarrow & & \Downarrow & & \\ Y_0 & \longrightarrow & Y_1 & \longrightarrow & Y_2 & \longrightarrow & \cdots \end{array}$$

Then the induced map $\operatorname{colim}_n X_n \Rightarrow \operatorname{colim}_n Y_n$ is a lax homotopy equivalence. □

We mentioned earlier, and it's easy to show, that $\iota: X \to LX$ is a lax homotopy equivalence if X is an ex-G-space. The same is true if X is a G-spectrum over B, but the proof is not as straightforward. (Note that it's easy to show that ι is a stable equivalence.)

Proposition 2.5.13 *If X is a G-spectrum over B, then $\iota: X \to LX$ is a lax homotopy equivalence. Its inverse is the lax map $\kappa: LX \Rightarrow X$ represented by the identity map $1: LX \to LX$.*

Proof The composite $\kappa\iota: X \Rightarrow X$ is the map represented by $\iota: X \to LX$, i.e., the identity map in $G\mathscr{P}_B^\lambda$.

The composite $\iota\kappa: LX \Rightarrow LX$ is represented by the map $L\iota: LX \to L^2X$. This is not the identity map, but we will show that it is (fiberwise) homotopic to ιL, which does represent the identity. To this end, for any ex-G-space K we define a homotopy $H: \iota L \to L\iota$,

$$H: LK \wedge_B [0, \infty]_+ \to L^2 K,$$

natural in K. Here, we use $[0, \infty]$ rather than the homeomorphic $[0, 1]$ to make the formulas defining H simpler:

$$H(x, \lambda, t) = \begin{cases} (x, \lambda|_0^{l-t}, \lambda|_{l-t}^{l}) & \text{if } t \le l \\ (x, \lambda|_0^0, \lambda|_0^l) & \text{if } t \ge l. \end{cases}$$

In order for H to define a map of spectra $LX \to L^2X$, we need the following diagram to commute (where we have in mind $K = X(V_i)$ and $V = W_{i+1}$):

$$\begin{array}{ccc} \Sigma_B^V LK \wedge_B I_+ & \xrightarrow{\Sigma^V H} & \Sigma_B^V L^2 K \\ {\scriptstyle \beta^V}\big\downarrow & & \big\downarrow{\scriptstyle (\beta^V)^2} \\ L\Sigma_B^V K \wedge_B I_+ & \xrightarrow[H]{} & L^2 \Sigma_B^V K \end{array}$$

To check that this diagram commutes is now a computation based on the formulas for β^V and H, which we will leave to the reader. □

We can now state the spectrum version of Lemma 2.3.1; the proof is exactly the same.

Lemma 2.5.14

(1) *If $f: X \to Y$ is a strict map that is also a lax homotopy equivalence, then it is a stable equivalence.*
(2) *If $f: X \Rightarrow Y$ is a lax homotopy equivalence, then its representing map $\hat{f}: X \to LY$ is a lax homotopy equivalence, hence a stable equivalence.* □

Corollary 2.5.15 *$\beta_Z^\infty: \Sigma_Z^\infty LK \to L\Sigma_Z^\infty K$ is a lax homotopy equivalence.*

Proof Write $\kappa: LK \Rightarrow K$ again for the lax map of ex-G-spaces represented by the identity $1: LK \to LK$. We know that κ is a lax homotopy equivalence (inverse to $\iota: K \Rightarrow LK$), hence $\Sigma_Z^\infty \kappa: \Sigma_Z^\infty LK \Rightarrow \Sigma_Z^\infty K$ is a lax homotopy equivalence. By the preceding lemma, $\widehat{\Sigma_Z^\infty \kappa}$ is also a lax homotopy equivalence, but it is easy to see from the definitions that $\widehat{\Sigma_Z^\infty \kappa} = \beta_Z^\infty$. □

We can now use lax maps to describe stable maps out of compact spaces. If C is a compact ex-G-space over B and Y is a G-spectrum over B, we want to describe $[\Sigma_Z^\infty C, Y]_{G,B}$ somewhat explicitly. May and Sigurdsson define a functor that takes G-spectra and produces so-called excellent spectra, as a composite of functors, the first of which is L. Analyzing the functors they use, we get the following, which we state without proof.

Proposition 2.5.16 *Let C be a compact ex-G-space over B of the homotopy type of a G-CW complex and let Y be a G-spectrum. Then*

$$[\Sigma_Z^\infty C, Y]_{G,B} \cong \operatorname*{colim}_{V \supset Z} \pi G\mathcal{K}_B^\lambda(\Sigma_B^{V-Z} C, Y(V)).$$

□

Corollary 2.5.17 *If C is a compact ex-G-space over B of the homotopy type of a G-CW complex and D is an ex-G-space over B, then the group of stable G-maps* *from C to D over B is given by*

$$[\Sigma_B^\infty C, \Sigma_B^\infty D]_{G,B} \cong \operatorname*{colim}_V \pi G\mathscr{K}_B^\lambda(\Sigma_B^V C, \Sigma_B^V D).$$

More generally,

$$[\Sigma_W^\infty C, \Sigma_Z^\infty D]_{G,B} \cong \operatorname*{colim}_{V\supset W,Z} \pi G\mathscr{K}_B^\lambda(\Sigma_B^{V-W} C, \Sigma_B^{V-Z} D).$$

□

Finally, some comments about computing the stable homotopy groups used to define stable equivalence of G-spectra in [51]. Specifically, the homotopy groups of a G-spectrum X are defined in [51, 12.3.4] to be the stable homotopy groups of the fibers of a levelwise fibrant approximation of X, for which we can use LX. So, for each $b \in B$ and $H \subset G_b$, we have

$$\begin{aligned}
\pi_q^H(X_b) &= \operatorname*{colim}_i \pi_q^H(\Omega^{V_i} LX(V_i)_b) \\
&\cong \operatorname*{colim}_i \pi H\mathscr{K}_*(S^{q+V_i}, LX(V_i)_b) \\
&\cong \operatorname*{colim}_i \pi H\mathscr{K}_B(b_! S^{q+V_i}, LX(V_i)) \\
&= \operatorname*{colim}_i \pi H\mathscr{K}_B^\lambda(b_! S^{q+V_i}, X(V_i)) \\
&\cong [b_! S^q, X]_{H,B}
\end{aligned}$$

as one would want.

We also have the following result that simplifies computing the stable homotopy groups of an ex-G-space: If K is an ex-G-space, $b \in B$, and $H \subset G_b$, then

$$\begin{aligned}
\pi_q^H((\Sigma_B^\infty K)_b) &\cong \pi_q^H((L\Sigma_B^\infty K)_b) \\
&\cong \pi_q^H((\Sigma_B^\infty LK)_b) \\
&\cong \pi_q^H(\Sigma_H^\infty LK_b) \\
&\cong \operatorname*{colim}_V \pi H\mathscr{K}_*(S^{q+V}, \Sigma^V LK_b)
\end{aligned}$$

is just the stable homotopy group of the (non-parametrized) H-space LK_b. This also follows from [51, 13.7.4].

2.6 The Stable Fundamental Groupoid

As mentioned earlier, we need a stable version of the fundamental groupoid. First, we need some notation.

Definition 2.6.1 Let $x\colon G/H \to B$ be a G-map and let V be a representation of H. We have the ex-G-space $G \times_H S^V$ over G/H and we let

$$G_+ \wedge_H S^{V,x} = x_!(G \times_H S^V)$$

over B. If α is a virtual representation of H, then, by [51, 11.5.4 & 13.7.9], corresponding to the H-spectrum S^α there is a G-spectrum over G/H we write as $G \times_H S^\alpha$. We then let

$$G_+ \wedge_H S^{\alpha,x} = x_!(G \times_H S^\alpha)$$

over B.

In one important case, we shall be much more specific:

Definition 2.6.2 Let δ be a dimension function for G. For each orbit G/H, choose an embedding $G/H \subset V$ in a G-representation V; note that any two embeddings are isotopic if V is sufficiently large. Then

$$G/H \times V \cong G \times_H ((V - \mathscr{L}(G/H)) \oplus \mathscr{L}(G/H)),$$

identifying $\mathscr{L}(G/H)$ with the vectors in V tangent to G/H at eH. Using this decomposition allows us to consider $G \times_H \delta(G/H)$ as a subbundle of $G/H \times V$ and to write $G \times_H (V - \delta(G/H))$ as its orthogonal complement. We then write

$$G \times_H S^{-\delta(G/H)} = G \times_H \Sigma_V^\infty S^{V-\delta(G/H)},$$

a G-spectrum over G/H, and, if $x\colon G/H \to B$ as in the preceding definition, we let

$$G_+ \wedge_H S^{-\delta(G/H),x} = x_!(G \times_H S^{-\delta(G/H)}).$$

Definition 2.6.3

(1) The *stable fundamental groupoid* of a G-space B is the category $\widehat{\Pi}_G B$ whose objects are the G-maps $x\colon G/H \to B$ (the same objects as the unstable fundamental groupoid) and whose maps from $(G/H, x)$ to $(G/K, y)$ are the stable G-maps $\{(G/H, x)_+, (G/K, y)_+\}_{B,G}$.
(2) More generally, suppose that δ is an additive dimension function for G. The δ-*stable fundamental groupoid* of B is the category $\widehat{\Pi}_{G,\delta} B = \widehat{\Pi}_\delta B$ with the same

objects as above, but whose maps from x to y are the stable G-maps

$$[G_+ \wedge_H S^{-\delta(G/H),x}, G_+ \wedge_K S^{-\delta(G/K),y}]_{B,G}.$$

(3) Even more generally, if γ is a virtual orthogonal representation of ΠB, the δ-γ-*stable fundamental groupoid* of B is the category $\widehat{\Pi}_{\delta,\gamma}B$ with the same objects as above but whose maps from x to y are the stable G-maps

$$[G_+ \wedge_H S^{\gamma_0(x)-\delta(G/H),x}, G_+ \wedge_K S^{\gamma_0(y)-\delta(G/K),y}]_{B,G}.$$

Here, $\gamma_0(x)$ is the fiber of $\gamma(x)$ over eH and $G_+ \wedge_H S^{\gamma_0(x)-\delta(G/H),x} = x_!(G \times_H \Sigma^{\gamma_0(x)} \Sigma_V^\infty S^{V-\delta(G/H)})$ for a chosen embedding of G/H in a sufficiently large V.

The purpose of this section is to prove the following result.

Theorem 2.6.4 *For any δ and γ, $\widehat{\Pi}_\delta B$ is naturally isomorphic to $\widehat{\Pi}_{\delta,\gamma}B$. Moreover, if $x: G/H \to B$ and $y: G/K \to B$ are orbits in B, then $\widehat{\Pi}_\delta B(x,y)$ is isomorphic to a free abelian group on equivalence classes of diagrams of maps of orbits over B of the form $[x \leftarrow z \Rightarrow y]$ where $z: G/L \to B$. We may assume that $L \leq H$ and $L \leq K^g$ for some g, and then we have the conditions that $\mathscr{L}(N_H L/L) - \delta(N_H L/L) = 0$ and $\delta(N_{K^g} L/L) = 0$. The diagram $x \leftarrow z \Rightarrow y$ is equivalent to $x \leftarrow z' \Rightarrow y$, where $z': G/L' \to B$, if there is a G-homeomorphism $z \to z'$ such that the left triangle below commutes and the right one lax homotopy commutes:*

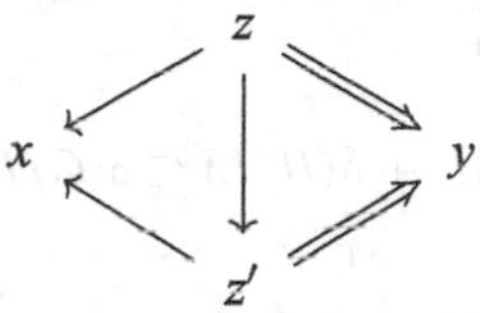

Remark 2.6.5 We conjecture that the preceding theorem is true as stated if we allow γ to be a virtual spherical representation rather than a virtual orthogonal representation. If true, that would allow us to grade ordinary homology and cohomology on virtual spherical representations throughout this work.

The calculation of $\widehat{\Pi}_\delta B(x,y)$ generalizes the calculation of stable maps between orbits in [42, V.9.4]. The case where B is a point and $\delta = 0$ gives that result, because the condition in the theorem amounts to saying that $N_H L/L$ is finite, as in [42, V.9.4]. At the other extreme, when $\delta = \mathscr{L}$, we get a calculation of the stable maps between the duals of orbits, which, as it should, simply has the condition reversed to say that $N_{K^g} L/L$ is finite.

A crucial ingredient in the proof is a general position result. This result is in the same vein as our results on "pseudotransversality" in [19]. In the following we shall mention several times that V should be "sufficiently large" that any two embeddings $G/L \subset V$ are isotopic and that any two such isotopies are themselves homotopic through isotopies. It suffices that V satisfy the *gap hypothesis* that, if $L \leq J$ are two isotropy groups of V, then either $V^J = V^L$ or

$$|V^J| \leq |V^L| - |G| - 3.$$

This is true, for example, if each irreducible representation contained in V appears at least $|G| + 3$ times.

Consider, then, the following context. Let

$$f\colon M = G \times_H N \to G_+ \wedge_K S^{V-\delta(G/K)}$$

be a G-map where N is a compact H-manifold with boundary whose tangent bundle is trivial and equal to $V - \delta(G/H) + n$ for some integer n. (As earlier, we assume chosen an embedding of G/H in V so that $\delta(G/H) \subset \mathscr{L}(G/H) \subset V$.) Suppose $x = [e, x'] \in G \times_H N$ is a point mapping to eH in the projection to G/H. Let L be the isotropy subgroup of x, so $L \leq H$. Consider the tangent plane to x' in N, which is $V - \delta(G/H) + n$ by assumption. The tangent vectors to the H-orbit of x' then give us an inclusion $\mathscr{L}(H/L) \subset V - \delta(G/H) + n$ and the vectors perpendicular to $\mathscr{L}(H/L)$, i.e., in $V - \delta(G/H) + n - \mathscr{L}(H/L)$, are the vectors normal to the orbit of x'. We can then write the space of tangent vectors to x in M normal to the orbit of x as

$$\tau_x - \mathscr{L}(G/L) = V - \delta(G/H) + n - \mathscr{L}(H/L).$$

Note that we have the inclusion

$$\delta(G/L) = \delta(G/H) + \delta(H/L) \subset \delta(G/H) + \mathscr{L}(H/L),$$

which leads to the inclusion

$$V - \delta(G/H) - \mathscr{L}(H/L) \subset V - \delta(G/L).$$

This copy of $V - \delta(G/L)$ can be thought of as coming from an embedding of G/L in V close to the given embedding of G/H in V; if V is sufficiently large, then there is a canonical identification (up to homotopy through orthogonal L-maps) of this copy of $V - \delta(G/L)$ with the one we get from any other embedding of G/L in V.

Now suppose that $f(x) \in G/K \times 0$, say, $f(x) = gK \times 0$. Then we have $L \leq K^g = gKg^{-1}$. We can write the space of tangent vectors at $f(x)$ normal to $G/K \times 0$ as

$$\tau_{f(x)} - \mathscr{L}(G/K^g) = V - \delta(G/K^g).$$

Note that we have an injection

$$i\colon V - \delta(G/H) - \mathscr{L}(H/L) \subset V - \delta(G/L) \hookrightarrow V - \delta(G/K^g).$$

The last map we should define using an isotropy from an embedding of G/L near G/H to one near G/K^g; this is well-defined up to homotopy through orthogonal maps if V is sufficiently large.

At a point x with $f(x) \in G/K \times 0$, we know that f must take the whole orbit of x to $G/K \times 0$. To specify what we mean by f being in general position, we are left with specifying the behaviour of the derivative of f on the space of tangent vectors at x normal to the orbit of x. We decompose this space into two pieces, the L-fixed vectors and the vectors perpendicular to those:

$$\begin{aligned}\tau_x - \mathscr{L}(G/L) &= [\tau_x - \mathscr{L}(G/L)]^L + [\tau_x - \mathscr{L}(G/L)]_L \\ &= [V - \delta(G/H) + n - \mathscr{L}(H/L)]^L + [V - \delta(G/H) - \mathscr{L}(H/L)]_L\end{aligned}$$

Definition 2.6.6 A map $f\colon M = G \times_H N \to G_+ \wedge_K S^{V-\delta(G/K)}$, with N as above, is said to be in *general position* if it is smooth in a neighborhood of $f^{-1}(G/K \times 0)$ and, for each point $x \in M$ over eH with $f(x) \in G/K \times 0$, we have, writing L for the isotropy subgroup of x and $Df\colon \tau_x \to \tau_{f(x)}$ for the induced map of tangent planes,

(1) the following map induced by Df^L is an epimorphism:

$$[\tau_x - \mathscr{L}(G/L)]^L \to \tau^L_{f(x)} \twoheadrightarrow [\tau_{f(x)} - \mathscr{L}(G/K^g)]^L,$$

where the second map is the orthogonal projection, and

(2) on the piece $[V - \delta(G/H) - \mathscr{L}(H/L)]_L$, Df is the injection i_L with i as above.

We also have the following definition of general position of a vector field from [19], where it was called canonical pseudotransversality.

Definition 2.6.7 Let M be a smooth G-manifold and let $\tau\colon TM \to M$ be its tangent bundle. We say that a smooth section $s\colon M \to TM$ is in *general position* if, for each $x \in M$ with $s(x) = x$ (i.e., equal to $0 \in \tau_x$), the composite

$$\tau_x \xrightarrow{Ds} \tau_x \oplus \tau_x \xrightarrow{p_2} \tau_x$$

is the identity on $[\tau_x - \mathscr{L}(G/G_x)]_{G_x}$ and takes $[\tau_x - \mathscr{L}(G/G_x)]^{G_x}$ isomorphically onto itself.

Proposition 2.6.8 (General Position) *Let $M = G \times_H N$ where N is a compact H-manifold with boundary whose tangent bundle is trivial and equal to $V - \delta(G/H) + n$ for some integer n. Let $f\colon M \to G_+ \wedge_K S^{V-\delta(G/K)}$ be a G-map that is in general position on a neighborhood of ∂M. Then, if V is sufficiently large in the sense above, f is G-homotopic rel ∂M to a map in general position.*

Proof The argument is similar to those given by Wasserman [70] and by Hauschild [29, 30]. The proof is by induction on the isotropy groups appearing in M. Let L be a maximal isotropy group appearing in M; we may assume that $L \le H$. Using the argument in [29, I.4], for example, we can homotope f rel boundary to a G-map transverse on the L-fixed sets. We can then G-homotope f rel boundary and L-fixed sets so that, at each L-fixed point mapping to $G/K \times 0$, in the directions normal to the fixed sets and the orbit of the point, f agrees with i_L as above.

Then f is in general position in a neighborhood of GM^L with no points on the boundary of the neighborhood mapping to $G/K \times 0$. We can then excise this neighborhood and, by induction, put f in regular position on the remainder of M. □

Proof of Theorem 2.6.4 We first define an auxiliary category $\mathscr{D}_\delta$. Its objects are the same as those of $\widehat{\Pi}_\delta B$, that is, maps $x: G/H \to B$. The set of maps from x to $y: G/K \to B$ is the Grothendieck group on the set of equivalence classes of diagrams $x \leftarrow M \Rightarrow y$ where M is a closed G-manifold thought of as a manifold over B via $M \to G/H \xrightarrow{x} B$. M has to satisfy a certain condition to be explained, and the equivalence relation must also be explained.

Name the maps from M to the orbits $\alpha: M \to G/H$ and $\beta: M \to G/K$. Let N be the fiber of α, so that $M = G \times_H N$. We can then decompose the tangent bundle of M as

$$\tau_M \cong \alpha^*(G \times_H \mathscr{L}(G/H)) \oplus (G \times_H \tau_N).$$

In other words, we identify $\alpha^*(G \times_H \mathscr{L}(G/H))$ with the subbundle of tangent vectors normal to the fiber of α. Similarly, we can identify $\beta^*(G \times_K \mathscr{L}(G/K))$ with the subbundle of tangent vectors normal to the fiber of β, and within this bundle we have its subbundle $\beta^*(G \times_K \delta(G/K))$. The condition we impose on M is then that

$$\beta^*(G \times_K \delta(G/K)) \subset \alpha^*(G \times_H \delta(G/H)) \oplus (G \times_H \tau_N).$$

Another way of putting this is that α carries $\beta^*(G \times_K \delta(G/K))$ into $G \times_H \delta(G/H)$. This is automatically true if M is an orbit of G, by the additivity of δ.

Now for the equivalence relation. Given the diagram $x \leftarrow M \Rightarrow y$, we define a stable map as follows. Choose a G-representation V so large that we can embed G/H, so that V has a distinguished H-subspace isomorphic to $\mathscr{L}(G/H)$. We assume V is also large enough that we can embed N in $V - \delta(G/H)$, which we extend to an embedding of M in $G \times_H (V - \delta(G/H))$. The normal bundle ν to this embedding will then be

$$\nu \cong V - \alpha^*(G \times_H \delta(G/H)) - G \times_H \tau_N.$$

By the condition imposed on M, this bundle is a subbundle of $V - \beta^*(G \times_K \delta(G/K))$. Hence, we can write down the following map:

$$G \times_H S^{V-\delta(G/H)} \xrightarrow{c} S^\nu \to S^{V-\beta^*(G\times_K\delta(G/K))} \xrightarrow{\beta} G \times_K S^{V-\delta(G/K)},$$

where S^ν denotes the sphere bundle over M obtained by taking the one-point compactification of each fiber of ν, and similarly for $S^{V-\beta^*(G\times_K\delta(G/K))}$. Now, using the homotopy involved in the lax map $M \Rightarrow y$, we get a lax stable map

$$\chi(x \leftarrow M \Rightarrow y)\colon G_+ \wedge_H S^{-\delta(G/H),x} \Rightarrow G_+ \wedge_K S^{-\delta(G/K),y},$$

an element of $\widehat{\Pi}_\delta B(x, y)$, which we call the *Euler characteristic* of the diagram. We consider two diagrams to be equivalent if they have the same Euler characteristic. (Note that, if G is trivial and B is a point, then $\widehat{\Pi}_\delta B(x, y) \cong \mathbb{Z}$ and $\chi(M)$ is the classical Euler characteristic of M.)

We define the sum of two such diagrams $[x \leftarrow M \Rightarrow y]$ and $[x \leftarrow N \Rightarrow y]$ to be $[x \leftarrow M \sqcup N \Rightarrow y]$ and then, as mentioned above, $\mathscr{D}_\delta(x, y)$ is defined to be the Grothendieck group of the resulting monoid.

Composition is defined by taking pullbacks. Precisely, if given two diagrams $[x \leftarrow M \Rightarrow y]$ and $[y \leftarrow N \Rightarrow z]$, we let P be the pullback in the following diagram:

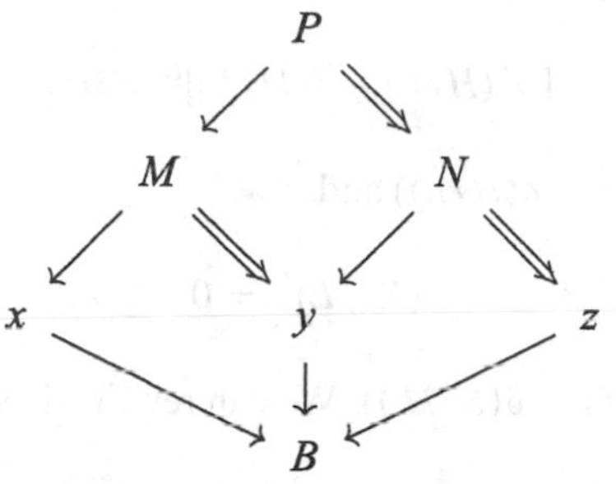

The homotopy involved in the lax map $P \Rightarrow N$ comes from the homotopy underlying $M \Rightarrow y$. We then define the composite

$$[y \leftarrow N \Rightarrow z] \circ [x \leftarrow M \Rightarrow y] = [x \leftarrow P \Rightarrow z].$$

It's a formal check that $x \leftarrow P \Rightarrow z$ satisfies the condition required of it. It follows from the construction that

$$\chi(x \leftarrow P \Rightarrow z) = \chi(y \leftarrow N \Rightarrow z) \circ \chi(x \leftarrow M \Rightarrow y),$$

from which it follows that composition respects the equivalence relation, hence we have a well-defined category $\mathscr{D}_\delta$.

In the process of defining $\mathscr{D}_\delta$, note that we've also defined a functor $\chi\colon \mathscr{D}_\delta \to \widehat{\Pi}_\delta B$, which, by definition, is a monomorphism on each mapping group.

Now suppose that we have element of $\widehat{\Pi}_\delta B(x, y)$. By Corollary 2.5.17 we can represent this element by a lax map $f\colon G_+ \wedge_H S^{V-\delta(G/H),x} \Rightarrow G_+ \wedge_K S^{V-\delta(G/K),y}$. By Proposition 2.6.8, we may assume that f is in general position (away from the basepoint). The inverse image of $G/K \times 0$ is then a disjoint union of orbits of G: If $m \in S^{V-\delta(G/H)}$ is a point such that $f(m) \in G/K \times 0$, with $Gm \cong G/L$, then the tangent plane at m normal to the orbit of m is isomorphic to $V - \delta(G/H) - \mathscr{L}(H/L)$. If $f(m) = gK \times 0$, then we can write the codimension of $G/K \times 0$ in the target of f

as $V - \delta(G/K^g)$. As noted while defining general position, we have an inclusion

$$V - \delta(G/H) - \mathscr{L}(H/L) \subset V - \delta(G/L) \subset V - \delta(G/K^g).$$

On L-fixed sets, this means that the dimension $|[V - \delta(G/H) - \mathscr{L}(H/L)]^L|$ is no more than $|[V - \delta(G/K^g)]^L|$. In order to satisfy the first condition of being in general position, these dimensions must therefore be the same and Df^L must be an isomorphism. Together with the second condition, this implies that the only tangent directions at m that can be taken into $G/K \times 0$ are those along the orbit of m, hence the orbit of m is isolated in $f^{-1}(G/K \times 0)$. Thus, $f^{-1}(G/K \times 0)$ is a disjoint union of orbits as claimed.

The inclusions above also tell us exactly which orbits may appear in the inverse image, namely those for which the inclusion are equalities on fixed sets: We need to have both

$$[\mathscr{L}(H/L) - \delta(H/L)]^L = 0$$

(using $\delta(G/L) = \delta(G/H) + \delta(H/L)$) and

$$\delta(K^g/L)^L = 0$$

(using $\delta(G/L) = \delta(G/K^g) + \delta(K^g/L)$). We can rewrite these as

$$\mathscr{L}(N_H L/L) - \delta(N_H L/L) = 0$$

and

$$\delta(N_{K^g} L/L) = 0$$

as in the statement of the theorem.

Each orbit in the inverse image determines a diagram $x \leftarrow G/L \Rightarrow y$; we write $z_i = [x \leftarrow G/L \Rightarrow y]$ for the class of this diagram in $\mathscr{D}_\delta(x, y)$. Also, for each orbit, the map Df^L above is an isomorphism, so determines a sign $\epsilon_i = \pm 1$, depending on whether Df^L preserves orientation or not, using the canonical identification of $|[V - \delta(G/H) - \mathscr{L}(H/L)]^L|$ with $|[V - \delta(G/K^g)]^L|$ for V sufficiently large. It then follows that

$$[f] = \chi\left(\sum_i \epsilon_i z_i\right)$$

where the sum runs over the orbits in the inverse image of $G/K \times 0$. This shows that χ is an epimorphism on mapping groups, hence is an isomorphism of categories.

Moreover, the argument above shows that $\mathscr{D}_\delta(x,y) \cong \widehat{\Pi}_\delta B(x,y)$ is generated by the classes $[x \leftarrow G/L \Rightarrow y]$ for orbits G/L satisfying the conditions in the statement of the theorem. To show that these classes freely generate, suppose that we have a sum $\sum_i n_i z_i$ such that $\chi(\sum_i n_i z_i) = 0$, where the z_i are all distinct generators. Let f be the map constructed above to represent $\chi(\sum_i n_i z_i)$, so that f is in general position and the inverse image of $G/K \times 0$ consists of n_i copies of each orbit underlying z_i. We are assuming that f is null homotopic and we may assume we have a null homotopy h in general position. For each orbit G/L that appears in a z_i, we know that

$$|(V - \delta(G/H) - \mathscr{L}(H/L))^L| = |(V - \delta(G/K^g))^L|,$$

hence that

$$|(V - \delta(G/H) - \mathscr{L}(H/L) + 1)^L| = |(V - \delta(G/K^g))^L| + 1.$$

So, each of these orbits appearing in $f^{-1}(G/K \times 0)$ will be one end of a manifold of the form $G/L \times I \subset h^{-1}(G/K \times 0)$. Because h is a null homotopy, the other end of this manifold must be on the same face of $G_+ \wedge_H S^{V-\delta(G/)} \wedge I_+$, so will have opposite orientation. The only way this can happen without contradiction is to have each $n_i = 0$. Thus, $\mathscr{D}_\delta(x,y)$ is free as claimed in the statement of the theorem.

We now define a functor $\psi\colon \mathscr{D}_\delta \to \widehat{\Pi}_{\delta,\gamma}B$. On objects it is the same as χ; on morphisms we define what it does on a generator: Given $x \leftarrow G/L \Rightarrow y$, let $\alpha\colon G/L \to G/H$ and $\beta\colon G/L \to G/K$ be the corresponding maps of orbits. Writing $z = x \circ \alpha$, we define $\psi(x \leftarrow G/L \Rightarrow y)$ to be the lax stable composite

$$\begin{aligned}
G_+ \wedge_H S^{\gamma(x)-\delta(G/H),x} &\xrightarrow{c} G_+ \wedge_L S^{\gamma(x)-\delta(G/H)-\mathscr{L}(H/L),z} \\
&\hookrightarrow G_+ \wedge_L S^{\gamma(x)-\delta(G/L),z} \\
&\xrightarrow{\gamma} G_+ \wedge_L S^{\gamma(y)-\delta(G/L),z} \\
&\hookrightarrow G_+ \wedge_L S^{\gamma(y)-\delta(G/K),z} \\
&\Rightarrow G_+ \wedge_K S^{\gamma(y)-\delta(G/K),y}.
\end{aligned}$$

The collapse map c is the same as used in the definition of χ. The map labeled γ is the one obtained by applying γ to the lax map $G/L \Rightarrow y$. The inclusions come from the fact that δ is a dimension function, while the last map comes from the projection $G/L \to G/K$.

We next have to show that ψ is a functor, i.e., that it respects composition. Let $x\colon G/H \to B$, $y\colon G/K \to B$, and $z\colon G/J \to B$. Consider two generators $x \leftarrow G/L \Rightarrow$

y and $y \leftarrow G/I \Rightarrow z$. Let P be the pullback in the following diagram:

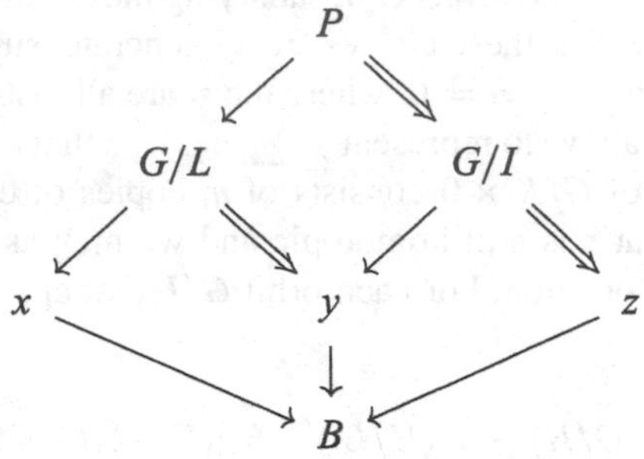

In order to understand what ψ does to the composite, we need to describe the composite $x \leftarrow P \Rightarrow z$ in terms of the generators. In [19], we showed that an equivariant manifold like P has a vector field in general position. In particular, its set of zeros is a disjoint union of orbits in P and to each of the orbits we can assign a sign ± 1 determined by the behaviour of the vector field in a neighborhood of the orbit. This determines a signed sum of generators. If we apply χ to $x \leftarrow P \Rightarrow z$, we can use the vector field on P to homotope the resulting map to the image under χ of that sum of generators. Therefore, $x \leftarrow P \Rightarrow z$ equals this sum of generators in $\mathscr{D}_\delta$. Now, we can use the vector field in the same way to homotope $\psi(y \leftarrow G/I \Rightarrow z) \circ \psi(x \leftarrow G/L \Rightarrow y)$ to the image under ψ of the same sum of generators (choosing a representative isomorphism in γ for each component of P/G), showing that

$$\psi(y \leftarrow G/I \Rightarrow z) \circ \psi(x \leftarrow G/L \Rightarrow y) = \psi(x \leftarrow P \Rightarrow z),$$

i.e., that ψ respects composition.

Suppose that $f: G_+ \wedge_H S^{V+\gamma_0(x)-\delta(G/H),x} \Rightarrow G_+ \wedge_K S^{V+\gamma_0(y)-\delta(G/K),y}$ represents an element of $\Pi_{\delta,\gamma} B$. Let $U \subset G_+ \wedge_K S^{V+\gamma_0(y)-\delta(G/K),y}$ be a G-invariant open neighborhood of $G_+ \wedge_K 0$ not including the base section. Then $f^{-1}(U)$ is an open subspace of $G_+ \wedge_H S^{V+\gamma_0(x)-\delta(G/H),x}$ not including the base section. Let $G \times_J A$ be the orbit of a component of $f^{-1}(U)$. Each point in A determines, via f, a homotopy in B^J, and any two points in A, being connected by a path, determine the same homotopy, up to path homotopy rel endpoints. Therefore, A determines a class of paths, hence, via γ, a homotopy class of isomorphisms of $\gamma_0(x)$ with $\gamma_0(y)$. Pick an isomorphism in that class. Using that isomorphism to identify $V + \gamma_0(x)$ with $V + \gamma_0(y)$, we can put f in general position inside $G \times_J A$. Doing this for each component of $f^{-1}(U)$, we put f in general position. It follows that ψ is an epimorphism. A similar argument applied to homotopies shows that ψ is a monomorphism. Hence, ψ is an isomorphism of categories.

We have therefore shown that have the following isomorphisms:

$$\widehat{\Pi}_\delta B \cong \mathscr{D}_\delta \cong \widehat{\Pi}_{\delta,\gamma} B,$$

which proves the theorem. □

2.7 Parametrized Homology and Cohomology Theories

The cellular theories we shall define in the following chapter can be thought of as consisting of a series of related "$RO(G)$-graded parametrized homology and cohomology theories." We define exactly what we mean by that, concentrating on reduced theories.

In [21] Dold gave axioms for nonequivariant homology and cohomology theories defined on spaces parametrized by B. Several authors, including Wirthmüller [71], have given axioms for $RO(G)$-graded theories defined on G-spaces; we follow those given by May in [47, XIII.1]. We combine the two approaches to give axioms for equivariant theories defined on parametrized spaces; this treatment is essentially the same as that of May and Sigurdsson [51, §21.1].

Let $\mathscr{R}O(G)$ be the category of finite-dimensional linear representations of G and G-linear isometric isomorphisms between them. Say that two maps $V \to W$ are homotopic if the associated G-maps $S^V \to S^W$ are stably homotopic. Let $h\mathscr{R}O(G)$ denote the resulting homotopy category. (Comparing with [47], we are assuming that we are using a complete G-universe $\mathscr{U}$. Variants for smaller universes are possible, but we are interested in duality, which at present requires a complete universe.)

Definition 2.7.1 An *$RO(G)$-graded parametrized (reduced) homology theory* consists of a functor

$$\tilde{h}^G_*: h\mathscr{R}O(G)^{\mathrm{op}} \times \mathrm{Ho}\, G\mathscr{K}_B \to Ab,$$

written $(V,X) \mapsto \tilde{h}^G_V(X)$ on objects. Similarly, an *$RO(G)$-graded parametrized (reduced) cohomology theory* consists of a functor

$$\tilde{h}^*_G: h\mathscr{R}O(G) \times \mathrm{Ho}\, G\mathscr{K}_B{}^{\mathrm{op}} \to Ab.$$

These functors are required to satisfy the following axioms.

(1) For each V, the functor $\tilde{h}^G_V$ is exact on cofiber sequences and sends wedges to direct sums. The functor $\tilde{h}^V_G$ is exact on cofiber sequences and sends wedges to products.
(2) For each W there are natural isomorphisms

$$\sigma^W: \tilde{h}^G_V(X) \to \tilde{h}^G_{V\oplus W}(\Sigma^W_B X)$$

and

$$\sigma^W : \tilde{h}^V_G(X) \to \tilde{h}^{V\oplus W}_G(\Sigma^W_B X).$$

$\sigma^0 = \mathrm{id}$ and $\sigma^Z \circ \sigma^W = \sigma^{W\oplus Z}$ for every pair of representations W and Z.

(3) The σ are natural in W in the following sense. If $\alpha : W \to W'$ is a map in $h\mathscr{R}O(G)$, then the following diagrams commute.

$$\begin{array}{ccc} \tilde{h}^G_V(X) & \xrightarrow{\sigma^{W'}} & \tilde{h}^G_{V\oplus W'}(\Sigma^{W'}_B X) \\ \Big\downarrow{\scriptstyle \sigma^W} & & \Big\downarrow{\scriptstyle \tilde{h}^G_{\mathrm{id}\oplus\alpha}(\mathrm{id})} \\ \tilde{h}^G_{V\oplus W}(\Sigma^W_B X) & \xrightarrow[(\Sigma^\alpha_B \mathrm{id})_*]{} & \tilde{h}^G_{V\oplus W}(\Sigma^{W'}_B X) \end{array}$$

$$\begin{array}{ccc} \tilde{h}^V_G(X) & \xrightarrow{\sigma^{W}} & \tilde{h}^{V\oplus W}_G(\Sigma^{W}_B X) \\ \Big\downarrow{\scriptstyle \sigma^{W'}} & & \Big\downarrow{\scriptstyle \tilde{h}^{\mathrm{id}\oplus\alpha}_G(\mathrm{id})} \\ \tilde{h}^{V\oplus W'}_G(\Sigma^{W'}_B X) & \xrightarrow[(\Sigma^\alpha_B \mathrm{id})^*]{} & \tilde{h}^{V\oplus W'}_G(\Sigma^{W}_B X) \end{array}$$

As in [47] we extend the grading to "formal differences" $V \ominus W$ by setting

$$\tilde{h}^G_{V\ominus W}(X) = \tilde{h}^G_V(\Sigma^W_B X)$$

and similarly in cohomology. Rigorously, we extend the grading from $h\mathscr{R}O(G)^{\mathrm{op}}$ to $h\mathscr{R}O(G)^{\mathrm{op}} \times h\mathscr{R}O(G)$. Now, $RO(G)$ can be obtained from $\mathscr{R}O(G)$ by considering $V \ominus W$ to be equivalent to $V' \ominus W'$ when there is a G-linear isometric isomorphism

$$\alpha : V \oplus W' \to V' \oplus W.$$

Given such an isomorphism we get the following diagram.

$$\begin{array}{ccc} \tilde{h}^G_{V'}(\Sigma^{W'}_B X) & \xrightarrow{\sigma^{W}} & \tilde{h}^G_{V'\oplus W}(\Sigma^{W'\oplus W}_B X) \\ \Big\downarrow & & \Big\downarrow{\scriptstyle \tilde{h}^G_\alpha(\Sigma^\tau \mathrm{id})} \\ \tilde{h}^G_V(\Sigma^W_B X) & \xrightarrow[\sigma^{W'}]{} & \tilde{h}^G_{V\oplus W'}(\Sigma^{W\oplus W'}_B X) \end{array}$$

Here, $\tau\colon W \oplus W' \to W' \oplus W$ is the transposition map. Thus,

$$\tilde{h}^G_{V\ominus W}(X) \cong \tilde{h}^G_{V'\ominus W'}(X).$$

(There is a similar isomorphism in cohomology.) It is in this sense that we can think of grading on $RO(G)$, but note that the isomorphism depends on the choice of α, so is not canonical.

Dold pointed out in the nonequivariant case that $\tilde{h}_*(-)$ and $\tilde{h}^*(-)$ restrict to functors on ΠB. Actually, the suspension isomorphisms imply that homology and cohomology theories extend to stable maps, and so we get restrictions to functors on the stable fundamental groupoid $\widehat{\Pi}B$. More generally, we can make the following definitions.

Definition 2.7.2 If $\tilde{h}^G_*(-)$ is an $RO(G)$-graded parametrized homology theory on ex-G-spaces over B and $H \leq G$ is a subgroup, define an $RO(H)$-graded parametrized homology theory $\tilde{h}^H_*(-)$ on ex-H-spaces over B by setting

$$\tilde{h}^H_V(X) = \tilde{h}^G_W(G_+ \wedge_H \Sigma^{W-V}X)$$

where W is any representation of G containing V as a sub-H-representation. Similarly, if $\tilde{h}^*_G(-)$ is an $RO(G)$-graded parametrized cohomology theory, define an $RO(H)$-graded parametrized cohomology theory $\tilde{h}^*_H(-)$ by setting

$$\tilde{h}^V_H(X) = \tilde{h}^W_G(G_+ \wedge_H \Sigma^{W-V}X).$$

Definition 2.7.3 If $\tilde{h}^G_*(-)$ is an $RO(G)$-graded parametrized homology theory, γ is a virtual representation of ΠB, and δ is a dimension function for G, we define a covariant $\widehat{\Pi}_\delta B$-module-valued homology theory $\underline{h}^{G,\delta}_\gamma(-)$ by

$$\begin{aligned}\underline{h}^{G,\delta}_\gamma(X)(b\colon G/H \to B) &= \tilde{h}^H_{\gamma_0(b)}(X \wedge_B S^{-\delta(G/H),b}) \\ &= \tilde{h}^H_{\gamma_0(b)+V}(X \wedge_B S^{V-\delta(G/H),b}),\end{aligned}$$

where V is a representation of H so large that $\delta(G/H) \subset V$. (Here, so that we actually get a homology theory, we should understand $\wedge_B$ to mean that we first turn $S^{V-\delta(G/H),b} \to B$ into a fibration and then take the smash product over B.) This is a functor on $\widehat{\Pi}_{\delta,-\gamma}B$, so we make it a functor on $\widehat{\Pi}_\delta B$ using Theorem 2.6.4. In particular, we write

$$\underline{h}^{G,\delta}_\gamma = \underline{h}^{G,\delta}_\gamma(B_+)$$

for the *coefficient system* of $\tilde{h}^G_*(-)$. If $\tilde{h}^*_G(-)$ is an $RO(G)$-graded parametrized cohomology theory, we define $\overline{h}^*_{G,\delta}(-)$, a contravariant $\widehat{\Pi}_\delta B$-module-valued homology theory, similarly, and write $\overline{h}^*_{G,\delta} = \overline{h}^*_{G,\delta}(B_+)$ for its coefficient system.

As usual, if X is an unbased space over B, or (X,A) is a pair of based or unbased G-spaces over B, we define

$$h_*^G(X) = \tilde{h}_*^G(X_+)$$

and

$$h_*^G(X,A) = \tilde{h}_*^G(C_B i)$$

where $i\colon A \to X$ is the inclusion and $C_B i$ is its cofiber over B. We use a similar convention for cohomology.

2.8 Representing Parametrized Homology and Cohomology Theories

We are interested in how parametrized homology and cohomology theories are represented. It is well-known how to represent parametrized cohomology theories using spectra over B (see, for example, [7] and [8]). The following is part of Theorem 21.2.3 in [51], and follows from Brown's representability theorem in the usual way.

Theorem 2.8.1 *If $\tilde{h}_G^*(-)$ is an $RO(G)$-graded cohomology theory on* $\operatorname{Ho} G\mathscr{K}_B$*, then it is represented by an Ω-G-spectrum E over B, in the sense that, for each V, there is a natural isomorphism*

$$\tilde{h}_G^V(X) \cong [\Sigma_B^\infty X, \Sigma_B^V E]_{G,B}.$$

The spectrum E is determined by h up to non-unique equivalence. □

We now consider homology theories. Curiously, previously to [51] and earlier versions of this book, there appears to have been no discussion in the literature of representing homology theories on parametrized spaces. The next proposition gives a way of defining a homology theory using a parametrized spectrum. Before that we should say a few words about smash products. (We summarize the more detailed discussion [51].)

The fiberwise smash product behaves poorly from the point of view of homotopy theory; it does not pass directly to homotopy categories. However, the *external* fiberwise smash product does behave well. If D is a G-spectrum over A and E is a G-spectrum over B we can form the external smash product $D \barwedge E$ over $A \times B$. (Since we are using LMS spectra and not orthogonal spectra, the smash product we use is either "external" with respect to indexing spaces as well, or is a "handicrafted" smash product [51, 13.7.1].) The internal smash product of spectra D and E in $\operatorname{Ho} G\mathscr{P}_B$ is then defined by $D \wedge_B E = \Delta^*(D \barwedge E)$, where $\Delta\colon B \to B \times B$ is the diagonal. As usual, this notation is convenient but misleading. In the homotopy category, Δ^* and $\barwedge$ must be understood as derived functors of their point-set level precursors. In particular, Δ^* is defined by first applying a fibrant approximation functor and then applying the point-set level Δ^*. Fortunately, May and Sigurdsson show that the familiar compatibility relations among the smash product and the change of base space functors remain true in the homotopy category.

Proposition 2.8.2 *If E is a G-spectrum over B then the groups*

$$\tilde{E}_V^G(X) = [S^V, \rho_!(E \wedge_B X)]_G$$

define an $RO(G)$-graded homology theory on $\mathrm{Ho}\, G\mathscr{K}_B$. *Here, $\rho: B \to *$ denotes projection to a point.* □

Proof (See also [51, 21.2.22].) The main problem is to see that $\tilde{E}_V^G(X)$ is exact on cofiber sequences and takes wedges to direct sums. Write $\rho_!(E \wedge_B X) = \rho_! \Delta^*(E \barwedge X)$. As in [51], each of $\rho_!$, Δ^*, and $E \barwedge -$ is exact on $\mathrm{Ho}\, G\mathscr{P}_B$ and also takes wedges to wedges. The remainder of the proof is as in [47, XIII.2.6]. □

For a discussion of the representability of general $RO(G)$-graded parametrized homology theories, see [51, 21.5]. Unfortunately, the story is not completely straightforward. May and Sigurdsson show that homology theories over certain nice base spaces are representable and sketch an argument for how to extend that result to all base spaces. Fortunately, this will not be a problem when we want to discuss the spectra representing cellular homology, which we will construct explicitly in Chap. 3.

At first glance, the construction of Proposition 2.8.2 is not an obvious way to define a homology theory using a spectrum. In particular, the use of $\rho_!$ on the right might not be expected, given that $\rho_!$ is a left adjoint. One justification is that it works: We do get a homology theory. Here is, perhaps, a better justification. The homology and cohomology theories determined by a parametrized spectrum E should, locally, reflect the nonparametrized theories given by the fibers of E. To make this precise, we first record a change of base space result.

Proposition 2.8.3 *Let $\alpha: A \to B$ be a G-map, let X be an ex-G-space over A, and let E be a G-spectrum over B. Then*

$$\tilde{E}_V^G(\alpha_! X) \cong (\widetilde{\alpha^* E})_V^G(X)$$

and

$$\tilde{E}_G^V(\alpha_! X) \cong (\widetilde{\alpha^* E})_G^V(X).$$

Proof We begin with cohomology, which is a little more obvious:

$$\begin{aligned}\tilde{E}^V_G(\alpha_!X) &= [\alpha_!X, \Sigma^V_B E]_{G,B}\\ &\cong [X, \alpha^*(\Sigma^V_B E)]_{G,A}\\ &\cong [X, \Sigma^V_A \alpha^* E]_{G,A}\\ &= (\widetilde{\alpha^* E})^V_G(X).\end{aligned}$$

For homology we have the following, using one of the stable equivalences listed in Sect. 2.4:

$$\begin{aligned}\tilde{E}^G_V(\alpha_!X) &= [S^V, \rho_!(E \wedge_B \alpha_!X)]_G\\ &\cong [S^V, \rho_!\alpha_!(\alpha^* E \wedge_A X)]_G\\ &= [S^V, \rho_!(\alpha^* E \wedge_A X)]_G\\ &= (\widetilde{\alpha^* E})^G_V(X).\end{aligned}$$

□

We can now say that these theories behave correctly over individual points of B:

Corollary 2.8.4 *Let E be a G-spectrum over B, let $b: G/H \to B$, and let X be a based H-space. Then*

$$\tilde{E}^G_V(b_!(G \times_H X)) \cong (\tilde{E}_b)^H_V(X)$$

and

$$\tilde{E}^V_G(b_!(G \times_H X)) \cong (\tilde{E}_b)^V_H(X)$$

where $b^*E = G \times_H E_b$. □

Thus, the homology and cohomology theories defined by E restrict over each point b of B to the nonparametrized theories defined by the fiber E_b.

Specializing further, we can examine the coefficient systems of the homology and cohomology theories determined by E.

Corollary 2.8.5 *Let γ be a virtual representation of ΠB and let δ be a dimension function for G. If E is a G-spectrum over B, then the coefficient systems of its associated homology and cohomology theories are given by*

$$\underline{E}^{G,\delta}_\gamma(b: G/H \to B) \cong \pi^H_{\gamma_0(b)-(\mathscr{L}-\delta)(G/H)}(E_b)$$

and

$$\overline{E}^{\gamma}_{G,\delta}(b) \cong \pi^{H}_{-\gamma_0(b)-\delta(b)}(E_b).$$

Proof The calculation for cohomology follows directly from the definitions, including Definition 2.7.3, and the preceding corollary. The calculation for homology uses the isomorphism of [42, II.6.5]. □

We can reframe this result as follows. For a fixed $b: G/H \to B$, define a covariant H-δ-Mackey functor $\underline{E}^{G,\delta}_{\gamma}|b$ by

$$(\underline{E}^{G,\delta}_{\gamma}|b)(H/K) = \underline{E}^{G,\delta}_{\gamma}(G/K \to G/H \to B).$$

Then we have

$$\underline{E}^{G,\delta}_{\gamma}|b \cong \overline{\pi}^{H,\mathscr{L}-\delta}_{\gamma_0(b)}(E_b).$$

Defining $\overline{E}^{\gamma}_{G,\delta}|b$ similarly, we get

$$\overline{E}^{\gamma}_{G,\delta}|b \cong \overline{\pi}^{H,\delta}_{-\gamma_0(b)}(E_b).$$

Finally, we give one more useful application of Proposition 2.8.3: The groups $\tilde{E}^G_*(X,p,\sigma)$ and $\tilde{E}^*_G(X,p,\sigma)$ depend only on the spectrum p^*E parametrized by X. To see this, note first that $(X,p,\sigma) = p_!(X',q,\tau)$ where $X' = X \cup_{\sigma(B)} X$, $q: X' \to X$ is the evident map, and $\tau: X \to X'$ takes X onto the second copy of X in X'. From this it follows that

$$\tilde{E}^G_V(X) = \tilde{E}^G_V(p_! X') \cong \widetilde{(p^*E)}^G_V(X')$$

and similarly for cohomology. In particular, if p^*E is trivial, in the sense that it is equivalent to $\rho^*D = X \times D$ for a nonparametrized G-spectrum D, then (assuming that X is well-sectioned)

$$\tilde{E}^G_V(X) \cong \tilde{D}^G_V(\rho_! X') \cong \tilde{D}^G_V(X/\sigma(B))$$

is just the usual D-homology of $X/\sigma(B)$, and

$$\tilde{E}^V_G(X) \cong \tilde{D}^V_G(X/\sigma(B))$$

is just the usual D-cohomology of $X/\sigma(B)$.

2.9 Duality

Let E be a G-spectrum parametrized by B. We would like to relate the homology and cohomology theories determined by E by finding, for each suitably "finite" G-spectrum X over B, a dual spectrum $\bar{D}_B X$ over B such that $E^*_G(X) \cong E^G_{-*}(\bar{D}_B X)$. We call this kind of duality *homological duality* to distinguish it from the better-known fiberwise duality (see, for example, [7] or [8] as well as [51]). What we call homological duality, May and Sigurdsson have seen fit to call Costenoble-Waner duality, for which we thank them.

Homological duality does not fit the usual formalism of duality in closed symmetric monoidal categories—fiberwise duality does. However, we can set up a very similar formalism for our purposes. May and Sigurdsson [51] have developed a more general notion of duality, in what they call closed symmetric bicategories, that includes both fiberwise duality and homological duality as special cases. Even though we recommend their treatment, we think it worth presenting the one we gave before [51] appeared.

It will be convenient to generalize as follows. Let A and B be base spaces, let W be a G-spectrum over A, let X be a G-spectrum over B, and let E be a G-spectrum over $A \times B$. Writing in terms of mapping sets, we would like a natural isomorphism of the form

$$[W \barwedge X, E]_{G,A\times B} \cong [W, (1 \times \rho)_!(1 \times \Delta)^*(E \barwedge \bar{D}_B X)]_{G,A}.$$

Here, $\barwedge$ denotes the external smash product while, as is our custom, ρ denotes projection to a point, so

$$1 \times \Delta : A \times B \to A \times B \times B$$

and

$$1 \times \rho : A \times B \to A.$$

We are most interested in the case $A = *$, but the extra generality comes at no cost and gives a considerably more satisfying result.

Recall from [51] that there is an external function spectrum functor $\bar{F}$: If Y is a G-spectrum over B and Z is a G-spectrum over $A \times B$, then $\bar{F}(Y, Z)$ is a G-spectrum over A. $\bar{F}(Y, -)$ is right adjoint to $- \barwedge Y$.

Definition 2.9.1 If X is a G-spectrum over B, let $\bar{D}_B X = \bar{F}(X, \Delta_! S_B)$ where S_B is the unit spectrum over B (i.e., $\Sigma^\infty_B B_+$) and $\Delta : B \to B \times B$ is the diagonal.

We shall justify this definition over the course of this section. It's interesting to note that the fiberwise dual of X can be written as $\bar{F}(X, \Delta_* S_B)$ where Δ_* is the right adjoint of Δ^*. We should reiterate our standard warning at this point: All of our constructions are done in the stable category. The point-set definition of $\bar{F}(X, \Delta_! S_B)$

gives a misleading picture of $\bar{D}_B X$, so we shall rely heavily on the formal properties of the functors we use, as given in [51].

Evaluation, the counit of the $\barwedge$-$\bar{F}$ adjunction, gives a map

$$\epsilon\colon \bar{D}_B X \barwedge X \to \Delta_! S_B.$$

The following innocuous-looking lemma is a key technical point in this circle of ideas.

Lemma 2.9.2 $(1\times\rho\times 1)_!(1\times\Delta\times 1)^*(E \barwedge \Delta_! S_B) \cong E$ *in the stable category, for every G-spectrum E over* $A \times B$.

Proof Here and elsewhere we use the commutation relation $j^*f_! \cong g_! i^*$ that holds when the following is a pullback diagram and at least one of f or j is a fibration:

$$\begin{array}{ccc} C & \xrightarrow{\;g\;} & D \\ {\scriptstyle i}\big\downarrow & & \big\downarrow{\scriptstyle j} \\ A & \xrightarrow[\;f\;]{} & B \end{array}$$

(This is [51, 13.7.7].) The proof of the present lemma would be considerably easier if the fibration condition were not required, but the result if false in general without it. The following chain of isomorphisms proves the lemma.

$$\begin{aligned}
&(1\times\rho\times 1)_!(1\times\Delta\times 1)^*(E \barwedge \Delta_! S_B)\\
&\qquad \cong (1\times\rho\times 1)_!\Delta^*_{A\times B\times B}[(1\times 1\times\rho)^*E \barwedge (\rho\times 1\times 1)^*\Delta_! S_B]\\
&\qquad = (1\times\rho\times 1)_![(1\times 1\times\rho)^*E \wedge_{A\times B\times B} (\rho\times 1\times 1)^*\Delta_! S_B]\\
&\qquad \cong (1\times\rho\times 1)_![(1\times 1\times\rho)^*E \wedge_{A\times B\times B} (1\times\Delta_!)(\rho\times 1)^* S_B]\\
&\qquad \cong (1\times\rho\times 1)_!(1\times\Delta)_![(1\times\Delta)^*(1\times 1\times\rho)^*E \wedge_{A\times B} (\rho\times 1)^* S_B]\\
&\qquad \cong E \wedge_{A\times B} S_{A\times B}\\
&\qquad \cong E
\end{aligned}$$

Here, Δ generally denotes the diagonal map $B \to B\times B$, but $\Delta_{A\times B\times B}$ denotes the diagonal map on $A\times B\times B$. □

Definition 2.9.3 Let X and Y be G-spectra over B, let W be a G-spectrum over A, and let E be a G-spectrum over $A\times B$.

(1) If $\epsilon\colon Y \barwedge X \to \Delta_! S_B$ is a G-map over $B\times B$, let

$$\epsilon_\sharp\colon [W, (1\times\rho)_!(1\times\Delta)^*(E \barwedge Y)]_{G,A} \to [W \barwedge X, E]_{G,A\times B}$$

be defined by letting $\epsilon_\sharp(f)$ be the composite

$$\begin{aligned} W \barwedge X &\xrightarrow{f\wedge 1} (1\times\rho)_!(1\times\Delta)^*(E\barwedge Y)\barwedge X \\ &\cong (1\times\rho\times 1)_!(1\times\Delta\times 1)^*(E\barwedge Y\barwedge X) \\ &\xrightarrow{1\wedge\epsilon} (1\times\rho\times 1)_!(1\times\Delta\times 1)^*(E\barwedge \Delta_! S_B) \\ &\cong E. \end{aligned}$$

(2) If $\eta\colon S \to \rho_!\Delta^*(X\barwedge Y) = \rho_!(X\wedge_B Y)$ is a G-map, let

$$\eta_\sharp\colon [W\barwedge X, E]_{G,A\times B} \to [W,(1\times\rho)_!(1\times\Delta)^*(E\barwedge Y)]_{G,A}$$

be defined by letting $\eta_\sharp(f)$ be the composite

$$\begin{aligned} W \cong W\barwedge S &\xrightarrow{1\wedge\eta} W\barwedge \rho_!\Delta^*(X\barwedge Y) \\ &\cong (1\times\rho)_!(1\times\Delta)^*(W\barwedge X\barwedge Y) \\ &\xrightarrow{f\wedge 1} (1\times\rho)_!(1\times\Delta)^*(E\barwedge Y). \end{aligned}$$

Definition 2.9.4 A G-spectrum X over B is *homologically finite* if

$$\epsilon_\sharp\colon [W,(1\times\rho)_!(1\times\Delta)^*(E\barwedge \bar{D}_B X)]_{G,A} \to [W\barwedge X, E]_{G,A\times B}$$

is an isomorphism for all W and E, where $\epsilon\colon \bar{D}_B X\barwedge X \to \Delta_! S_B$ is the evaluation map.

We have the following characterizations of the dual, similar to [42, III.1.6].

Theorem 2.9.5 *Let X and Y be G-spectra over B. The following are equivalent.*

(1) *X is homologically finite and $Y \cong \bar{D}_B X$ in the stable category.*
(2) *There exists a map $\epsilon\colon Y\barwedge X \to \Delta_! S_B$ such that*

$$\epsilon_\sharp\colon [W,(1\times\rho)_!(1\times\Delta)^*(E\barwedge \bar{Y})]_{G,A} \to [W\barwedge X, E]_{G,A\times B}$$

is an isomorphism for all W and E.
(3) *There exists a map $\eta\colon S \to \rho_!\Delta^*(X\barwedge Y)$ such that*

$$\eta_\sharp\colon [W\barwedge X, E]_{G,A\times B} \to [W,(1\times\rho)_!(1\times\Delta)^*(E\barwedge Y)]_{G,A}$$

is an isomorphism for all W and E.

(4) *There exist maps* $\epsilon\colon Y \barwedge X \to \Delta_! S_B$ *and* $\eta\colon S \to \rho_! \Delta^*(X \barwedge Y)$ *such that the following two composites are each the identity in the stable category:*

$$\begin{aligned} X &\xrightarrow{\eta\wedge 1} \rho_!\Delta^*(X \barwedge Y) \barwedge X \\ &\cong (\rho \times 1)_!(\Delta \times 1)^*(X \barwedge Y \barwedge X) \\ &\xrightarrow{1\wedge\epsilon} (\rho \times 1)_!(\Delta \times 1)^*(X \barwedge \Delta_! S_B) \\ &\cong X \end{aligned}$$

and

$$\begin{aligned} Y &\xrightarrow{1\wedge\eta} Y \barwedge \rho_!\Delta^*(X \barwedge Y) \\ &\cong (1 \times \rho)_!(1 \times \Delta)^*(Y \barwedge X \barwedge Y) \\ &\xrightarrow{\epsilon\wedge 1} (1 \times \rho)_!(1 \times \Delta)^*(\Delta_! S_B \wedge Y) \\ &\cong Y \end{aligned}$$

Moreover, when these conditions are satisfied, the following are also true: The maps $\epsilon_\sharp$ *and* $\eta_\sharp$ *are inverse isomorphisms; the adjoint of* $\epsilon\colon Y \barwedge X \to \Delta_! S_B$ *is an equivalence* $\tilde{\epsilon}\colon Y \to \bar{D}_B X$*; and the maps* $\epsilon\gamma$ *and* $\gamma\eta$ *satisfy the same conditions with the roles of* X *and* Y *reversed, where* γ *is the twist map.*

Proof (1) implies (2) trivially.

Suppose that (4) is true. The following diagram shows that $\epsilon_\sharp\eta_\sharp(f) = f$ for any $f\colon W \barwedge X \to E$:

$$\begin{array}{ccc} W\barwedge X & & \\ \big\downarrow{\scriptstyle 1\wedge\eta\wedge 1} & & \\ W\barwedge\rho_!\Delta^*(X\barwedge Y)\barwedge X & & \\ \big\downarrow{\scriptstyle\cong} & & \\ (1\times\rho\times 1)_!(1\times\Delta\times 1)^*(W\barwedge X\barwedge Y\barwedge X) & \xrightarrow{f\wedge 1} & (1\times\rho\times 1)_!(1\times\Delta\times 1)^*(E\barwedge Y\barwedge X) \\ \big\downarrow{\scriptstyle 1\wedge 1\wedge\epsilon} & & \big\downarrow{\scriptstyle 1\wedge\epsilon} \\ (1\times\rho\times 1)_!(1\times\Delta\times 1)^*(W\barwedge X\barwedge\Delta_! S_B) & \xrightarrow{f\wedge 1} & (1\times\rho\times 1)_!(1\times\Delta\times 1)^*(E\barwedge\Delta_! S_B) \\ \big\downarrow{\scriptstyle\cong} & & \big\downarrow{\scriptstyle\cong} \\ W\barwedge X & \xrightarrow{f} & E \end{array}$$

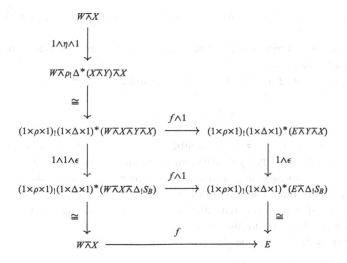

On the other hand, the following diagram shows that $\eta_\sharp\epsilon_\sharp(g) = g$ for any $g\colon W \to (1\times\rho)_!(1\times\Delta^*)(E \barwedge Y)$:

$$\begin{array}{ccc}
W & \xrightarrow{\quad g \quad} & (1\times\rho)_!(1\times\Delta)^*(E\barwedge Y) \\
\Big\downarrow{\scriptstyle 1\wedge\eta} & & \Big\downarrow{\scriptstyle 1\wedge\eta} \\
W\barwedge\rho_!\Delta^*(X\barwedge Y) & \xrightarrow{\;g\wedge 1\;} & (1\times\rho)_!(1\times\Delta)^*(E\barwedge Y)\barwedge\rho_!\Delta^*(X\barwedge Y) \\
\Big\downarrow{\scriptstyle\cong} & & \Big\downarrow{\scriptstyle\cong} \\
(1\times\rho)_!(1\times\Delta)^*(W\barwedge X\barwedge Y) & \xrightarrow{\;g\wedge 1\;} & (1\times\rho\times\rho)_!(1\times\Delta\times\Delta)^*(E\barwedge Y\barwedge X\barwedge Y) \\
& & \Big\downarrow{\scriptstyle 1\wedge\epsilon\wedge 1} \\
& & (1\times\rho\times\rho)_!(1\times\Delta\times\Delta)^*(E\barwedge \Delta_!S_B\barwedge Y) \\
& & \Big\downarrow{\scriptstyle\cong} \\
& & (1\times\rho)_!(1\times\Delta)^*(E\barwedge Y)
\end{array}$$

Here we use the natural isomorphism

$$(1\times\rho)_!(1\times\Delta)^*(1\times\rho\times 1\times 1)_!(1\times\Delta\times 1\times 1)^* \cong (1\times\rho\times\rho)_!(1\times\Delta\times\Delta)^*,$$

which follows from the isomorphism $(1\times\Delta)^*(1\times\rho\times 1\times 1)_! \cong (1\times\rho\times 1)_!(1\times 1\times\Delta)^*$. We also use the isomorphism

$$(1\times\rho)_!(1\times\Delta)^*(1\times 1\times 1\times\rho)_!(1\times 1\times 1\times\Delta)^* \cong (1\times\rho\times\rho)_!(1\times\Delta\times\Delta)^*,$$

which is true for a similar reason. Therefore, $\epsilon_\sharp$ and $\eta_\sharp$ are inverse isomorphisms, so (4) implies both (2) and (3).

Assume now that (2) is true. Taking $W = S$ and $E = X$, let

$$\eta\colon S \to \rho_!\Delta^*(X\barwedge Y)$$

be the map such that $\epsilon_\sharp(\eta) = 1$, the identity map of X. The first composite in (4) is just $\epsilon_\sharp(\eta)$, so is the identity by construction. The first of the diagrams above shows that $\epsilon_\sharp\circ\eta_\sharp = 1$, so $\eta_\sharp$ is the inverse isomorphism to $\epsilon_\sharp$. Now take $W = Y$ and $E = \Delta_!S_B$. Using the identification $(1\times\rho)_!(1\times\Delta)^*(\Delta_!S_B\barwedge Y) \cong Y$, we have $\epsilon_\sharp(1) = \epsilon$, so $\eta_\sharp(\epsilon) = 1$. Written out, this says that the second composite in (4) is also the identity. Thus, (2) implies (4).

Now suppose that (3) is true. Take $W = Y$ and $E = \Delta_!S_B$ and let

$$\epsilon\colon Y\barwedge X \to \Delta_!S_B$$

be the map such that $\eta_\sharp(\epsilon) = 1$. The remainder of the proof that (3) implies (4) is now similar to the preceding argument.

Assuming (4), hence (2), let $\tilde{\epsilon}\colon Y \to \bar{D}_B X$ be the adjoint of $\epsilon\colon Y \barwedge X \to \Delta_! S_B$. A reasonably straightforward diagram chase shows that $\tilde{\epsilon}$ is an equivalence with inverse the composite

$$\begin{aligned}\bar{D}_B X &\xrightarrow{1\wedge\eta} \bar{D}_B X \barwedge \rho_! \Delta^*(X \barwedge Y)\\ &\cong (1\times\rho)_!(1\times\Delta)^*(\bar{D}_B X \barwedge X \barwedge Y)\\ &\xrightarrow{\epsilon\wedge 1} (1\times\rho)_!(1\times\Delta)^*(\Delta_! S_B \barwedge Y)\\ &\cong Y.\end{aligned}$$

(1) now follows.

The claim in the theorem about $\epsilon\gamma$ and $\gamma\eta$ follows from (4) by symmetry. The other claims in the last part of the theorem have been shown along the way. □

Corollary 2.9.6 *If X is homologically finite then so is $\bar{D}_B X$, and $\bar{D}_B\bar{D}_B X \cong X$.* □

We have the following useful compatibility result.

Proposition 2.9.7 *If X is a homologically finite G-spectrum over B and $\beta\colon B \to B'$ is a G-map, then $\beta_! X$ is homologically finite and $\beta_!\bar{D}_B X \cong \bar{D}_{B'}\beta_! X$.*

Proof Let E be a G-spectrum over $A \times B'$. We first show that, for any G-spectrum Y over B, we have an isomorphism

$$(1\times\rho)_!(1\times\Delta)^*(E \barwedge \beta_! Y) \cong (1\times\rho)_!(1\times\Delta)^*[(1\times\beta)^* E \barwedge Y].$$

Once more, we need to detour carefully around commutation relations we wish were true but need not be in general. The following chain of isomorphisms works:

$$\begin{aligned}(1\times\rho)_!(1\times\Delta)^*(E \barwedge \beta_! Y) &\cong (1\times\rho)_![E \wedge_{A\times B'} (\rho\times 1)^*\beta_! Y]\\ &\cong (1\times\rho)_![E \wedge_{A\times B'} (1\times\beta)_!(\rho\times 1)^* Y]\\ &\cong (1\times\rho)_!(1\times\beta)_![(1\times\beta)^* E \wedge_{A\times B} (\rho\times 1)^* Y]\\ &\cong (1\times\rho)_!(1\times\Delta)^*[(1\times\beta)^* E \barwedge Y]\end{aligned}$$

Let W be a G-spectrum over A. We now have the following chain of isomorphisms:

$$\begin{aligned}&[W, (1\times\rho)_!(1\times\Delta)^*(E \barwedge \beta_!\bar{D}_B X)]_{G,A}\\ &\qquad\cong [W, (1\times\rho)_!(1\times\Delta)^*[(1\times\beta)^* E \barwedge \bar{D}_B X]]_{G,A}\\ &\qquad\cong [W \barwedge X, (1\times\beta)^* E]_{G,A\times B}\\ &\qquad\cong [W \barwedge \beta_! X, E]_{G,A\times B'}\end{aligned}$$

Tracing through the adjunctions, we can see that this is $\epsilon'_\sharp$ where ϵ' is the composite

$$\begin{aligned}\beta_! Y \barwedge \beta_! X &\cong (\beta \times \beta)_!(Y \barwedge X)\\ &\xrightarrow{\epsilon} (\beta \times \beta)_! \Delta_! S_B\\ &\cong \Delta_! \beta_! S_B\\ &\to \Delta_! S_{B'},\end{aligned}$$

where the last map is induced by the adjoint of the isomorphism $S_B \xrightarrow{\cong} \beta^* S_{B'}$. The proposition now follows from Theorem 2.9.5. □

In particular, if X is homologically finite over B, then $\rho_! X$ is a finite nonparametrized spectrum and $D\rho_! X \cong \rho_! \bar{D}_B X$. Thus, the homological dual of X is essentially the ordinary dual of the underlying nonparametrized spectrum $\rho_! X$, except that it is constructed so as to be parametrized by B. For example, we have the following sequence of results, giving examples of duals of many spaces. (Our original proofs were modeled on [42, §III.4] and were rather involved. Peter informs us that he no longer trusts the argument of that section and we're not confident that our arguments were correct, either. Given that May and Sigurdsson have since provided better proofs, we shall defer to them for the proofs of the following results.)

We first introduce the following notion for spaces.

Definition 2.9.8 Let X and Y be ex-G-spaces over B. We say that X and Y are V-*dual* if there exist weak maps

$$\epsilon\colon Y \barwedge X \to \Delta_! S_B \wedge S^V$$

and

$$\eta\colon S^V \to \rho_!(X \wedge_B Y)$$

such that the following diagrams stably commute:

$$\begin{array}{ccc} S^V \wedge X & \xrightarrow{\eta\wedge 1} & (\rho \times 1)_![(1 \times \rho)^* X \wedge_{B\times B} (Y \barwedge X)] \\ \gamma \big\downarrow & & \big\downarrow 1\wedge\epsilon \\ X \wedge S^V & \underset{\cong}{\longleftarrow} & (\rho \times 1)_![(1 \times \rho)^* X \wedge_{B\times B} \Delta_! S_B \wedge S^V] \end{array}$$

and

$$\begin{array}{ccc}
Y \wedge S^V & \xrightarrow{1\wedge\eta} & (1\times\rho)_![(Y \,\bar{\wedge}\, X) \wedge_{B\times B} (\rho\times 1)^* Y] \\
\gamma \downarrow & & \\
S^V \wedge Y & & \downarrow \epsilon\wedge 1 \\
\sigma\wedge 1 \downarrow & & \\
S^V \wedge Y & \underset{\cong}{\longleftarrow} & (1\times\rho)_![(\Delta_! S_B \wedge S^V) \wedge_{B\times B} (\rho\times 1)^* Y]
\end{array}$$

Here, γ is transposition and $\sigma(v) = -v$.

Proposition 2.9.9 ([51, 18.3.2]) *If X and Y are V-dual, then $\Sigma_B^\infty X$ is homologically finite and $\bar{D}_B X \cong \Sigma_B^{-V} \Sigma_B^\infty Y$.* □

If $(X, A) \to B$ is any pair of (unbased) spaces over B, let $C_B(X, A)$ denote the fiberwise mapping cone of $A \to X$ over B, with section given by the conepoints. In particular, $C_B(X, \emptyset) = X_+$.

Let X be a compact G-ENR and let $p: X \to B$. Take an embedding of X as an equivariant neighborhood retract in a G-representation V. Let $r: N \to X$ be a G-retraction of an open neighborhood of X in V. Via r, we think of N as a space over X, hence over B.

Theorem 2.9.10 ([51, 18.5.2]) *Let (X, A) be a compact G-ENR pair over B and let V and N be as above. Then $C_B(X, A)$ is V-dual to $C_B(N - A, N - X)$, hence $\Sigma_B^\infty C_B(X, A)$ is homologically finite with*

$$\bar{D}_B \Sigma_B^\infty C_B(X, A) \cong \Sigma^{-V} \Sigma_B^\infty C_B(N - A, N - X).$$

□

As usual, we can specialize to smooth compact manifolds. Let M be a smooth compact G-manifold and let $p: M \to B$ be a G-map. Take a proper embedding of M in $V' \times [0, \infty)$ for some G-representation V', so ∂M embeds in $V' \times 0$ and M meets $V' \times 0$ transversely. Write $V = V' \oplus \mathbb{R}$. Write ν for the normal bundle of the embedding of M in $V' \times [0, \infty)$ and write $\partial\nu$ for the normal bundle of the embedding of ∂M in V', so $\partial\nu = \nu | \partial M$. Write S^ν for the sectioned bundle over M formed by taking the one-point compactification of each fiber of ν, the section being given by the compactification points. Let $S_B^\nu = p_! S^\nu$. In particular, $S_*^\nu = \rho_! S^\nu = T\nu$, the Thom space of ν. We have the following version of Atiyah duality.

Theorem 2.9.11 ([51, 18.7.2]) *Let M be a compact G-manifold and let $p: M \to B$ be a G-map. Embed M in a G-representation V as above. Then $\Sigma_B^\infty M_+$ is homologically finite, with dual $\Sigma^{-V} S_B^\nu /_B S_B^{\partial\nu}$, and $\Sigma_B^\infty M /_B \partial M$ is homologically finite* *with dual* $\Sigma^{-V} S^{\nu}_B$. *The cofibration sequence*

$$\partial M_+ \to M_+ \to M/_B\, \partial M \to \Sigma_B \partial M_+$$

is V*-dual to the cofibration sequence*

$$\Sigma_B S^{\partial\nu}_B \leftarrow S^{\nu}_B/_B\, S^{\partial\nu}_B \leftarrow S^{\nu}_B \leftarrow S^{\partial\nu}_B.$$

□

Note: In [51], Atiyah duality is shown first and used to prove the more general duality theorem for G-ENRs.

In particular, if $b\colon G/H \to B$ and W is a representation of H, for example, $W = \delta(G/H)$ for a dimension function δ, then we have

$$\bar{D}_B(G_+ \wedge_H S^{-W,b}) \cong G_+ \wedge_H S^{-(\mathscr{L}(G/H)-W),b}.$$

One last observation:

Proposition 2.9.12 *If* X *and* Y *are homologically finite spectra over* B*, then*

$$[X, Y]_{G,B} \cong [\bar{D}_B Y, \bar{D}_B X]_{G,B}.$$

Proof We have

$$\begin{aligned}[X, Y]_{G,B} &\cong [S, \rho_!(Y \wedge_B \bar{D}_B X)]_G \\ &\cong [S, \rho_!(\bar{D}_B X \wedge_B Y)]_G \\ &\cong [\bar{D}_B Y, D_B X]_{G,B}.\end{aligned}$$

□

This allows us to identify the opposites of the stable orbit categories.

Corollary 2.9.13 *If* δ *is a dimension function for* G *then*

$$(\widehat{\Pi}_\delta B)^{\mathrm{op}} \cong \widehat{\Pi}_{\mathscr{L}-\delta} B.$$

□

Chapter 3
$RO(\Pi B)$-Graded Ordinary Homology and Cohomology

As we've already mentioned, $RO(G)$-graded ordinary homology is not adequate to give Poincaré duality for G-manifolds except in the case of manifolds modeled on a single representation. To get Poincaré duality for general G-manifolds, we need to extend to a theory indexed on representations of ΠX. (For simplicity of notation we shall now write ΠX for $\Pi_G X$.) That is, the homology and cohomology of X should be graded on representations of ΠX. A construction of the $RO(\Pi X)$-graded theory for finite G was given in [14], and the theory was used in [15] and [17] to obtain π–π theorems for equivariant Poincaré duality spaces and equivariant simple Poincaré duality spaces. In this chapter we give the construction for all compact Lie groups.

Our construction uses CW spaces in which the cells are modeled on disks of varying representations, as specified by a virtual representation of ΠX. It will be convenient (and, to represent the theories, necessary) to have the representation (and the coefficient system) carried by a parametrizing space. Thus, we work in the category of spaces parametrized by a fixed base space B and grade our theories on $RO(\Pi B)$.

When deciding what to use as the grading group, there are several possibilities. We've chosen to work with the most general possible, the group of virtual orthogonal representations of ΠB. Another possibility would be to work with the Grothendieck group constructed from the orthogonal representations, which is isomorphic when B is compact. However, for noncompact B, there can be virtual representations that cannot be expressed as differences of actual representations.

As much as possible, this chapter parallels Chap. 1. In many cases we will refer to results from that chapter or point out how proofs given there can be generalized to the present context, rather than repeating them in full.

S.R. Costenoble, S. Waner, *Equivariant Ordinary Homology and Cohomology*, Lecture Notes in Mathematics 2178, DOI 10.1007/978-3-319-50448-3_3

3.1 Examples of Parametrized Cell Complexes

We give examples of the kinds of parametrized cell complexes that will be used to define $RO(\Pi B)$-graded ordinary homology and cohomology. As in Chap. 1, these will be special cases of the general definition given in the next section.

3.1.1 G-CW(γ) Complexes

Fix a compactly generated G-space B of the homotopy type of a G-CW complex. If γ is a virtual representation of ΠB and $b\colon G/H \to B$ is an orbit over B, write $\gamma_0(b)$ for the virtual representation of H given by restricting $\gamma(b)$ to eH. Recall that we write $\bar{D}(Z)$ for the pair $(D(Z), S(Z))$.

Definition 3.1.1 Let γ be a virtual representation of ΠB.

(1) A *γ-cell* is a pair of objects $c = (G \times_H \bar{D}(Z), p)$ in $G\mathscr{K}/B$ where Z is an actual representation of H such that Z is stably equivalent to

$$\gamma_0(p|G \times_H 0) + n$$

for some integer n. The *dimension* of c is $\gamma + n$. The *boundary* of c is $\partial c = (G \times_H S(Z), p)$.

(2) A *G-CW(γ) complex* is a G-space (X, p) over B together with a decomposition

$$(X, p) = \operatorname*{colim}_n (X^n, p^n)$$

in $G\mathscr{K}/B$, where

(a) X^0 is a union of $(\gamma - |\gamma|)$-dimensional γ-cells—which is to say, a union of orbits $(G/H, p)$ over B such that H acts trivially on $\gamma_0(G/H, p)$—and
(b) each (X^n, p^n), $n > 0$, is obtained from (X^{n-1}, p^{n-1}) by attaching $(\gamma - |\gamma| + n)$-dimensional γ-cells.

For notational convenience and to remind ourselves of the role of γ, we shall also write $X^{\gamma+n}$ for $X^{|\gamma|+n}$.

We define *relative complexes* in the obvious way, and ex-G-CW(γ) complexes as complexes relative to the base section.

Example 3.1.2 We give a simple example, a projective space. We shall give this space three different G-CW structures, the first being one you might think of, being similar to the structure often used to calculate the nonequivariant homology, the second being a G-CW(τ) structure, where τ is the tangent representation, and the third being a G-CW(λ) structure, where λ is the representation associated with the canonical line bundle.

Let $G = \mathbb{Z}/2$ and write Λ for $\mathbb{R}$ with the nontrivial action of G. Let $X = \mathbb{R}P(\mathbb{R}^2 \oplus \Lambda)$, the real projective space of lines in $\mathbb{R}^2 \oplus \Lambda$, constructed as the quotient of $S(\mathbb{R}^2 \oplus \Lambda)$ by the antipodal map. We think of $S(\mathbb{R}^2)$ as the equator and $S(\Lambda)$ as the north and south poles in $S(\mathbb{R}^2 \oplus \Lambda)$. We consider X as parametrized over itself so that we can look at G-CW(γ) structures on X for various representations γ of the fundamental groupoid ΠX. This fundamental groupoid is equivalent to a category with three objects: one object R over G/e and two, P and E, over G/G, where P is the image of the poles and E is a point in the image of the equator. We can picture ΠX as follows:

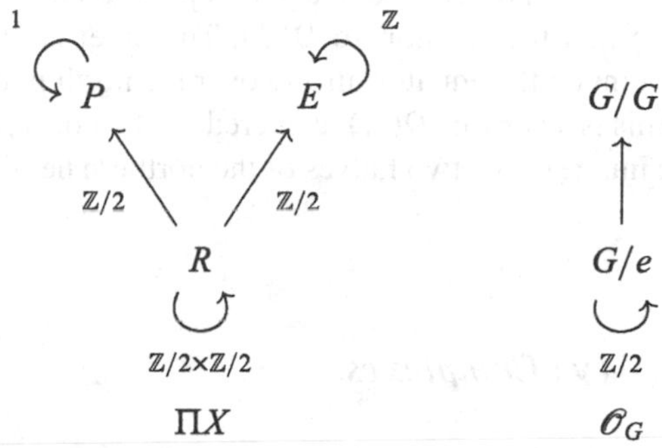

The self-maps of E are a copy of the fundamental groupoid of the image of the equator, which is a circle. The self-maps of R map to the two self-maps of G/e; over the identity of G/e we see $\mathbb{Z}/2$, the fundamental groupoid of the nonequivariant $\mathbb{R}P^2$. There is a unique map $G/e \to G/G$, and over that are two maps from R to P and two maps from R to E, coming again from the nonequivariant fundamental group of $\mathbb{R}P^2$.

(1) Let γ be the following representation of ΠX: $\gamma(P) = G \times_G \Lambda^2$, $\gamma(E) = G \times_G \mathbb{R}^2$, and $\gamma(R) = G \times_e \mathbb{R}^2$. All the maps in ΠX act essentially trivially, that is, by identities.

X has a natural G-CW(γ) structure: It has one fixed 0-cell, the image of a point x on the equator. Because x is on the equator, $\gamma(x) \cong \gamma(E)$ satisfies the requirement that G act trivially on $\gamma_0(x)$. There is one fixed 1-cell of the form $G \times_G D(\mathbb{R})$, which forms the equator when attached to the 0-cell. If y is the center of this 1-cell, it again is a point on the equator, so $G \times_G D(\mathbb{R})$ is a γ-cell. Finally, there is one 2-cell of the form $G \times_G D(\Lambda^2)$, whose center is P; this is the image of the northern (or southern) hemisphere.

Note that this cell structure is not a G-CW(V) structure for any V. It has different representations of G at the poles vs at the equator.

(2) $X = \mathbb{R}P(\mathbb{R}^2 \oplus \Lambda)$ is a G-manifold. Its tangent representation τ has $\tau(P) = \Lambda^2$ and $\tau(E) = \mathbb{R} \oplus \Lambda$. We can give X a G-CW(τ) structure as follows. It has one free 0-cell, the orbit of a chosen point at 45° north latitude. It has two 1-cells: One is free, represented by half the line of latitude at 45° north, connecting one point of the 0-cell to the other point in its orbit. The other 1-cell has the form

$G \times_G D(\Lambda)$ and is represented by a line starting at the chosen point at 45° north and descending due south until it meets the other point in the orbit of the 0-cell, at 45° south. Its center is on the equator, so this is a valid τ-cell. There are two 2-cells: One has the form $G \times_G D(\Lambda^2)$ and is represented by a cap centered at the north pole, descending to the line of latitude at 45° north. The other 2-cell fills in the remaining square outlined by the two 1-cells. Its center is again on the equator and it has the form $G \times_G D(\mathbb{R} \oplus \Lambda)$. Its image in X is a Möbius strip.

(3) As a projective space, X has a canonical line bundle. The associated representation λ has $\lambda(P) = \Lambda$ and $\lambda(E) = \mathbb{R}$. We can give X the following G-CW(λ) structure: It has one fixed 0-cell, the image of a point on the equator. It has two 1-cells, one being half the equator, so $D(\mathbb{R})$. The other is the image of a line going from the point on the equator up and over the north pole, and back down to the equator. This is a copy of $D(\Lambda)$, centered on the north pole. There is then a free 2-cell, the image of the two halves of the northern hemisphere carved out by the 1-cells.

3.1.2 Dual G-CW(γ) Complexes

Definition 3.1.3 Let γ be a virtual representation of ΠB.

(1) A *dual γ-cell* is a pair of objects $c = (G \times_H \bar{D}(Z), p)$ in $G\mathscr{K}/B$ where Z is an actual representation of H such that Z is stably equivalent to

$$\gamma_0(p|G \times_H 0) - \mathscr{L}(G/H) + n$$

for some integer n. The *dimension* of c is $\gamma + n$. The *boundary* of c is $\partial c = (G \times_H S(Z), p)$.

(2) A *dual G-CW(γ) complex* is a G-space (X, p) over B together with a decomposition

$$(X, p) = \operatorname*{colim}_n (X^n, p^n)$$

in $G\mathscr{K}/B$, where

(a) X^0 is a union of $(\gamma - |\gamma|)$-dimensional dual γ-cells—which is to say, a union of orbits $(G/H, p)$ over B such that $\mathscr{L}(G/H) = 0$ and H acts trivially on $\gamma_0(G/H, p)$—and

(b) each (X^n, p^n), $n > 0$, is obtained from (X^{n-1}, p^{n-1}) by attaching $(\gamma - |\gamma| + n)$-dimensional dual γ-cells.

For notational convenience and to remind ourselves of the role of γ, we shall also write $X^{\gamma+n}$ for $X^{|\gamma|+n}$.

We define *relative dual complexes* in the obvious way, and dual ex-G-CW(γ) complexes as complexes relative to the base section.

Examples 3.1.4

(1) Let $G = S^1$ and consider the action of G on S^2 by rotation, with the G-triangulation considered in Example 1.1.3(6). Let τ be the tangent representation on the sphere. Dualizing the triangulation gives a dual G-CW(τ) structure, which is the same as the dual G-CW(V) structure given in Example 1.1.8(4). However, as mentioned in Example 2.1.5(3), even though S^2 is a V-manifold, τ is not the constant representation $\mathbb{V}$, which leads to different boundary homomorphisms in the resulting chain complexes, as defined later in this chapter.
(2) With $G = S^1$ acting on S^2 as above, consider the induced action on the projective plane, as considered in Example 1.1.3(7). Let τ be the tangent representation. The cell structure from the preceding example gives the following dual G-CW(τ) structure on the projective plane: It has one free dual $(\tau-1)$-cell, the image of the line of latitude at 45° north; this has the form $G\times_e D(\tau(L)-2)$, where L is a point on that line of latitude. It has two τ-cells. One is the image of the cap over the north pole, and has the form $G \times_G D(\tau(P))$, where P is the image of the pole. The other is the image of the band around the equator between the lines of latitude at $\pm 45°$, and has the form $G \times_{Z/2} D(\tau(E) - \mathbb{R})$, where E is a point on the equator.

   Note that the projective plane is not a V-manifold, for any V, and this is not a dual G-CW(V) structure: the representation of $\mathbb{Z}/2$ that occurs on the equator is not the restriction of the representation of G that occurs at the pole.
(3) If you take the projective space from the preceding example and remove the τ-cell centered at the pole, you get a dual G-CW(τ) structure on the Möbius strip, with one $(\tau - 1)$-cell and one τ-cell. This also gives a relative dual cell structure on the Möbius strip, relative to its boundary, with but a single τ-cell.
(4) If M is any smooth G-manifold, let τ be the tangent representation of ΠM. Then the dual of a smooth G-triangulation (as in Definition 1.1.2 and the discussion following it) gives an explicit dual G-CW(τ) structure on M, considering M as a space over itself. This is the geometry underlying Poincaré duality.

3.2 δ-G-CW(γ) Complexes

We now give a definition that generalizes the two special cases we looked at in the preceding section, and is also the parametrized version of the definition in Sect. 1.4.

Fix a compactly generated G-space B of the homotopy type of a G-CW complex, a virtual representation γ of ΠB, and a dimension function δ for G. We describe a theory of CW complexes in $G\mathscr{K}/B$, the cells being locally modeled on γ, with cells having dimensions given by δ.

Recall that, if $b\colon G/H \to B$ is an orbit over B, we write $\gamma_0(b)$ for the virtual representation of H given by restricting $\gamma(b)$ to eH.

Definition 3.2.1 Let γ be a virtual representation of ΠB and let δ be a dimension function for G.

(1) An orbit $b: G/H \to B$ is *δ-γ-admissible*, or simply *admissible*, if $H \in \mathscr{F}(\delta)$ and $\gamma_0(b) - \delta(G/H) + n$ is stably equivalent to an actual H-representation for some integer n.
(2) A *δ-γ-cell* is a pair of objects $c = (G \times_H \bar{D}(Z), p)$ in $G\mathscr{K}/B$ where Z is an actual representation of H such that Z is stably equivalent to

$$\gamma_0(p|G \times_H 0) - \delta(G/H) + n$$

for some integer n. The *dimension* of c is $\gamma + n$. The *boundary* of c is $\partial c = (G \times_H S(Z), p)$.

Note that $p: G \times_H D(Z) \to B$ can be any G-map in the definition above. By definition, $p|G \times_H 0$ must be admissible.

Because $G\mathscr{K}/B$ has pushouts, we can now speak of attaching a cell c to a space X over B along an attaching map $\partial c \to X$ over B.

Definition 3.2.2

(1) A *δ-G-CW(γ) complex* is a G-space (X, p) over B together with a decomposition

$$(X, p) = \operatorname{colim}_n (X^n, p^n)$$

in $G\mathscr{K}/B$, where

(a) X^0 is a union of $(\gamma - |\gamma|)$-dimensional δ-γ-cells—which is to say, a union of orbits $(G/H, p)$ over B such that $\delta(G/H) = 0$ and H acts trivially on $\gamma_0(G/H, p)$—and
(b) each (X^n, p^n), $n > 0$, is obtained from (X^{n-1}, p^{n-1}) by attaching $(\gamma - |\gamma| + n)$-dimensional δ-γ-cells.

For notational convenience and to remind ourselves of the role of γ, we shall also write $X^{\gamma+n}$ for $X^{|\gamma|+n}$.

(2) A *relative* δ-G-CW(γ) complex is a pair (X, A, p) over B where $(X, p) = \operatorname{colim}_n(X^n, p^n)$, X^0 is the disjoint union of A with $(\gamma - |\gamma|)$-dimensional γ-cells as in (a) above, and cells are attached as in (b).
(3) An *ex-δ-G-CW(γ) complex* is an ex-G-space (X, p, σ) with a relative δ-G-CW(γ) structure on $(X, \sigma(B), p)$.
(4) If we allow cells of any dimension to be attached at each stage, we get the weaker notions of absolute, relative, or ex- *δ-γ-cell complex*.
(5) If X is a δ-G-CW(γ) complex or δ-γ-cell complex with cells only of dimension less than or equal to $\gamma + n$, we say that X is *$(\gamma + n)$-dimensional*.

Examples 3.2.3

(1) As in Sect. 1.4, an (ordinary) G-CW(γ) complex is a 0-G-CW(γ) complex, i.e., one with $\delta = 0$, while a dual G-CW(γ) complex is an $\mathscr{L}$-G-CW(γ) complex.
(2) Again as in Sect. 1.4, the product of a δ-G-CW(α) complex and an ϵ-H-CW(β) complex has a natural structure as a $(\delta \times \epsilon)$-$(G \times H)$-CW$(\alpha + \beta)$ complex.
(3) If α is a virtual representation of G and X is a (nonparametrized) δ-G-CW(α) complex, as in Chap. 1, then any map $X \to B$ gives X the structure of a δ-G-CW($\bar{\alpha}$) complex over B, where $\bar{\alpha}$ is the constant representation of ΠB at α.
(4) If $p: E \to B$ is a G-vector bundle, let ρ be the associated representation of ΠB as in 2.1.3. Let $T_B(p)$ be the fiberwise one-point compactification of the bundle, with basepoint section $\sigma: B \to T_B(p)$, so $(T_B(p), p, \sigma)$ is an ex-G-space. If B is an ordinary G-CW complex, then $T_B(p)$ has an evident structure as an ex-G-CW(ρ) complex (taking $\delta = 0$), with cells in one-to-one correspondence with the cells of B. More generally, if B has a δ-G-CW(γ) structure over itself, then $T_B(p)$ is an ex-δ-G-CW$(\gamma + \rho)$ complex over B. This is the geometry underlying the Thom isomorphism. We use $T_B(p)$ as our model for the Thom space of p in this context because, unlike the actual Thom space (which is $T(p) = T_B(p)/\sigma(B)$), it comes with a natural parametrization, hence a sensible way to consider it as a δ-G-CW$(\gamma + \rho)$ complex.

Definition 3.2.4 If n is an integer, a lax map $f: X \Rightarrow Y$ is a *δ-$(\gamma + n)$-equivalence* if, for every δ-γ-cell c of dimension $\gamma + i$ with $i \leq n$, every diagram of the following form is lax homotopic to one in which there exists a lift $c \Rightarrow X$:

$$\begin{array}{ccc} \partial c & \Longrightarrow & X \\ \big\downarrow & & \big\Downarrow{\scriptstyle f} \\ c & \Longrightarrow & Y. \end{array}$$

We say that f is a *δ-weak$_\gamma$ equivalence* if it a δ-$(\gamma + n)$-equivalence for all n.

Note that, because $\partial c \to c$ is a lax cofibration by Theorem 2.2.7, we can say that f is a δ-$(\gamma + n)$-equivalence if every diagram as above is homotopic *rel* ∂c to one in which we can find a lift.

We insert the following reassuring result.

Theorem 3.2.5 *Let $f: X \Rightarrow Y$ be a lax map. Then f is a δ-weak$_\gamma$ equivalence if and only if, for each $b \in B$, $f_b: LX_b \to LY_b$ is a δ-weak$_{\gamma_0(b)}$ equivalence.*

Proof A δ-γ-cell $c = (G \times_H D(Z), p)$ with $p(eH \times 0) = b$ is lax homotopy equivalent to the cell

$$c' = (G \times_H \bar{D}(Z), p')$$

of the same dimension with $p'(G \times_H \bar{D}(Z)) = Gb$. It follows that f is a δ-weak$_\gamma$ equivalence if and only if each map of fibers $f_b\colon LX_b \to LY_b$ is a δ-weak$_{\gamma_0(b)}$ equivalence. □

We have the following variant of the "homotopy extension and lifting property".

Lemma 3.2.6 (H.E.L.P.) *Let $r\colon Y \Rightarrow Z$ be a δ-$(\gamma + n)$-equivalence. Let (X, A) be a relative δ-γ-cell complex of dimension $\gamma + n$. If the following diagram commutes in $G\mathscr{K}^\lambda/B$ without the dashed arrows, then there exist lax maps $\tilde{g}$ and $\tilde{h}$ making the diagram commute.*

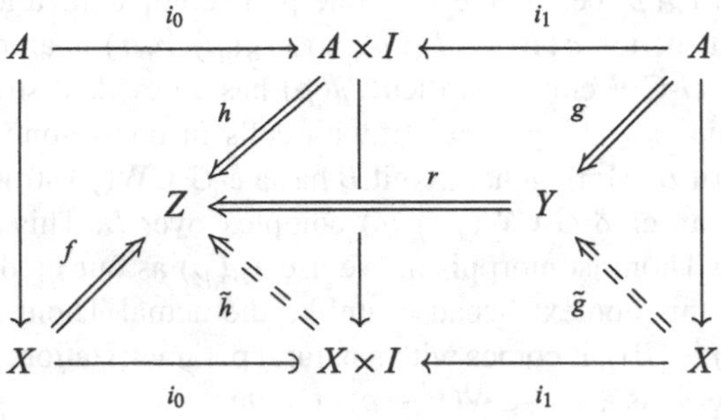

The result remains true when $n = \infty$.

Proof We can use the same proof as that given for Lemma 1.4.8, understanding all maps to be in the lax category. □

To state the Whitehead theorem, we need the following notations. Write $\pi(G\mathscr{K}^\lambda/B)(X, Y)$ for the set of lax homotopy classes of maps $X \Rightarrow Y$. If A is a space over B, write $A/G\mathscr{K}^\lambda/B$ for the category of spaces under A in $G\mathscr{K}^\lambda/B$, meaning spaces Y over B equipped with lax maps $A \Rightarrow Y$. We then write $\pi(A/G\mathscr{K}^\lambda/B)(X, Y)$ for the set of lax homotopy classes of maps under A and over B.

Theorem 3.2.7 (Whitehead)

(1) *If $f\colon Y \Rightarrow Z$ is a δ-$(\gamma + n)$-equivalence and X is a $(\gamma + n - 1)$-dimensional δ-γ-cell complex, then*

$$f_*\colon \pi(G\mathscr{K}^\lambda/B)(X, Y) \to \pi(G\mathscr{K}^\lambda/B)(X, Z)$$

is an isomorphism. It is an epimorphism if X is $(\gamma + n)$-dimensional.

(2) *If $f\colon Y \Rightarrow Z$ is a δ-weak$_\gamma$ equivalence and X is a δ-γ-cell complex, then*

$$f_*\colon \pi(G\mathscr{K}^\lambda/B)(X, Y) \to \pi(G\mathscr{K}^\lambda/B)(X, Z)$$

is an isomorphism. In particular any δ-weak$_\gamma$ equivalence of δ-γ-cell complexes is a lax homotopy equivalence.

(3) *If* $f: Y \Rightarrow Z$ *is a* δ-$(\gamma + n)$*-equivalence of spaces under* A *and* (X, A) *is a* $(\gamma + n - 1)$*-dimensional relative* δ-γ*-cell complex, then*

$$f_*: \pi(A/G\mathscr{K}^\lambda/B)(X, Y) \to \pi(A/G\mathscr{K}^\lambda/B)(X, Z)$$

is an isomorphism. It is an epimorphism if (X, A) *is* $(\gamma + n)$*-dimensional.*

(4) *If* $f: Y \Rightarrow Z$ *is a* δ*-weak*$_\gamma$ *equivalence of spaces under* A *and* (X, A) *is a relative* δ-γ*-cell complex, then*

$$f_*: \pi(A/G\mathscr{K}^\lambda/B)(X, Y) \to \pi(A/G\mathscr{K}^\lambda/B)(X, Z)$$

is an isomorphism.

Proof This follows from the H.E.L.P. lemma in exactly the same way as in Theorems 1.4.9 and 1.4.10. □

Note that, when we take $A = B$ in the relative Whitehead theorem, we get the special case of ex-G-spaces and ex-δ-γ-cell complexes.

Although δ-weak$_\gamma$ equivalence of parametrized spaces is, in general, weaker than weak G-equivalence, for δ-γ-cell complexes these notions coincide and are equivalent to lax homotopy equivalence. The point is that δ-γ-cell complexes are limited in what cells they can use, and this limitation allows δ-weak$_\gamma$ equivalence to detect weak G-equivalence of δ-γ-cell complexes.

As in the nonparametrized case, we have the following example of a δ-$(\gamma + n)$-equivalence.

Proposition 3.2.8 *Let* δ *be a familial dimension function and let* ϵ *be another dimension function (not necessarily familial) with* $\delta \succcurlyeq \epsilon$*. If* (X, A) *is a relative* ϵ-G*-CW(*γ*) complex, then the inclusion* $X^{\gamma+n} \to X$ *is a* δ-$(\gamma + n)$*-equivalence.*

Proof By the usual induction, this reduces to showing that, if Y is obtained from C by attaching a $(\gamma + k)$-dimensional ϵ-γ-cell of the form $(G \times_K \bar{D}(W), q)$, and $(G \times_H \bar{D}(V), p)$ is a $(\gamma + i)$-dimensional δ-γ-cell, with $i < k$, then any lax G-map of pairs $G \times_H \bar{D}(V) \to (Y, C)$ is homotopic rel boundary to a map into B. For this we take an ordinary G-triangulation of $G \times_H D(V)$ and show by induction on the cells that we can (lax) homotope the map to miss the orbit $G/K \times 0 = (G/K, b')$ in the attached cell. We use the fact that the fixed-set dimensions of V must equal those of $\gamma_0(b) - \delta(G/H) + i$, where $(G/H, b)$ is the center of $G \times_H D(V)$, and the fixed set dimensions of W equal those of $\gamma_0(b') - \epsilon(G/K) + k$.

The only simplices that might hit the orbit have the form $G/J \times \Delta^j$ where $J \leq H$ and J is subconjugate to K; by replacing K with a conjugate we may assume $J \leq K$ also. If such a simplex meets the orbit, then we get a lax map $(G/J, b) \Rightarrow (G/K, b')$, where G/J is any orbit in $G/J \times \Delta^j$. Applying γ we get a virtual equivalence of J-representations $\gamma_0(b) \simeq \gamma_0(b')$, the import of which is that the J-fixed points of these representations have the same dimension. Because such a simplex is embedded in

$G \times_H D(V)$, we now have

$$\begin{aligned} j &\le |[\gamma_0(b) - \delta(G/H) - \mathscr{L}(H/J)]^J| + i \\ &\le |[\gamma_0(b) - \delta(G/H) - \delta(H/J)]^J| + i \\ &= |[\gamma_0(b) - \delta(G/J)]^J| + i \\ &\le |[\gamma_0(b') - \epsilon(G/K)]^J| + i \\ &< |[\gamma_0(b') - \epsilon(G/K)]^J| + k \end{aligned}$$

as in the proof of Proposition 1.4.11. Finally, note that $|[\gamma_0(b') - \epsilon(G/K)]^J| + k$ is the codimension of $(G/K \times 0)^J$ in Y^J, so the desired homotopy exists for dimensional reasons. □

Definition 3.2.9 A lax map $f: (X,A) \Rightarrow (Y,C)$ from a relative δ-G-CW(γ) complex to a relative ϵ-G-CW(γ) complex is *cellular* if $f(X^{\gamma+n}) \subset Y^{\gamma+n}$ for each n.

Theorem 3.2.10 (Cellular Approximation of Maps) *Let δ be a familial dimension function and let ϵ be any dimension function such that $\delta \succcurlyeq \epsilon$. Let (X,A) be a relative δ-G-CW(γ) complex and let (Y,C) be a relative ϵ-G-CW(γ) complex. Let $f: (X,A) \Rightarrow (Y,C)$ be a lax map and suppose given a subcomplex $(Z,A) \subset (X,A)$ and a lax homotopy h of $f|Z$ to a cellular map. Then h can be extended to a lax homotopy of f to a cellular map.*

Proof This follows by induction on skeleta, using the H.E.L.P. lemma and Proposition 3.2.8 applied to the inclusion $Y^{\gamma+n} \to Y$. □

Theorem 3.2.11 *Let δ be a familial dimension function, let (A,P) be a relative δ-G-CW(γ) complex, let X be a G-space over B, and let $f: A \to X$ be a (strict) map. Then there exists a relative δ-G-CW(γ) complex (Y,P), containing (A,P) as a subcomplex, and a δ-weak$_\gamma$ equivalence $g: Y \to X$ extending f.*

Proof The proof is similar to that of Theorem 1.4.21. We start by letting

$$Y^{\gamma-|\gamma|} = A^{\gamma-|\gamma|} \sqcup \coprod (G/K, b),$$

where the coproduct runs over all $b: G/K \to B$ such that $\delta(G/K) = 0$ and K acts trivially on $\gamma(b)$, and all maps $G/K \to X$ over B. The map $g: Y^{\gamma-|\gamma|} \to X$ is the one induced by those maps of orbits. $(Y^{\gamma-|\gamma|}, P)$ is then a relative δ-G-CW(γ) complex of dimension $\gamma - |\gamma|$, containing $(A^{\gamma-|\gamma|}, P)$, and $Y^{\gamma-|\gamma|} \to X$ is a δ-$(\gamma - |\gamma|)$-equivalence.

Inductively, suppose that we have constructed $(Y^{\gamma+n-1}, P)$, a relative δ-G-CW(γ) complex of dimension $\gamma + n - 1$ containing $(A^{\gamma+n-1}, P)$ as a subcomplex, and a δ-$(\gamma + n - 1)$-equivalence $g: Y^{\gamma+n-1} \to X$ extending f on $A^{\gamma+n-1}$. Let

$$Y^{\gamma+n} = Y^{\gamma+n-1} \cup A^{\gamma+n} \cup \coprod G \times_K D(V),$$

where the coproduct runs over all $(|\gamma| + n)$-dimensional δ-γ-cells $G \times_K D(V)$ and all diagrams of the form

$$\begin{array}{ccc} G \times_K S(V) & \longrightarrow & Y^{\gamma+n-1} \\ \downarrow & & \downarrow \\ G \times_K D(V) & \longrightarrow & X \end{array}$$

over B. The union that defines $Y^{\gamma+n}$ is along $A^{\gamma+n-1} \to Y^{\gamma+n-1}$ and the maps $G \times_K S(V) \to Y^{\gamma+n-1}$ displayed above. By construction, $(Y^{\gamma+n}, P)$ is a relative δ-G-CW(γ) complex of dimension $\gamma + n$ containing $(A^{\gamma+n}, P)$ as a subcomplex. We let $g: Y^{\gamma+n} \to X$ be the induced map and we claim that g is a δ-$(\gamma+n)$-equivalence. To see this, consider any diagram of the following form, with $G \times_K D(V)$ being a δ-γ-cell of dimension $\leq \gamma + n$:

$$\begin{array}{ccc} G \times_K S(V) & \overset{\alpha}{\Longrightarrow} & Y^{\gamma+n} \\ \downarrow & & \downarrow \\ G \times_K D(V) & \underset{\beta}{\Longrightarrow} & X \end{array}$$

Let $b: G/K \to B$ be the map of the center of the cell. By Lemma 1.4.13, $S(V)$ is K-homotopy equivalent to a δ-K-CW($\gamma_0(b)$) complex of dimension strictly less than $\gamma_0(b) + n$. Taking $G \times_K -$ and composing with the map of the cell into B, we can consider this a δ-G-CW(γ) structure. So, by cellular approximation of maps, the diagram above is lax homotopic to one in which α maps the sphere into $Y^{\gamma+n-1}$. We can then find a lift of β up to lax homotopy using the inductive hypothesis if $|V| < |\gamma| + n$ or the construction of $Y^{\gamma+n}$ if $|V| = |\gamma| + n$.

Finally, $Y = \operatorname{colim}_n Y^{\gamma+n}$ satisfies the claim of the theorem. □

Theorem 3.2.12 (Approximation by δ-G-CW(γ) Complexes) *Let δ be a familial dimension function and let X be a G-space over B. Then there exists a δ-G-CW(γ) complex ΓX and a δ-weak$_\gamma$ equivalence $g: \Gamma X \to X$ over B. If $f: X \Rightarrow Y$ is a lax G-map and $g: \Gamma Y \to Y$ is an approximation of Y by a δ-G-CW(γ) complex, then there exists a lax G-map $\Gamma f: \Gamma X \Rightarrow \Gamma Y$, unique up to lax G-homotopy, such that the following diagram commutes up to lax G-homotopy:*

$$\begin{array}{ccc} \Gamma X & \overset{\Gamma f}{\Longrightarrow} & \Gamma Y \\ {\scriptstyle g}\downarrow & & \downarrow{\scriptstyle g} \\ X & \underset{f}{\Longrightarrow} & Y \end{array}$$

Proof The existence of $g\colon \Gamma X \to X$ is the special case of Theorem 3.2.11 in which we take $A = P = \emptyset$. The existence and uniqueness of Γf follows from Whitehead's theorem. □

Theorem 3.2.13 (Approximation of Ex-Spaces) *Let δ be a familial dimension function and let X be an ex-G-space over B. Then there exists an ex-δ-G-CW(γ) complex ΓX and a δ-weak$_\gamma$ equivalence $g\colon \Gamma X \to X$ over and under B. Further, Γ is functorial up to lax homotopy under B.*

Proof The existence of $g\colon \Gamma X \to X$ is the special case of Theorem 3.2.11 in which we take $A = P = B$.

Given $f\colon X \Rightarrow Y$, the existence and uniqueness of $\Gamma f\colon \Gamma X \Rightarrow \Gamma Y$ follows from the relative part of Whitehead's theorem, which tells us that

$$\pi(B/G\mathscr{K}^\lambda/B)(\Gamma X, \Gamma Y) \cong \pi(B/G\mathscr{K}^\lambda/B)(\Gamma X, Y).$$ □

Theorem 3.2.14 (Approximation of Pairs) *Let δ be a familial dimension function and Let (X, A) be a pair of G-spaces over B. Then there exists a pair of δ-G-CW(γ) complexes $(\Gamma X, \Gamma A)$ and a pair of δ-weak$_\gamma$ equivalences $g\colon (\Gamma X, \Gamma A) \to (X, A)$ over B. Further, Γ is functorial on lax maps of pairs up to lax homotopy.*

Proof Take any approximation $g\colon \Gamma A \to A$, then apply Theorem 3.2.11 to $\Gamma A \to X$ (taking $P = \emptyset$ in that theorem) to get ΓX with ΓA as a subcomplex.

Given $f\colon (X, A) \Rightarrow (Y, C)$, we first construct $\Gamma f\colon \Gamma A \Rightarrow \Gamma C$ using the Whitehead theorem and then extend to $\Gamma X \Rightarrow \Gamma Y$ using the relative Whitehead theorem (considering the category of spaces under ΓA and lax maps over B). □

We say that a triad $(X; A, C)$ over B is excisive if it is excisive as a triad of G-spaces, ignoring B.

Theorem 3.2.15 (Approximation of Triads) *Let δ be a familial dimension function and let $(X; A, C)$ be an excisive triad over B. Then there exists a δ-G-CW(γ) triad $(\Gamma X; \Gamma A, \Gamma C)$ and a map of triads*

$$g\colon (\Gamma X; \Gamma A, \Gamma C) \to (X; A, C)$$

over B such that each of the maps $\Gamma A \cap \Gamma C \to A \cap C$, $\Gamma A \to A$, $\Gamma C \to C$, and $\Gamma X \to X$ is a δ-weak$_\gamma$ equivalence. Γ is functorial on lax maps of excisive triads up to lax homotopy.

Proof Let $D = A \cap C$. Take a δ-G-CW(γ) approximation $g\colon \Gamma D \to D$. Using Theorem 3.2.11, extend to approximations

$$g\colon (\Gamma A, \Gamma D) \to (A, D)$$

and

$$g\colon (\Gamma C, \Gamma D) \to (C, D).$$

Let $\Gamma X = \Gamma A \cup_{\Gamma D} \Gamma C$. All the statements of the theorem are clear except that the map $g: \Gamma X \to X$ is a δ-weak$_\gamma$ equivalence. By Theorem 3.2.5, it suffices to show that g induces a δ-weak$_{\gamma_0(b)}$ equivalence on each homotopy fiber. This follows from the observation that, for example, $L(A \cup_D C)_b = LA_b \cup_{LD_b} LC_b$ and an application of Theorem 3.2.5 to say that $L\Gamma A_b \to LA_b$ is a δ-weak$_{\gamma_0(b)}$ equivalence, and similarly for the other approximations. □

3.3 Homology and Cohomology of Parametrized Cell Complexes

We shall now construct the cellular chains of δ-G-CW(γ) spaces. As usual, we work with a fixed base space B, a virtual representation γ of ΠB, and a dimension function δ for G. Referring to Definition 1.6.5, our coefficient systems will be $\widehat{\Pi}_\delta B$-modules. (We might call these "G-δ-Mackey functors over B," but that's a bit long to repeat as many times as we will need to refer to them. Recall that we write $\widehat{\Pi}_{G,\delta} B$ when we need to specify the group.) Such modules can be either contravariant or covariant and, as previously, we adopt the convention that contravariant modules are written with a bar on top and covariant ones with a bar underneath.

We have the following generalization of Corollary 1.6.4.

Corollary 3.3.1 *If δ is a dimension function for G, then the category of contravariant $\widehat{\Pi}_\delta B$-modules and natural transformations between them is isomorphic to the category of covariant $\widehat{\Pi}_{\mathscr{L}-\delta} B$-modules.*

Proof This is an immediate consequence of Corollary 2.9.13. □

If (X, A) is a pair of G-spaces over B, we write $X/_B A$ for the ex-G-space over B obtained by taking the fiberwise quotient. This is the same ex-G-space as $X_+/_B A_+$. Recall also the notation of Definition 2.6.1.

Lemma 3.3.2 *Let (X, A, p) be a relative δ-G-CW(γ) complex. Then*

$$X^{\gamma+n}/_B X^{\gamma+n-1} \simeq_\lambda \bigvee_x G_+ \wedge_H S^{\gamma_0(px)-\delta(G/H)+n,px}$$

as ex-G-spaces, where the wedge runs through the centers x of the $(\gamma+n)$-cells of X and the equivalence is lax homotopy equivalence. (Here we think of $x: G/H \to X$ and px denotes the composite $px: G/H \to B$. When we write $\gamma_0(px) - \delta(G/H) + n$, we mean an actual representation of H stably equivalent to $\gamma_0(px) - \delta(G/H) + n$.)

Proof $X^{\gamma+n}/_B X^{\gamma+n-1}$ consists of a copy of B, the image of the section, with a copy of each $(\gamma+n)$-cell of X adjoined via the composite of its attaching map and the projection to B. The attaching map for a cell, as a map into B, may be deformed to be the constant map at the image of the center of the cell, and then the assertion is clear. □

Definition 3.3.3 Let (X,A) be a relative δ-G-CW(γ) complex. The *cellular chain complex of* (X,A), $\overline{C}_{\gamma+*}(X,A)$ (which we also write as $\overline{C}^{G,\delta}_{\gamma+*}(X,A)$ when we need to emphasize the group involved or δ), is the chain complex of contravariant $\widehat{\Pi}_\delta B$-modules defined by

$$\begin{aligned}\overline{C}_{\gamma+n}(X,A)(b) &= [\Sigma^\infty_B G_+ \wedge_H S^{\gamma_0(b)-\delta(G/H)+n,b}, \Sigma^\infty_B(X^{\gamma+n}/_B X^{\gamma+n-1})]_{G,B} \\ &\cong \pi^H_{\gamma_0(b)-\delta(G/H)+n}(\Sigma^\infty_B(X^{\gamma+n}/_B X^{\gamma+n-1})_b)\end{aligned}$$

if $b\colon G/H \to B$. (We are using shorthand notation—we should really be as careful as in Definition 1.7.1.) This is obviously a contravariant functor on $\widehat{\Pi}_{\delta,\gamma}B$, and we use the isomorphism of Theorem 2.6.4 to consider it as a functor on $\widehat{\Pi}_\delta B$. Let $d\colon \overline{C}_{\gamma+n}(X,A) \to \overline{C}_{\gamma+n-1}(X,A)$ be the natural transformation induced by the composite

$$X^{\gamma+n}/_B X^{\gamma+n-1} \to \Sigma_B(X^{\gamma+n-1}/_B A) \to \Sigma_B(X^{\gamma+n-1}/_B X^{\gamma+n-2}).$$

The analogues of Remarks 1.7.2 hold here: $\overline{C}_{\gamma+*}(X,A)$ is covariant in cellular maps $(X,A) \to (Y,B)$ and contravariant in virtual maps $\gamma \to \gamma'$.

If $\underline{S}$ is a covariant $\widehat{\Pi}_\delta B$-module, and $\overline{T}$ and $\overline{U}$ are contravariant $\widehat{\Pi}_\delta B$-modules, then Definition 1.6.5 defines for us the groups

$$\mathrm{Hom}_{\widehat{\Pi}_\delta B}(\overline{T}, \overline{U})$$

and

$$\overline{T} \otimes_{\widehat{\Pi}_\delta B} \underline{S}.$$

Recall that a $\widehat{\Pi}_\delta B$-module is said to be *free* if it is a sum of functors of the form $\widehat{\Pi}_\delta B(-,b)$. The following follows from Lemma 3.3.2.

Proposition 3.3.4 *$\overline{C}_{\gamma+*}(X)$ is a chain complex of free $\widehat{\Pi}_\delta B$-modules. If $\overline{T}$ is a contravariant $\widehat{\Pi}_\delta B$-module then*

$$\mathrm{Hom}_{\widehat{\Pi}_\delta B}(\overline{C}_{\gamma+n}(X), \overline{T}) \cong \prod_x \overline{T}(px)$$

where the product runs over the centers x of the $(\gamma+n)$-cells of X. Similarly, if $\underline{S}$ is a covariant $\widehat{\Pi}_\delta B$-module then

$$\overline{C}_{\gamma+n}(X) \otimes_{\widehat{\Pi}_\delta B} \underline{S} \cong \bigoplus_x \underline{S}(px)$$

where again x runs over the centers of the $(\gamma+n)$-cells of X. □

Definition 3.3.5 Let $\overline{T}$ be a contravariant $\widehat{\Pi}_\delta B$-module and let $\underline{S}$ be a covariant $\widehat{\Pi}_\delta B$-module.

(1) Let (X, A) be a relative δ-G-CW(γ) complex. We define the $(\gamma + n)$*th cellular homology* of (X, A), with coefficients in $\underline{S}$, to be

$$H^{G,\delta}_{\gamma+n}(X, A; \underline{S}) = H_{\gamma+n}(\overline{C}^{G,\delta}_{\gamma+*}(X, A) \otimes_{\widehat{\Pi}_\delta B} \underline{S}).$$

and we define the $(\gamma + n)$*th cellular cohomology* of (X, A), with coefficients in $\overline{T}$, to be

$$H^{\gamma+n}_{G,\delta}(X, A; \overline{T}) = H^{\gamma+n}(\mathrm{Hom}_{\widehat{\Pi}_\delta B}(\overline{C}^{G,\delta}_{\gamma+*}(X, A), \overline{T})),$$

where we introduce the same sign as in Definition 1.7.3. Homology is covariant in cellular maps of (X, A) while cohomology is contravariant in cellular maps of (X, A).

(2) If X is a δ-G-CW(γ) complex, so $(X, \emptyset)$ is a relative δ-G-CW(γ) complex, we define

$$H^{G,\delta}_{\gamma+n}(X; \underline{S}) = H^{G,\delta}_{\gamma+n}(X, \emptyset; \underline{S})$$

and

$$H^{\gamma+n}_{G,\delta}(X; \overline{T}) = H^{\gamma+n}_{G,\delta}(X, \emptyset; \overline{T}).$$

(3) If X is an ex-δ-G-CW(γ) complex, so (X, B) is a relative δ-G-CW(γ) complex, we define the *reduced* homology and cohomology of X to be

$$\tilde{H}^{G,\delta}_{\gamma+n}(X; \underline{S}) = H^{G,\delta}_{\gamma+n}(X, B; \underline{S})$$

and

$$\tilde{H}^{\gamma+n}_{G,\delta}(X; \overline{T}) = H^{\gamma+n}_{G,\delta}(X, B; \overline{T}).$$

Theorem 3.3.6 (Reduced Homology and Cohomology of Complexes) *Let δ be a dimension function for G, let γ be a virtual representation of $\Pi_G B$, and let $\underline{S}$ and $\overline{T}$ be respectively a covariant and a contravariant $\widehat{\Pi}_\delta B$-module. Then the abelian groups $\tilde{H}^{G,\delta}_\gamma(X; \underline{S})$ and $\tilde{H}^\gamma_{G,\delta}(X; \overline{T})$ are respectively covariant and contravariant functors on the homotopy category of ex-δ-G-CW(γ) complexes and cellular maps and homotopies. They are also respectively contravariant and covariant functors of γ. These functors satisfy the following properties.*

(1) *(Exactness) If A is a subcomplex of X, then the following sequences are exact:*

$$\tilde{H}^{G,\delta}_\gamma(A; \underline{S}) \to \tilde{H}^{G,\delta}_\gamma(X; \underline{S}) \to \tilde{H}^{G,\delta}_\gamma(X/_B A; \underline{S})$$

and

$$\tilde{H}^{\gamma}_{G,\delta}(X/_BA;\overline{T}) \to \tilde{H}^{\gamma}_{G,\delta}(X;\overline{T}) \to \tilde{H}^{\gamma}_{G,\delta}(A;\overline{T}).$$

(2) *(Additivity) If $X = \bigvee_i X_i$ is a fiberwise wedge of ex-δ-G-CW(γ) complexes, then the inclusions of the wedge summands induce isomorphisms*

$$\bigoplus_i \tilde{H}^{G,\delta}_{\gamma}(X_i;\underline{S}) \cong \tilde{H}^{G,\delta}_{\gamma}(X;\underline{S})$$

and

$$\tilde{H}^{\gamma}_{G,\delta}(X;\overline{T}) \cong \prod_i \tilde{H}^{\gamma}_{G,\delta}(X_i;\overline{T}).$$

(3) *(Suspension) There are suspension isomorphisms*

$$\sigma^V\colon \tilde{H}^{G,\delta}_{\alpha}(X;\underline{S}) \xrightarrow{\cong} \tilde{H}^{G,\delta}_{\alpha+V}(\Sigma^V_B X;\underline{S})$$

and

$$\sigma^V\colon \tilde{H}^{\gamma}_{G,\delta}(X;\overline{T}) \xrightarrow{\cong} \tilde{H}^{\gamma+V}_{G,\delta}(\Sigma^V_B X;\overline{T}).$$

These isomorphisms satisfy $\sigma^0 = \mathrm{id}$, $\sigma^W \circ \sigma^V = \sigma^{V\oplus W}$, and the following naturality condition: If $\zeta\colon V \to V'$ is an isomorphism, then the following diagrams commute:

$$\begin{array}{ccc}
\tilde{H}^{G,\delta}_{\gamma}(X;\underline{S}) & \xrightarrow{\sigma^V} & \tilde{H}^{G,\delta}_{\gamma+V}(\Sigma^V_B X;\underline{S}) \\
{\scriptstyle \sigma^{V'}}\downarrow & & \downarrow{\scriptstyle \tilde{H}_{\mathrm{id}}(\mathrm{id}\wedge\zeta)} \\
\tilde{H}^{G,\delta}_{\gamma+V'}(\Sigma^{V'}_B X;\underline{S}) & \xrightarrow[\tilde{H}_{\mathrm{id}+\zeta}(\mathrm{id})]{} & \tilde{H}^{G,\delta}_{\gamma+V}(\Sigma^{V'}_B X;\underline{S})
\end{array}$$

$$\begin{array}{ccc}
\tilde{H}^{\gamma}_{G,\delta}(X;\overline{T}) & \xrightarrow{\sigma^V} & \tilde{H}^{\gamma+V}_{G,\delta}(\Sigma^V_B X;\overline{T}) \\
{\scriptstyle \sigma^{V'}}\downarrow & & \downarrow{\scriptstyle \tilde{H}^{\mathrm{id}+\zeta}(\mathrm{id})} \\
\tilde{H}^{\gamma+V'}_{G,\delta}(\Sigma^{V'}_B X;\overline{T}) & \xrightarrow[\tilde{H}^{\mathrm{id}}(\mathrm{id}\wedge\zeta)]{} & \tilde{H}^{\gamma+V'}_{G,\delta}(\Sigma^V_B X;\overline{T})
\end{array}$$

(4) *(Dimension Axiom) If $H \in \mathscr{F}(\delta)$, $b\colon G/H \to B$, and V is a representation of H so large that $\gamma_0(b) + V - \delta(G/H) + n$ is an actual representation, then there*

are natural isomorphisms

$$\tilde{H}^{G,\delta}_{\gamma+V+k}(G_+ \wedge_H S^{\gamma_0(b)+V-\delta(G/H)+n}; \underline{S}) \cong \begin{cases} \underline{S}(b) & \textit{if } k = n \\ 0 & \textit{if } k \neq n \end{cases}$$

and

$$\tilde{H}_{G,\delta}^{\gamma+V+k}(G_+ \wedge_H S^{\gamma_0(b)+V-\delta(G/H)+n}; \overline{T}) \cong \begin{cases} \overline{T}(b) & \textit{if } k = n \\ 0 & \textit{if } k \neq n. \end{cases}$$

The proof is the same as Theorem 1.7.4 with appropriate changes.

3.4 Stable G-CW Approximation of Parametrized Spaces

As in Chap. 1, we will have a technical problem extending the definition of cellular homology and cohomology from G-CW complexes to arbitrary G-spaces, stemming from Example 1.4.29. In the parametrized case we have no analogue of Corollary 1.4.28, so the use of the CW spectra we now describe is essential.

If $\mathscr{V}$ is an indexing sequence in a G-universe $\mathscr{U}$, recall from Definition 2.5.2 that $G\mathscr{P}^\lambda_B$ is the category of G-spectra over B indexed on $\mathscr{V}$ and lax maps. We write $G\mathscr{P}^\lambda\mathscr{V}_B$ if we want to make the indexing sequence explicit. We have the following analogue of Definition 1.9.2.

Definition 3.4.1 Let $\mathscr{V}$ be an indexing sequence in a universe $\mathscr{U}$, let δ be a dimension function for G, and let γ be a virtual representation of ΠB.

(1) A lax map $f: D \Rightarrow E$ in $G\mathscr{P}^\lambda\mathscr{V}_B$ is a *δ-weak$_\gamma$ equivalence* if, for each i, $f_i: D(V_i) \Rightarrow E(V_i)$ is a δ-weak$_{\gamma+V_i}$ equivalence of G-spaces over B.
(2) A G-spectrum D in $G\mathscr{P}^\lambda\mathscr{V}_B$ is a *δ-G-CW(γ) spectrum* if, for each i, $D(V_i)$ is an ex-δ-G-CW$(\gamma + V_i)$ complex and if each structure map

$$\Sigma^{V_i - V_{i-1}} D(V_{i-1}) \to D(V_i)$$

is the inclusion of a subcomplex.
(3) A (strict) map $D \to E$ of δ-G-CW(γ) spectra is *the inclusion of a subcomplex* if, for each i, the map $D(V_i) \to E(V_i)$ is the inclusion of a subcomplex. We also say simply that D is a subcomplex of E.
(4) A lax map $D \Rightarrow E$ of δ-G-CW(γ) spectra is *cellular* if, for each i, the map $D(V_i) \to E(V_i)$ is cellular.
(5) If D is a G-spectrum in $G\mathscr{P}^\lambda\mathscr{V}_B$, a *$\delta$-$G$-CW$(\gamma)$ approximation* of D is a δ-G-CW(γ) spectrum $\Gamma^\delta_\alpha D$ and a strict δ-weak$_\gamma$ equivalence $\Gamma^\delta_\alpha D \to D$.

Our results on G-CW parametrized spaces give quick proofs of the following results.

Lemma 3.4.2 (H.E.L.P.) *Let $r: E \Rightarrow F$ be a δ-weak$_\gamma$ equivalence of G-spectra over B and let D be a δ-G-CW(γ) spectrum over B with subcomplex C. If the following diagram commutes without the dashed arrows, then there exist maps $\tilde{g}$ and $\tilde{h}$ making the diagram commute.*

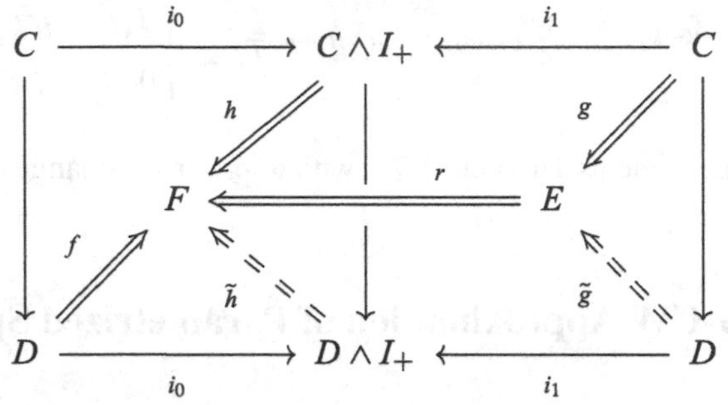

Proof We construct $\tilde{g}$ and $\tilde{h}$ inductively on the indexing space V_i. For $i = 1$ we simply quote the space-level H.E.L.P. lemma to find $\tilde{g}_1$ and $\tilde{h}_1$.

For the inductive step, we assume that we've constructed $\tilde{g}_{i-1}$ and $\tilde{h}_{i-1}$. We then apply the space-level H.E.L.P. lemma with (using the notation of Lemma 3.2.6) $Y = E_i$, $Z = F_i$, $X = D_i$, and $A = \Sigma X_{i-1} \cup C_i$. □

Proposition 3.4.3 (Whitehead) *Suppose that D is a δ-G-CW(γ) spectrum over B and that $f: E \Rightarrow F$ is a δ-weak$_\gamma$ equivalence. Then*

$$f_*: \pi G\mathscr{P}^\lambda \mathscr{V}_B(D, E) \to \pi G\mathscr{P}^\lambda \mathscr{V}_B(D, F)$$

is an isomorphism, where $\pi G\mathscr{P}^\lambda \mathscr{V}_B(-,-)$ denotes lax homotopy classes of lax G-maps. Therefore, any δ-weak$_\gamma$ equivalence of δ-G-CW(γ) spectra over B is a lax G-homotopy equivalence.

Proof We get surjectivity by applying the H.E.L.P. lemma to D and its subcomplex $*$. We get injectivity by applying it to $D \wedge I_+$ and its subcomplex $D \wedge \partial I_+$. □

Proposition 3.4.4 (Cellular Approximation of Maps) *Suppose that $f: D \Rightarrow E$ is a lax map of δ-G-CW(γ) spectra over B, C is a subcomplex of D, and h is a lax G-homotopy of $f|C$ to a cellular map. Then h can be extended to a lax G-homotopy of f to a cellular map.*

Proof This follows by induction on the indexing space V_i. The first case to consider is $f_1: D(V_1) \Rightarrow E(V_1)$, and we know from the space-level result that we can extend h_1 to a lax G-homotopy k_1 from f_1 to a cellular map g_1. For the inductive step, assume we have a lax homotopy k_{i-1}, extending h_{i-1}, from f_{i-1} to a cellular map g_{i-1}. Then $\Sigma^{V_i - V_{i-1}} k_{i-1} \cup h_i$ is a lax homotopy on the subcomplex $\Sigma^{V_i - V_{i-1}} D(V_{i-1}) \cup C_i$ of

$D(V_i)$. By the space-level result again, we can extend to a lax homotopy k_i on $D(V_i)$ from f_i to a cellular map g_i. □

Proposition 3.4.5 (Cellular Approximation of Spectra) *If D is a G-spectrum in $G\mathscr{P}^\lambda\mathscr{V}_B$, then there exists a δ-G-CW(γ) approximation $\Gamma D \to D$. If $f\colon D \Rightarrow E$ is a lax map of G-spectra over B and $\Gamma E \to E$ is an approximation of E, then there exists a lax cellular map $\Gamma f\colon \Gamma D \Rightarrow \Gamma E$, unique up to lax cellular homotopy, making the following diagram lax homotopy commute:*

$$\begin{array}{ccc} \Gamma D & \overset{\Gamma f}{\Longrightarrow} & \Gamma E \\ \downarrow & & \downarrow \\ D & \underset{f}{\Longrightarrow} & E \end{array}$$

Proof We construct ΓD recursively on the indexing space V_i. For $i = 1$, we take $(\Gamma D)(V_1) = \Gamma(D(V_1))$ to be any δ-G-CW$(\gamma + V_1)$ approximation of $D(V_1)$. Suppose that we have constructed $\Gamma D(V_{i-1}) \to D(V_{i-1})$, a δ-G-CW$(\gamma + V_{i-1})$ approximation. Then $\Sigma^{V_i - V_{i-1}}\Gamma D(V_{i-1})$ is a δ-G-CW$(\gamma + V_i)$ complex and, by Theorem 3.2.11, we can find a δ-G-CW$(\gamma + V_i)$ approximation $\Gamma D(V_i) \to D(V_i)$ making the following diagram commute, in which the map σ at the top is the inclusion of a subcomplex:

$$\begin{array}{ccc} \Sigma^{V_i - V_{i-1}}\Gamma D(V_{i-1}) & \overset{\sigma}{\longrightarrow} & \Gamma D(V_i) \\ \downarrow & & \downarrow \\ \Sigma^{V_i - V_{i-1}} D(V_{i-1}) & \underset{\sigma}{\longrightarrow} & D(V_i) \end{array}$$

Thus, ΓD is a δ-G-CW(γ) spectrum and the map $\Gamma D \to D$ so constructed is a δ-weak$_\gamma$ equivalence.

The existence and uniqueness of Γf follow from the Whitehead theorem and cellular approximation of maps and homotopies. □

As usual, these results imply that we can invert the δ-weak$_\gamma$ equivalences of spectra and that the result is equivalent to the ordinary lax homotopy category of δ-G-CW(γ) spectra. We emphasize, once again, that this is not the parametrized stable category.

3.5 Homology and Cohomology of Parametrized Spaces

We now extend the definition of ordinary homology and cohomology to arbitrary ex-G-spaces over B. As in the nonparametrized case, we do so by approximating by G-CW spectra.

Definition 3.5.1 Let E be a δ-G-CW(γ) spectrum over B. We define the *cellular chain complex of* E to be the colimit

$$\overline{C}^{G,\delta}_{\gamma+*}(E) = \operatorname*{colim}_i \overline{C}_{\gamma+V_i+*}(E(V_i)).$$

If F is an arbitrary spectrum over B and δ is familial, we define

$$\overline{C}^{G,\delta}_{\gamma+*}(F) = \overline{C}^{G,\delta}_{\gamma+*}(\Gamma F)$$

where $\Gamma F \to F$ is a δ-G-CW(γ) approximation of F. If X is an ex-G-space over B and $\Sigma^\infty_B X$ denotes its suspension spectrum, we define

$$\overline{C}^{G,\delta}_{\gamma+*}(X) = \overline{C}^{G,\delta}_{\gamma+*}(\Sigma^\infty_B X) = \overline{C}^{G,\delta}_{\gamma+*}(\Gamma\Sigma^\infty_B X).$$

The analogues of Propositions 1.10.11 through 1.10.13 are true, so that the chain complex $\overline{C}^{G,\delta}_{\gamma+*}(X)$ is well-defined up to chain homotopy equivalence and functorial up to chain homotopy, independent of the choices of universe, indexing sequence, and CW approximation.

We also have the analogue of Definition 1.10.4:

Definition 3.5.2

(1) Let $f: A \to X$ be a map of G-spaces over B The *(unreduced) mapping cylinder*, Mf, is the pushout in the following diagram:

$$\begin{array}{ccc} A & \xrightarrow{f} & X \\ {\scriptstyle i_0}\downarrow & & \downarrow \\ A \times I & \longrightarrow & Mf \end{array}$$

The *(unreduced) mapping cone* is $Cf = Mf/(A\times 1)$ with basepoint the image of $A \times 1$.

(2) Let $f\colon A \to X$ be a map of ex-G-spaces. The *reduced mapping cylinder*, $\tilde{M}f$, is the pushout in the following diagram:

$$\begin{array}{ccc} A & \xrightarrow{f} & X \\ {\scriptstyle i_0}\downarrow & & \downarrow \\ A \wedge I_+ & \longrightarrow & \tilde{M}f \end{array}$$

The *reduced mapping cone*, $\tilde{C}f$, is the pushout in the following diagram, in which I has basepoint 1:

$$\begin{array}{ccc} A & \xrightarrow{f} & X \\ {\scriptstyle i_0}\downarrow & & \downarrow \\ A \wedge I & \longrightarrow & \tilde{C}f \end{array}$$

In the special case in which $\sigma\colon B \to X$ is the section, we call $M\sigma$ the *whiskering construction* (perhaps better, the *beard construction*) and write $X_w = M\sigma$ for X with a whisker attached.

Definition 3.5.3 Let δ be a familial dimension function for G, let γ be a virtual representation of G, let $\overline{T}$ be a contravariant $\widehat{\Pi}_\delta B$-module, and let $\underline{S}$ be a covariant $\widehat{\Pi}_\delta B$-module. If X is an ex-G-space, let

$$\tilde{H}^{G,\delta}_{\gamma+n}(X;\underline{S}) = H_{\gamma+n}(\overline{C}^{G,\delta}_{\gamma+*}(X_w) \otimes_{\widehat{\Pi}_\delta B} \underline{S})$$

and

$$\tilde{H}^{\gamma+n}_{G,\delta}(X;\overline{T}) = H^{\gamma+n}(\mathrm{Hom}_{\widehat{\Pi}_\delta B}(\overline{C}^{G,\delta}_{\gamma+*}(X_w), \overline{T})).$$

The following proposition follows just as in the non-parametrized case.

Proposition 3.5.4 *$\tilde{H}^{G,\delta}_{\gamma+n}(X;\underline{S})$ is a well-defined covariant homotopy functor of X, while $\tilde{H}^{\gamma+n}_{G,\delta}(X;\overline{T})$ is a well-defined contravariant homotopy functor of X. If X is a δ-G-CW(γ) complex, these groups are naturally isomorphic to those given by Definition 3.3.5.* □

Theorem 3.5.5 (Reduced Homology and Cohomology of Spaces Over B) *Let δ be a familial dimension function for G, let γ be a virtual representation of $\Pi_G B$, and let $\underline{S}$ and $\overline{T}$ be respectively a covariant and a contravariant $\widehat{\Pi}_\delta B$-module. Then the abelian groups $\tilde{H}^{G,\delta}_\gamma(X;\underline{S})$ and $\tilde{H}^\gamma_{G,\delta}(X;\overline{T})$ are respectively covariant and contravariant functors on the homotopy category of ex-G-spaces over B. They are* *also respectively contravariant and covariant functors of γ. These functors satisfy the following properties.*

(1) *(Weak Equivalence) If $f: X \to Y$ is a an $\mathscr{F}(\delta)$-equivalence of ex-G-spaces, then*

$$f_*: \tilde{H}^{G,\delta}_{\gamma}(X; \underline{S}) \to \tilde{H}^{G,\delta}_{\gamma}(Y; \underline{S})$$

and

$$f^*: \tilde{H}^{\gamma}_{G,\delta}(Y; \overline{T}) \to \tilde{H}^{\gamma}_{G,\delta}(X; \overline{T})$$

are isomorphisms.

(2) *(Exactness) If $A \to X$ is a cofibration, then the following sequences are exact:*

$$\tilde{H}^{G,\delta}_{\gamma}(A; \underline{S}) \to \tilde{H}^{G,\delta}_{\gamma}(X; \underline{S}) \to \tilde{H}^{G,\delta}_{\gamma}(X/_B A; \underline{S})$$

and

$$\tilde{H}^{\gamma}_{G,\delta}(X/_B A; \overline{T}) \to \tilde{H}^{\gamma}_{G,\delta}(X; \overline{T}) \to \tilde{H}^{\gamma}_{G,\delta}(A; \overline{T}).$$

(3) *(Additivity) If $X = \bigvee_i X_i$ is a fiberwise wedge of well-based ex-G-spaces, then the inclusions of the wedge summands induce isomorphisms*

$$\bigoplus_i \tilde{H}^{G,\delta}_{\gamma}(X_i; \underline{S}) \cong \tilde{H}^{G,\delta}_{\gamma}(X; \underline{S})$$

and

$$\tilde{H}^{\gamma}_{G,\delta}(X; \overline{T}) \cong \prod_i \tilde{H}^{\gamma}_{G,\delta}(X_i; \overline{T}).$$

(4) *(Suspension) If X is well-based, there are suspension isomorphisms*

$$\sigma^V: \tilde{H}^{G,\delta}_{\alpha}(X; \underline{S}) \xrightarrow{\cong} \tilde{H}^{G,\delta}_{\alpha+V}(\Sigma^V_B X; \underline{S})$$

and

$$\sigma^V: \tilde{H}^{\gamma}_{G,\delta}(X; \overline{T}) \xrightarrow{\cong} \tilde{H}^{\gamma+V}_{G,\delta}(\Sigma^V_B X; \overline{T}).$$

These isomorphisms satisfy $\sigma^0 = \mathrm{id}$, $\sigma^W \circ \sigma^V = \sigma^{V \oplus W}$, and the following naturality condition: If $\zeta: V \to V'$ is an isomorphism, then the following *diagrams commute:*

$$\begin{array}{ccc}
\tilde{H}^{G,\delta}_{\gamma}(X;\underline{S}) & \xrightarrow{\sigma^{V}} & \tilde{H}^{G,\delta}_{\gamma+V}(\Sigma^{V}_{B}X;\underline{S}) \\
\Big\downarrow{\scriptstyle \sigma^{V'}} & & \Big\downarrow{\scriptstyle \tilde{H}_{\mathrm{id}}(\mathrm{id}\wedge\zeta)} \\
\tilde{H}^{G,\delta}_{\gamma+V'}(\Sigma^{V'}_{B}X;\underline{S}) & \xrightarrow[\tilde{H}_{\mathrm{id}+\zeta}(\mathrm{id})]{} & \tilde{H}^{G,\delta}_{\gamma+V}(\Sigma^{V'}_{B}X;\underline{S})
\end{array}$$

$$\begin{array}{ccc}
\tilde{H}^{\gamma}_{G,\delta}(X;\overline{T}) & \xrightarrow{\sigma^{V}} & \tilde{H}^{\gamma+V}_{G,\delta}(\Sigma^{V}_{B}X;\overline{T}) \\
\Big\downarrow{\scriptstyle \sigma^{V'}} & & \Big\downarrow{\scriptstyle \tilde{H}^{\mathrm{id}+\zeta}(\mathrm{id})} \\
\tilde{H}^{\gamma+V'}_{G,\delta}(\Sigma^{V'}_{B}X;\overline{T}) & \xrightarrow[\tilde{H}^{\mathrm{id}}(\mathrm{id}\wedge\zeta)]{} & \tilde{H}^{\gamma+V'}_{G,\delta}(\Sigma^{V}_{B}X;\overline{T})
\end{array}$$

(5) *(Dimension Axiom) If $H \in \mathscr{F}(\delta)$, $b\colon G/H \to B$, and V is a representation of H so large that $\gamma_0(b) + V - \delta(G/H) + n$ is an actual representation, then there are natural isomorphisms*

$$\tilde{H}^{G,\delta}_{\gamma+V+k}(G_+ \wedge_H S^{\gamma_0(b)+V-\delta(G/H)+n,b};\underline{S}) \cong \begin{cases} \underline{S}(b) & \text{if } k = n \\ 0 & \text{if } k \neq n \end{cases}$$

and

$$\tilde{H}^{\gamma+V+k}_{G,\delta}(G_+ \wedge_H S^{\gamma_0(b)+V-\delta(G/H)+n,b};\overline{T}) \cong \begin{cases} \overline{T}(b) & \text{if } k = n \\ 0 & \text{if } k \neq n. \end{cases}$$

□

The proof is the same as that of Theorem 1.10.7 with appropriate changes. The following analogue of Corollary 1.10.8 follows.

Corollary 3.5.6

(1) *If $f\colon A \to X$ is a map of well-based ex-G-spaces, then the following sequences are exact:*

$$\tilde{H}^{G,\delta}_{\gamma}(A;\underline{S}) \to \tilde{H}^{G,\delta}_{\gamma}(X;\underline{S}) \to \tilde{H}^{G,\delta}_{\gamma}(\tilde{C}f;\underline{S})$$

and

$$\tilde{H}^{\gamma}_{G,\delta}(\tilde{C}f;\overline{T}) \to \tilde{H}^{\gamma}_{G,\delta}(X;\overline{T}) \to \tilde{H}^{\gamma}_{G,\delta}(A;\overline{T}).$$

(2) *If $f: A \to X$ is any map of ex-G-spaces, then the following sequences are exact, where $f_w: A_w \to X_w$ is the induced map:*

$$\tilde{H}_\gamma^{G,\delta}(A;\underline{S}) \to \tilde{H}_\gamma^{G,\delta}(X;\underline{S}) \to \tilde{H}_\gamma^{G,\delta}(\tilde{C}(f_w);\underline{S})$$

and

$$\tilde{H}^\gamma_{G,\delta}(\tilde{C}(f_w);\overline{T}) \to \tilde{H}^\gamma_{G,\delta}(X;\overline{T}) \to \tilde{H}^\gamma_{G,\delta}(A;\overline{T}).$$

□

Finally, we define unreduced homology and cohomology of pairs.

Definition 3.5.7 If (X, A) is a pair of G-spaces over B, write $i: A \to X$ for the inclusion and let

$$H_\gamma^{G,\delta}(X, A;\underline{S}) = \tilde{H}_\gamma^{G,\delta}(Ci;\underline{S})$$

and

$$H^\gamma_{G,\delta}(X, A;\overline{T}) = \tilde{H}^\gamma_{G,\delta}(Ci;\overline{T}).$$

In particular, we write

$$H_\gamma^{G,\delta}(X;\underline{S}) = H_\gamma^{G,\delta}(X,\emptyset;\underline{S}) = \tilde{H}_\gamma^{G,\delta}(X_+;\underline{S})$$

and

$$H^\gamma_{G,\delta}(X;\overline{T}) = H^\gamma_{G,\delta}(X,\emptyset;\overline{T}) = \tilde{H}^\gamma_{G,\delta}(X_+;\overline{T}).$$

Theorem 3.5.8 (Unreduced Homology and Cohomology of Spaces Over B) *Let δ be a familial dimension function for G, let γ be a virtual representation of $\Pi_G B$, and let $\underline{S}$ and $\overline{T}$ be respectively a covariant and a contravariant $\widehat{\Pi}_\delta B$-module. Then the abelian groups $H_\gamma^{G,\delta}(X, A;\underline{S})$ and $H^\gamma_{G,\delta}(X, A;\overline{T})$ are respectively covariant and contravariant functors on the homotopy category of pairs of G-spaces over B. They are also respectively contravariant and covariant functors of γ. These functors satisfy the following properties.*

(1) *(Weak Equivalence) If $f: (X, A) \to (Y, B)$ is an $\mathscr{F}(\delta)$-equivalence of pairs of G-spaces over B, then*

$$f_*: H_\gamma^{G,\delta}(X, A;\underline{S}) \to H_\gamma^{G,\delta}(Y, B;\underline{S})$$

and

$$f^*: H^\gamma_{G,\delta}(Y, B;\overline{T}) \to H^\gamma_{G,\delta}(X, A;\overline{T})$$

are isomorphisms.

(2) *(Exactness) If (X,A) is a pair of G-spaces over B, then there are natural homomorphisms*

$$\partial\colon H^{G,\delta}_{\gamma+n}(X,A;\underline{S}) \to H^{G,\delta}_{\gamma+n-1}(A;\underline{S})$$

and

$$d\colon H_{G,\delta}^{\gamma+n}(A;\overline{T}) \to H_{G,\delta}^{\gamma+n+1}(X,A;\overline{T})$$

and long exact sequences

$$\cdots \to H^{G,\delta}_{\gamma+n}(A;\underline{S}) \to H^{G,\delta}_{\gamma+n}(X;\underline{S}) \to H^{G,\delta}_{\gamma+n}(X,A;\underline{S}) \to H^{G,\delta}_{\gamma+n-1}(A;\underline{S}) \to \cdots$$

and

$$\cdots \to H_{G,\delta}^{\gamma+n-1}(A;\overline{T}) \to H_{G,\delta}^{\gamma+n}(X,A;\overline{T}) \to H_{G,\delta}^{\gamma+n}(X;\overline{T}) \to H_{G,\delta}^{\gamma+n}(A;\overline{T}) \to \cdots.$$

(3) *(Excision) If $(X;A,B)$ is an excisive triad, i.e., X is the union of the interiors of A and B, then the inclusion $(A, A\cap B) \to (X,B)$ induces isomorphisms*

$$H^{G,\delta}_{\gamma}(A,A\cap B;\underline{S}) \cong H^{G,\delta}_{\gamma}(X,B;\underline{S})$$

and

$$H_{G,\delta}^{\gamma}(X,B;\overline{T}) \cong H_{G,\delta}^{\gamma}(A,A\cap B;\overline{T}).$$

(4) *(Additivity) If $(X,A) = \coprod_k (X_k,A_k)$ is a disjoint union of pairs of G-spaces over B, then the inclusions $(X_k,A_k) \to (X,A)$ induce isomorphisms*

$$\bigoplus_k H^{G,\delta}_{\gamma}(X_k,A_k;\underline{S}) \cong H^{G,\delta}_{\gamma}(X,A;\underline{S})$$

and

$$H_{G,\delta}^{\gamma}(X,A;\overline{T}) \cong \prod_k H_{G,\delta}^{\gamma}(X_k,A_k;\overline{T}).$$

(5) *(Suspension) If $A \to X$ is a cofibration, there are suspension isomorphisms*

$$\sigma^V\colon H^{G,\delta}_{\gamma}(X,A;\underline{S}) \xrightarrow{\cong} H^{G,\delta}_{\gamma+V}((X,A)\times(D(V),S(V));\underline{S})$$

and

$$\sigma^V\colon H_{G,\delta}^{\gamma}(X,A;\overline{T}) \xrightarrow{\cong} H_{G,\delta}^{\gamma+V}((X,A)\times(D(V),S(V));\overline{T}),$$

where $(X,A)\times(D(V),S(V)) = (X\times D(V), X\times S(V)\cup A\times D(V))$. *These isomorphisms satisfy* $\sigma^0 = \mathrm{id}$ *and* $\sigma^W\circ\sigma^V = \sigma^{V\oplus W}$, *under the identification* $\bar D(V)\times\bar D(W)\approx \bar D(V\oplus W)$ *(here we use the notation* $\bar D(V) = (D(V),S(V))$*). They also satisfy the following naturality condition: If* $\zeta\colon V\to V'$ *is an isomorphism, then the following diagrams commute:*

$$\begin{array}{ccc}
H_\gamma^{G,\delta}(X,A;\underline{S}) & \xrightarrow{\sigma^V} & H^{G,\delta}_{\gamma+V}((X,A)\times\bar D(V);\underline{S}) \\
\downarrow{\scriptstyle\sigma^{V'}} & & \downarrow{\scriptstyle H^G_{\mathrm{id}}(\mathrm{id}\times\zeta)} \\
H^{G,\delta}_{\gamma+V'}((X,A)\times\bar D(V');\underline{S}) & \xrightarrow[H^G_{\mathrm{id}\oplus\zeta}(\mathrm{id})]{} & H^{G,\delta}_{\gamma+V}((X,A)\times\bar D(V');\underline{S})
\end{array}$$

$$\begin{array}{ccc}
H^\gamma_{G,\delta}(X,A;\overline{T}) & \xrightarrow{\sigma^V} & H^{\gamma+V}_{G,\delta}((X,A)\times\bar D(V);\overline{T}) \\
\downarrow{\scriptstyle\sigma^{V'}} & & \downarrow{\scriptstyle H_G^{\mathrm{id}\oplus\zeta}(\mathrm{id})} \\
H^{\gamma+V'}_{G,\delta}((X,A)\times\bar D(V');\overline{T}) & \xrightarrow[H_G^{\mathrm{id}}(\mathrm{id}\times\zeta)]{} & H^{\gamma+V'}_{G,\delta}((X,A)\times\bar D(V);\overline{T})
\end{array}$$

(6) *(Dimension Axiom) If* $H\in\mathscr{F}(\delta)$, $b\colon G/H\to B$, *and* V *is a representation of* H *so large that* $\gamma_0(b)+V-\delta(G/H)+n$ *is an actual representation, write*

$$\bar D_b(\gamma_0(b)+V-\delta(G/H)+n) = b_!(G\times_H\bar D(\gamma_0(b)+V-\delta(G/H)+n)),$$

i.e., the pair $G\times_H\bar D(\gamma_0(b)+V-\delta(G/H)+n)$ *mapping to* B *via* b. *Then there are natural isomorphisms*

$$H^{G,\delta}_{\gamma+V+k}(\bar D_b(\gamma_0(b)+V-\delta(G/H)+n);\underline{S}) \cong \begin{cases}\underline{S}(b) & \text{if } k=n\\ 0 & \text{if } k\neq n\end{cases}$$

and

$$H^{\gamma+V+k}_{G,\delta}(\bar D_b(\gamma_0(b)+V-\delta(G/H)+n);\overline{T}) \cong \begin{cases}\overline{T}(b) & \text{if } k=n\\ 0 & \text{if } k\neq n.\end{cases}$$

□

The proof is the same as Theorem 1.10.10 with appropriate changes.

Definition 2.7.3 gives us the covariant $\widehat{\Pi}_\delta B$-module $\underline{H}_\gamma^{G,\delta}(X;\underline{S})$ defined by

$$\begin{aligned}
\underline{H}_\gamma^{G,\delta}(X;\underline{S})(b\colon G/H\to B) &= \tilde H_\gamma^{G,\delta}(X\wedge(G_+\wedge_H S^{-\delta(G/H),b});\underline{S}) \\
&= \tilde H^{G,\delta}_{\gamma+V}(X\wedge(G_+\wedge_H S^{V-\delta(G/H),b});\underline{S})
\end{aligned}$$

for a V sufficiently large that $V - \delta(G/H)$ is an actual representation, and the contravariant $\widehat{\Pi}_\delta B$-module $\overline{H}^{\gamma}_{G,\delta}(X;\overline{T})$ defined similarly. In these terms, the dimension axioms take following form:

$$\underline{H}^{G,\delta}_{n}(B_+;\underline{S}) \cong \begin{cases} \underline{S} & \text{if } n = 0 \\ 0 & \text{if } n \neq 0 \end{cases}$$

and

$$\overline{H}^{n}_{G,\delta}(B_+;\overline{T}) \cong \begin{cases} \overline{T} & \text{if } n = 0 \\ 0 & \text{if } n \neq 0. \end{cases}$$

3.6 Atiyah-Hirzebruch Spectral Sequences and Uniqueness

The following theorem is proved in the same way as Theorem 1.11.2. Recall that, if $\tilde{h}^G_*(-)$ is a generalized $RO(G)$-graded homology theory on ex-G-spaces over B (as in Definition 2.7.1), γ is a virtual representation of ΠB, and δ is a dimension function for G, then Definition 2.7.3 defines the coefficient system $\underline{h}^{G,\delta}_{\gamma}$. Similarly, if $\tilde{h}^*_G(-)$ is a reduced $RO(G)$-graded cohomology theory, we have the coefficient system $\overline{h}^{\gamma}_{G,\delta}$.

Theorem 3.6.1 (Atiyah-Hirzebruch Spectral Sequence) *Suppose that $\tilde{h}^G_*(-)$ is an $RO(G)$-graded homology theory on ex-G-spaces over B. Let δ be a familial dimension function for G, let α be an element of $RO(G)$, and let γ be a virtual representation of ΠB. Assume that $\tilde{h}^G_*(-)$ takes $\mathscr{F}(\delta)$-equivalences to isomorphisms. Then there is a strongly convergent spectral sequence*

$$E^2_{p,q} = \tilde{H}^{G,\delta}_{\alpha-\gamma+p}(X;\underline{h}^{G,\delta}_{\gamma+q}) \Rightarrow \tilde{h}^G_{\alpha+p+q}(X).$$

Similarly, if $h^*_G(-)$ is an $RO(G)$-graded cohomology theory on ex-G-spaces over B, taking $\mathscr{F}(\delta)$-equivalences to isomorphisms, then there is a conditionally convergent spectral sequence

$$E^{p,q}_2 = \tilde{H}^{\alpha-\gamma+p}_{G,\delta}(X;\overline{h}^{\gamma+q}_{G,\delta}) \Rightarrow h^{\alpha+p+q}_G(X).$$ □

Uniqueness follows in the usual way.

Corollary 3.6.2 (Uniqueness of Ordinary $RO(\Pi B)$-Graded Homology) *Let δ be a familial dimension function for G. Let $\tilde{h}^G_*(-)$ be an $RO(G)$-graded homology theory on ex-G-spaces over B that takes $\mathscr{F}(\delta)$-equivalences to isomorphisms and* *let γ be a virtual representation of ΠB. Suppose that, for integers n,*

$$\underline{h}^{G,\delta}_{\gamma+n} = 0 \qquad \text{for } n \neq 0$$

Then, for $\alpha \in RO(G)$, there is a natural isomorphism

$$\tilde{h}^G_\alpha(-) \cong \tilde{H}^{G,\delta}_{\alpha-\gamma}(-; \underline{h}^{G,\delta}_\gamma).$$

There is a similar statement for cohomology theories. □

Finally, we have a universal coefficients spectral sequence, constructed in the usual way. Write $\overline{H}^{G,\delta}_\gamma(X)$ for the contravariant $\widehat{\Pi}_\delta B$-module given by

$$\overline{H}^{G,\delta}_\gamma(X)(b) = \tilde{H}^{G,\delta}_\gamma(X; \widehat{\Pi}_\delta B(b, -)).$$

$\mathrm{Tor}_*^{\widehat{\Pi}_\delta B}$ and $\mathrm{Ext}^*_{\widehat{\Pi}_\delta B}$ below are the derived functors of $\otimes_{\widehat{\Pi}_\delta B}$ and $\mathrm{Hom}_{\widehat{\Pi}_\delta B}$ respectively.

Theorem 3.6.3 (Universal Coefficients Spectral Sequence) *If $\underline{S}$ is a covariant $\widehat{\Pi}_\delta B$-module, $\overline{T}$ is a contravariant $\widehat{\Pi}_\delta B$-module, and γ is a virtual representation of ΠB, there are spectral sequences*

$$E^2_{p,q} = \mathrm{Tor}_p^{\widehat{\Pi}_\delta B}(\overline{H}^{G,\delta}_{\gamma+q}(X), \underline{S}) \Rightarrow \tilde{H}^{G,\delta}_{\gamma+p+q}(X; \underline{S})$$

and

$$E_2^{p,q} = \mathrm{Ext}^p_{\widehat{\Pi}_\delta B}(\overline{H}^{G,\delta}_{\gamma+q}(X), \overline{T}) \Rightarrow \tilde{H}^{\gamma+p+q}_{G,\delta}(X; \overline{T}).$$ □

3.7 The Representing Spectra

Let δ be a familial dimension function for G, let γ be a virtual representation of $\Pi_G B$, and let $\overline{T}$ be a contravariant $\widehat{\Pi}_\delta B$-module. By Theorem 2.8.1, there exists a parametrized spectrum representing the $RO(G)$-graded cohomology theory $\tilde{H}^{\gamma+*}_{G,\delta}(-; \overline{T})$, which we will call $H_\delta \overline{T}^\gamma$.

If E is any G-spectrum over B, define the contravariant $\widehat{\Pi}_\delta B$-module $\overline{\pi}^\delta_\gamma E$ by

$$\overline{\pi}^\delta_\gamma E(b\colon G/K \to B) = [G_+ \wedge_K S^{\gamma_0(b)-\delta(G/K),b}, E]_{G,B}.$$

This obviously defines a functor on $\widehat{\Pi}_{\delta,\gamma}B$, and we make it a functor on $\widehat{\Pi}_{\delta}B$ using the isomorphism of Theorem 2.6.4. From the dimension axiom for cellular cohomology, we have

$$\overline{\pi}^{\delta}_{\gamma+n}H_{\delta}\overline{T}^{\gamma} \cong \begin{cases} \overline{T} & n = 0 \\ 0 & n \neq 0 \end{cases}$$

and, from the uniqueness of ordinary cohomology, this together with the equivalence $H_{\delta}\overline{T}^{\gamma} \simeq F(E\mathscr{F}(\delta)_+, H_{\delta}\overline{T}^{\gamma})$ characterizes $H_{\delta}\overline{T}^{\gamma}$ up to nonunique homotopy equivalence. We call any spectrum satisfying these conditions a *parametrized Eilenberg-Mac Lane spectrum.* Using Corollary 2.8.5 and the discussion following it, we can rewrite the calculation above as

$$\overline{\pi}^{K,\delta}_{\gamma_0(b)+n}((H_{\delta}\overline{T}^{\gamma})_b) \cong \begin{cases} b^*\overline{T} & n = 0 \\ 0 & n \neq 0 \end{cases}$$

for $b\colon G/K \to B$. Thus, we can identify each fiber as the suspension of a nonparametrized Eilenberg-Mac Lane spectrum:

$$(H_{\delta}\overline{T}^{\gamma})_b \simeq \Sigma^{\gamma_0(b)} H_{\delta}(b^*\overline{T})$$

as K-spectra. Again, this characterizes $H_{\delta}\overline{T}^{\gamma}$.

What about cellular homology? It follows from [51, 18.6.7 & 18.1.5] that

$$[S^n, \rho_!(H_{\delta}\overline{T}^{\gamma} \wedge_B G_+ \wedge_H S^{-\gamma_0(b)-(\mathscr{L}-\delta)(G/H),b})]_{G,B}$$
$$\cong [G_+ \wedge_H S^{\gamma_0(b)-\delta(G/H)+n,b}, H_{\delta}\overline{T}^{\gamma}]_{G,B}.$$

It follows, as in Theorem 1.12.2, that

$$H^{\mathscr{L}-\delta}\overline{T}_{-\gamma} = H_{\delta}\overline{T}^{\gamma} \wedge E\mathscr{F}(\delta)_+$$

represents $\tilde{H}^{G,\mathscr{L}-\delta}_{-\gamma+*}(-;\overline{T})$, or that

$$H^{\delta}\underline{S}_{\gamma} = H_{\mathscr{L}-\delta}\underline{S}^{-\gamma} \wedge E\mathscr{F}(\delta)_+$$

represents $\tilde{H}^{G,\delta}_{\gamma+*}(-;\underline{S})$. The characterization of the homotopy groups then looks like

$$\overline{\pi}^{\mathscr{L}-\delta}_{-\gamma+n}H^{\delta}\underline{S}_{\gamma} \cong \begin{cases} \underline{S} & n = 0 \\ 0 & n \neq 0 \end{cases}$$

We call such a spectrum a *parametrized Mac Lane-Eilenberg spectrum.* The spectrum $H^{\delta}\underline{S}_{\gamma}$ can also be characterized by the fact that its fibers are nonparametrized Mac Lane-Eilenberg spectra representing ordinary homology theories.

It will be useful to have a simple, explicit construction of these spectra. We simply have to elaborate Construction 1.12.5.

Construction 3.7.1 Let δ be a familial dimension function for G, let γ be a virtual representation of $\Pi_G B$, and let $\overline{T}$ be a contravariant $\widehat{\Pi}_{G,\delta}B$-module. Define

$$F_{\delta}\overline{T}^{\gamma} = \bigvee_{\overline{T}(b)} G_+ \wedge_{H_b} S^{\gamma_0(b)-\delta(G/H),b},$$

where the wedge runs over all objects b in $\widehat{\Pi}_{G,\delta}B$ and all elements in $\overline{T}(b)$. Then $F_{\delta}(-)^{\gamma}$ is a functor taking contravariant $\widehat{\Pi}_{G,\delta}B$-modules to G-spectra and we have a natural epimorphism

$$\epsilon\colon \overline{\pi}_{\gamma}^{\delta} F_{\delta}\overline{T} \to \overline{T}.$$

Let

$$K = \{\kappa\colon G_+ \wedge_{H_\kappa} S^{\gamma_0(b_\kappa)-\delta(G/H_\kappa),b_\kappa} \to F_{\delta}\overline{T} \mid \epsilon(\kappa) = 0\},$$

and let

$$R_{\delta}\overline{T}^{\gamma} = \bigvee_{\kappa\in K} S^{\gamma_0(b_\kappa)-\delta(G/H_\kappa),b_\kappa}.$$

Then $R_{\delta}(-)^{\gamma}$ is also a functor, there is a natural transformation $R_{\delta}\overline{T}^{\gamma} \to F_{\delta}\overline{T}^{\gamma}$ (given by the maps κ), and we have an exact sequence

$$\overline{\pi}_{\gamma}^{\delta} R_{\delta}\overline{T}^{\gamma} \to \overline{\pi}_{\gamma}^{\delta} F_{\delta}\overline{T}^{\gamma} \to \overline{T} \to 0.$$

It follows that, if we let $C_{\delta}\overline{T}^{\gamma}$ be the cofiber of $R_{\delta}\overline{T}^{\gamma} \to F_{\delta}\overline{T}^{\gamma}$, then $\overline{\pi}_{\gamma}^{\delta} C_{\delta}\overline{T}^{\gamma} \cong \overline{T}$ and $\overline{\pi}_{\gamma+n}^{\delta} C_{\delta}\overline{T}^{\gamma} = 0$ for $n < 0$.

We can then functorially kill all the homotopy $\overline{\pi}_{\gamma+n}^{\delta} C_{\delta}\overline{T}^{\gamma}$ for $n > 0$, obtaining a functor $P_{\delta}(-)^{\gamma}$ with

$$\overline{\pi}_{\gamma+n}^{\delta} P_{\delta}\overline{T}^{\gamma} = \begin{cases} \overline{T} & \text{if } n = 0 \\ 0 & \text{if } n \neq 0. \end{cases}$$

Finally, we let

$$H_{\delta}\overline{T}^{\gamma} = F(E\mathscr{F}(\delta)_+, P_{\delta}\overline{T}^{\gamma})$$

and

$$H^\delta \underline{S}_\gamma = P_{\mathscr{L}-\delta}\underline{S}^{-\gamma} \wedge E\mathscr{F}(\delta)_+$$

if $\underline{S}$ is a covariant δ-Mackey functor. Then $H_\delta(-)^\gamma$ and $H^\delta(-)^\gamma$ are functors and produce spectra that satisfy the characterizations of the Eilenberg-Mac Lane spectra representing cohomology and homology, respectively. □

3.8 Change of Base Space

It may have occurred to the reader to ask how the homology or cohomology of a parametrized space depends on the choice of base space. In some examples, such as Example 3.1.2, we have considered spaces as parametrized over themselves. The answer, as we shall see, is that the homology and cohomology do not, in fact, depend on the base space.

Let $f: A \to B$ be a G-map. The main results of this section come from the observation that, if X is a CW complex over A, then $f_!X$ is a CW complex over B with corresponding cells. More precisely, if γ is a virtual representation of $\Pi_G B$, then, if X is a δ-G-CW$(f^*\gamma)$ complex over A, then $f_!X$ is a δ-G-CW(γ) complex over B.

For the algebra, we have

$$f_!: \widehat{\Pi}_{G,\delta}A \to \widehat{\Pi}_{G,\delta}B,$$

given by the change-of-base functor $f_!$ on parametrized spectra. We write $f_! = (f_!)_!$ and $f^* = (f_!)^*$ for the induced functors on modules.

This leads to the following isomorphism of chain complexes.

Proposition 3.8.1 *If γ is a virtual representation of $\Pi_G B$ and X is a δ-G-CW$(f^*\gamma)$ complex over A, then*

$$\overline{C}^{G,\delta}_{\gamma+*}(f_!X) \cong f_!\overline{C}^{G,\delta}_{f^*\gamma+*}(X).$$

Moreover, this isomorphism respects suspension isomorphisms.

Proof We define a map $f_!\overline{C}^{G,\delta}_{f^*\gamma+*}(X) \to \overline{C}^{G,\delta}_{\gamma+*}(f_!X)$ as follows. For simplicity of notation, we write γ again for $f^*\gamma$ and, for $a: G/H \to A$ an object of $\widehat{\Pi}_{G,\delta}A$, we write

$$S^{\gamma+n,a} = G_+ \wedge_H S^{\gamma_0(a)-\delta(G/H)+n,a}.$$

If $b \in \widehat{\Pi}_{G,\delta}B$, we consider the following map:

$$\begin{aligned}
f_! \overline{C}^{G,\delta}_{f^*\gamma+n}(X)(b) &= \int^{a\in\widehat{\Pi}_{G,\delta}A} \overline{C}^{G,\delta}_{f^*\gamma+n}(X)(a) \otimes \widehat{\Pi}_{G,\delta}B(b,f_!(a)) \\
&= [S^{\gamma+n,a}, \Sigma^\infty_A X^{\gamma+n}/X^{\gamma+n-1}]_{G,A} \otimes \widehat{\Pi}_{G,\delta}B(b,f_!(a)) \\
&\xrightarrow{f_!} [f_! S^{\gamma+n,a}, f_!\Sigma^\infty_A X^{\gamma+n}/X^{\gamma+n-1}]_{G,B} \otimes \widehat{\Pi}_{G,\delta}B(b,f_!(a)) \\
&\cong [S^{\gamma+n,f_!(a)}, \Sigma^\infty_B (f_!X)^{\gamma+n}/(f_!X)^{\gamma+n-1}]_{G,B} \otimes \widehat{\Pi}_{G,\delta}B(b,f_!(a)) \\
&\to [S^{\gamma+n,b}, \Sigma^\infty_B (f_!X)^{\gamma+n}/(f_!X)^{\gamma+n-1}]_{G,B} \\
&= \overline{C}^{G,\delta}_{\gamma+n}(f_!X)(b).
\end{aligned}$$

This description shows that the map is a chain map. Further, when we look at its effect on one cell in $X^{\gamma+n}/X^{\gamma+n-1}$, Proposition 1.6.6(5) implies that we have an isomorphism, so Lemma 3.3.2 shows that the map above is a chain isomorphism.

That this isomorphism respects suspension isomorphisms follows from the definition of the map above. □

This gives us the following isomorphisms in homology and cohomology.

Theorem 3.8.2 *Let δ be a dimension function for G, let $f: A \to B$ be a G-map, let γ be a virtual representation of $\Pi_G B$, let $\overline{T}$ be a contravariant $\widehat{\Pi}_{G,\delta}B$-module, and let $\underline{S}$ be a covariant $\widehat{\Pi}_{G,\delta}B$-module. If X is a δ-G-CW($f^*\gamma$) complex, then there are natural isomorphisms*

$$\tilde{H}^{G,\delta}_\gamma(f_!X;\underline{S}) \cong \tilde{H}^{G,\delta}_{f^*\gamma}(X;f^*\underline{S})$$

and

$$\tilde{H}^\gamma_{G,\delta}(f_!X;\overline{T}) \cong \tilde{H}^{f^*\gamma}_{G,\delta}(X;f^*\overline{T}).$$

If δ is familial, for X a well-based ex-G-space there are natural isomorphisms

$$\tilde{H}^{G,\delta}_\gamma(f_!X;\underline{S}) \cong \tilde{H}^{G,\delta}_{f^*\gamma}(X;f^*\underline{S})$$

and

$$\tilde{H}^\gamma_{G,\delta}(f_!X;\overline{T}) \cong \tilde{H}^{f^*\gamma}_{G,\delta}(X;f^*\overline{T}).$$

These isomorphisms all respect suspension isomorphisms.

Proof If X is a δ-G-CW$(f^*\gamma)$ complex, then the preceding proposition gives us

$$\begin{aligned}\overline{C}^{G,\delta}_{\gamma+*}(f_!X) \otimes_{\widehat{\Pi}_{G,\delta}B} \underline{S} &\cong f_!\overline{C}^{G,\delta}_{f^*\gamma+*}(X) \otimes_{\widehat{\Pi}_{G,\delta}B} \underline{S} \\ &\cong \overline{C}^{G,\delta}_{f^*\gamma+*}(X) \otimes_{\widehat{\Pi}_{G,\delta}A} f^*\underline{S}.\end{aligned}$$

On taking homology we get the first isomorphism of the theorem. Similarly, we have

$$\begin{aligned}\mathrm{Hom}_{\widehat{\Pi}_{G,\delta}B}(\overline{C}^{G,\delta}_{\gamma+*}(f_!X), \overline{T}) &\cong \mathrm{Hom}_{\widehat{\Pi}_{G,\delta}B}(f_!\overline{C}^{G,\delta}_{f^*\gamma+*}(X), \overline{T}) \\ &\cong \mathrm{Hom}_{\widehat{\Pi}_{G,\delta}A}(\overline{C}^{G,\delta}_{f^*\gamma+*}(X), f^*\overline{T}).\end{aligned}$$

On taking homology we get the second isomorphism of the theorem. That these isomorphisms respect suspension follows from the preceding proposition.

Now let X be a well-based ex-G-space over A. If $\Gamma X \to X$ is an approximation by a δ-G-CW$(f^*\gamma)$ complex, then $f_!\Gamma X \to f_!X$ is an approximation by a δ-G-CW(γ) complex. This follows from Theorems 3.2.5 and 1.4.16 and the fact that, nonequivariantly, $f_!$ preserves weak equivalence of well-based ex-spaces [51, 7.3.4]. From this it follows that $f_!$ takes an approximation $\Gamma\Sigma^\infty_A X \to \Sigma^\infty_A X$ by a G-CW$(f^*\gamma)$ spectrum to an approximation of $\Sigma^\infty_B f_!X$, which gives a chain isomorphism $\overline{C}^{G,\delta}_{\gamma+*}(f_!X) \cong f_!\overline{C}^{G,\delta}_{f^*\gamma+*}(X)$. The homology and cohomology isomorphisms now follow as for CW complexes. □

We can use these isomorphisms to get the following "push-forward" maps: Let $\underline{S}$ be a covariant $\widehat{\Pi}_{G,\delta}A$-module and let $\overline{T}$ be a contravariant $\widehat{\Pi}_{G,\delta}A$-module. Using the unit $\eta\colon \underline{S} \to f^*f_!\underline{S}$ of the adjunction, we can then define

$$f_!\colon \tilde{H}^{G,\delta}_{f^*\gamma}(X;\underline{S}) \to \tilde{H}^{G,\delta}_{\gamma}(f_!X;f_!\underline{S})$$

to be the composite

$$\tilde{H}^{G,\delta}_{f^*\gamma}(X;\underline{S}) \xrightarrow{\eta_*} \tilde{H}^{G,\delta}_{f^*\gamma}(X;f^*f_!\underline{S}) \cong \tilde{H}^{G,\delta}_{\gamma}(f_!X;f_!\underline{S}).$$

We define

$$f_!\colon \tilde{H}^{f^*\gamma}_{G,\delta}(X;\overline{T}) \to \tilde{H}^{\gamma}_{G,\delta}(f_!X;f_!\overline{T})$$

similarly. Note that, if $\underline{U}$ is a covariant $\widehat{\Pi}_{G,\delta}B$-module, then the composite

$$\tilde{H}^{G,\delta}_{f^*\gamma}(X;f^*\underline{U}) \xrightarrow{f_!} \tilde{H}^{G,\delta}_{\gamma}(f_!X;f_!f^*\underline{U}) \xrightarrow{\epsilon_*} \tilde{H}^{G,\delta}_{\gamma}(f_!X;\underline{U})$$

agrees with the isomorphism of the Theorem 3.8.2, where ϵ is the counit of the adjunction. Of course, the similar statement is true for cohomology.

We now look at how the isomorphisms of Theorem 3.8.2 are represented. The main result is the following.

Proposition 3.8.3 *Let $f: A \to B$ be a G-map. Let δ be a familial dimension function for G, let γ be a virtual representation of $\Pi_G B$, let $\underline{S}$ be a covariant $\widehat{\Pi}_{G,\delta}B$-module, and let $\overline{T}$ be a contravariant $\widehat{\Pi}_{G,\delta}B$-module. Then we have stable equivalences*

$$f^* H^\delta \underline{S}_\gamma \simeq H^\delta(f^*\underline{S})_{f^*\gamma}$$

and

$$f^* H_\delta \overline{T}^\gamma \simeq H_\delta(f^*\overline{T})^{f^*\gamma}$$

Proof If $a: G/K \to A$ is a G-map, then

$$a^* f^* H_\delta \overline{T}^\gamma \simeq (fa)^* H_\delta \overline{T}^\gamma$$

is a nonparametrized Eilenberg-Mac Lane spectrum by the characterization given in the preceding section, with

$$\overline{\pi}^{K,\delta}_{f^*\gamma_0(a)}((f^* H_\delta \overline{T}^\gamma)_a) \cong \overline{\pi}^{K,\delta}_{\gamma_0(fa)}((H_\delta \overline{T}^\gamma)_{fa}) \cong (fa)^* \overline{T},$$

hence

$$\overline{\pi}^\delta_{f^*\gamma} f^* H_\delta \overline{T}^\gamma \cong f^* \overline{T}.$$

Further,

$$F(E\mathscr{F}(\delta)_+, f^* H_\delta \overline{T}^\gamma) \simeq f^* F(E\mathscr{F}(\delta)_+, H_\delta \overline{T}^\gamma) \simeq f^* H_\delta \overline{T}^\gamma.$$

Hence, $f^* H_\delta \overline{T}^\gamma \simeq H_\delta(f^*\overline{T})^{f^*\gamma}$ as claimed. The proof for covariant modules is essentially the same. □

The isomorphisms of Theorem 3.8.2 are then represented as follows:

$$\begin{aligned}
\tilde{H}^{G,\delta}_\gamma(f_! X; \underline{S}) &\cong [S, \rho_!(H^\delta \underline{S}_\gamma \wedge_B f_! X]_G \\
&\cong [S, \rho_! f_!(f^* H^\delta \underline{S}_\gamma \wedge_A X]_G \\
&\cong [S, \rho_!(H^\delta(f^*\underline{S})_{f^*\gamma} \wedge_A X]_G \\
&\cong \tilde{H}^{G,\delta}_{f^*\gamma}(X; f^*\underline{S})
\end{aligned}$$

and

$$\begin{aligned}\tilde{H}^{\gamma}_{G,\delta}(f_!X;\overline{T}) &\cong [f_!X, H_\delta\overline{T}^{\gamma}]_{G,B}\\ &\cong [X, f^*H_\delta\overline{T}^{\gamma}]_{G,A}\\ &\cong [X, H_\delta(f^*\overline{T})^{f^*\gamma}]_{G,A}\\ &\cong \tilde{H}^{f^*\gamma}_{G,\delta}(X;f^*\overline{T}).\end{aligned}$$

As for the push-forward map, if $\underline{S}$ is a covariant $\widehat{\Pi}_{G,\delta}A$-module, the push-forward is represented as

$$\begin{aligned}\tilde{H}^{G,\delta}_{f^*\gamma}(X;\underline{S}) &\cong [S, \rho_!(H^\delta\underline{S}_{f^*\gamma}\wedge_A X)]_G\\ &\cong [S, \rho_!f_!(H^\delta\underline{S}_{f^*\gamma}\wedge_A X)]_G\\ &\to [S, \rho_!f_!(H^\delta(f^*f_!\underline{S})_{f^*\gamma}\wedge_A X)]_G\\ &\cong [S, \rho_!f_!(f^*H^\delta(f_!\underline{S})_\gamma\wedge_A X)]_G\\ &\cong [S, \rho_!(H^\delta(f_!\underline{S})_\gamma\wedge_B f_!X)]_G\\ &\cong \tilde{H}^{G,\delta}_{\gamma}(f_!X;f_!\underline{S}).\end{aligned}$$

We could write down a similar composite in cohomology, but it's interesting to write it in a slightly different way. If $\overline{T}$ is a contravariant $\widehat{\Pi}_{G,\delta}A$-module, we have a map

$$f_!H_\delta\overline{T}^{f^*\gamma} \to H_\delta(f_!\overline{T})^{\gamma}$$

adjoint to the map

$$H_\delta\overline{T}^{f^*\gamma} \xrightarrow{\eta} H_\delta(f^*f_!\overline{T})^{f^*\gamma} \xrightarrow{\simeq} f^*H_\delta(f_!\overline{T})^{\gamma}.$$

The push-forward map in cohomology is then

$$\begin{aligned}\tilde{H}^{f^*\gamma}_{G,\delta}(X;\overline{T}) &\cong [X, H_\delta\overline{T}^{f^*\gamma}]_{G,A}\\ &\to [f_!X, f_!H_\delta\overline{T}^{f^*\gamma}]_{G,B}\\ &\to [f_!X, H_\delta(f_!\overline{T})^{\gamma}]_{G,B}\\ &\cong \tilde{H}^{\gamma}_{G,\delta}(f_!X;f_!\overline{T}).\end{aligned}$$

For homology, we have a similar map $f_! H^\delta \underline{S}_{f^*\gamma} \to H^\delta(f_!\underline{S})_\gamma$. We also use the map of spectra $f_!(E \wedge_A F) \to f_!E \wedge_B f_!F$ given by the composite

$$f_!(E \wedge_A F) \to f_!(f^* f_! E \wedge_A F) \simeq f_! E \wedge_B f_! F.$$

We can then write the push-forward in homology as

$$\begin{aligned} \tilde{H}^{G,\delta}_{f^*\gamma}(X;\underline{S}) &\cong [S, \rho_!(H^\delta \underline{S}_{f^*\gamma} \wedge_A X)]_G \\ &\cong [S, \rho_! f_!(H^\delta \underline{S}_{f^*\gamma} \wedge_A X)]_G \\ &\to [S, \rho_!(f_! H^\delta \underline{S}_{f^*\gamma} \wedge_B f_! X)]_G \\ &\to [S, \rho_!(H^\delta(f_!\underline{S})_\gamma \wedge_B f_! X)]_G \\ &\cong \tilde{H}^{G,\delta}_\gamma(f_! X; f_!\underline{S}). \end{aligned}$$

An interesting fact about this way of writing the push-forward is the following.

Proposition 3.8.4 *Let $f: A \to B$ be a G-map. Let δ be a familial dimension function for G, let γ be a virtual representation of $\Pi_G B$, let $\underline{S}$ be a covariant $\widehat{\Pi}_{G,\delta} A$-module, and let $\overline{T}$ be a contravariant $\widehat{\Pi}_{G,\delta} A$-module. Then we have*

$$\pi^{\mathscr{L}-\delta}_{-\gamma+n} f_! H^\delta \underline{S}_{f^*\gamma} \cong \begin{cases} f_!\underline{S} & \text{if } n = 0 \\ 0 & \text{if } n < 0. \end{cases}$$

Further, the map $f_! H^\delta \underline{S}_{f^*\gamma} \to H^\delta(f_!\underline{S})_\gamma$ induces an isomorphism in $\pi^{\mathscr{L}-\delta}_{-\gamma}$. If δ is complete, then we have

$$\pi^\delta_{\gamma+n} f_! H_\delta \overline{T}^{f^*\gamma} \cong \begin{cases} f_!\overline{T} & \text{if } n = 0 \\ 0 & \text{if } n < 0. \end{cases}$$

Further, the map $f_! H_\delta \overline{T}^{f^*\gamma} \to H_\delta(f_!\overline{T})^\gamma$ induces an isomorphism in π^δ_γ.

Proof The approach to calculating the homotopy groups is similar to that used in the proof of Proposition 1.14.16. Consider the case of $H^\delta \underline{S}_{f^*\gamma}$ and assume that it has been constructed as in Construction 3.7.1, so that we have a cofibration

$$R^\delta \underline{S}_{f^*\gamma} \to F^\delta \underline{S}_{f^*\gamma} \to C^\delta \underline{S}_{f^*\gamma}$$

which induces an exact sequence

$$\pi^{\mathscr{L}-\delta}_{-f^*\gamma} R^\delta \underline{S}_{f^*\gamma} \to \pi^{\mathscr{L}-\delta}_{-f^*\gamma} F^\delta \underline{S}_{f^*\gamma} \to \underline{S} \to 0.$$

Now, $F^\delta \underline{S}_{f^*\gamma}$ and $R^\delta \underline{S}_{f^*\gamma}$ are wedges of spheres and we have

$$f_!(G_+ \wedge_K S^{-f^*\gamma_0(a)-(\mathscr{L}-\delta)(G/K),a}) = G_+ \wedge_K S^{-\gamma_0(fa)-(\mathscr{L}-\delta)(G/K),fa},$$

which implies that

$$\overline{\pi}_{-\gamma}^{\mathscr{L}-\delta} f_! F^\delta \underline{S}_{f^*\gamma} \cong f_! \overline{\pi}_{-f^*\gamma}^{\mathscr{L}-\delta} F^\delta \underline{S}_{f^*\gamma}$$

and similarly for R^δ. Because the functor $f_!$ on spectra preserves cofibrations, we have the cofibration sequence

$$f_! R^\delta \underline{S}_{f^*\gamma} \to f_! F^\delta \underline{S}_{f^*\gamma} \to f_! C^\delta \underline{S}_{f^*\gamma}$$

Taking homotopy and using the isomorphism above, we get exact sequences

$$f_! \overline{\pi}_{-f^*\gamma+n}^{\mathscr{L}-\delta} R^\delta \underline{S}_{f^*\gamma} \to f_! \overline{\pi}_{-f^*\gamma+n}^{\mathscr{L}-\delta} F^\delta \underline{S}_{f^*\gamma} \to \overline{\pi}_{-\gamma+n}^{\mathscr{L}-\delta} f_! C^\delta \underline{S}_{f^*\gamma} \to 0$$

for $n \leq 0$. Using the fact that the algebraic functor $f_!$ is right exact, this implies that

$$\overline{\pi}_{-\gamma+n}^{\mathscr{L}-\delta} f_! C^\delta \underline{S}_{f^*\gamma} \cong \begin{cases} f_! \underline{S} & \text{if } n = 0 \\ 0 & \text{if } n < 0. \end{cases}$$

The spectrum $P^\delta \underline{S}_{f^*\gamma}$ is obtained from $C^\delta \underline{S}_{f^*\gamma}$ by killing higher homotopy groups, so we have the same calculation of the lower homotopy groups of $f_! P^\delta \underline{S}_{f^*\gamma}$. This implies the same calculation for

$$f_! H^\delta \underline{S}_{f^*\gamma} = f_!(P^\delta \underline{S}_{f^*\gamma} \wedge E\mathscr{F}(\delta)_+) \simeq f_! P^\delta \underline{S}_{f^*\gamma} \wedge E\mathscr{F}(\delta)_+.$$

Now consider the map

$$\overline{\pi}_{-\gamma}^{\mathscr{L}-\delta} f_! H^\delta \underline{S}_{f^*\gamma} \to \overline{\pi}_{-\gamma}^{\mathscr{L}-\delta} H^\delta (f_! \underline{S})_\gamma.$$

By the calculation above, we can build a model of $H^\delta(f_!\underline{S})_\gamma$ by starting with $f_! H^\delta(\underline{S})_{f^*\gamma}$ and killing higher homotopy groups. This gives us a map $f_! H^\delta(\underline{S})_{f^*\gamma} \to H^\delta(f_!\underline{S})_\gamma$ that is an isomorphism in $\overline{\pi}_{-\gamma}^{\mathscr{L}-\delta}$, but is it the map we're interested in? It's adjoint can be written as

$$H^\delta(\underline{S})_{f^*\gamma} \to f^* f_! H^\delta(\underline{S})_{f^*\gamma} \to f^* H^\delta(f_!\underline{S})_\gamma.$$

On applying $\overline{\pi}_{f^*\gamma}^{G,\delta}$, the first map is the unit $\eta\colon \underline{S} \to f^* f_! \underline{S}$ and the second is the identity, so this map of spectra coincides with the map $H^\delta(\underline{S})_{f^*\gamma} \to H^\delta(f^* f_! \underline{S})_{f^*\gamma}$ induced by η. Therefore, we are looking at the correct map and conclude that $f_! H^\delta(\underline{S})_{f^*\gamma} \to H^\delta(f_!\underline{S})_\gamma$ does give us an isomorphism on applying $\overline{\pi}_{-\gamma}^{\mathscr{L}-\delta}$, as claimed.

When δ is complete, the results for cohomology follow from the identity $H_\delta \overline{T}^\gamma = H^{\mathscr{L}-\delta}\overline{T}_{-\gamma}$. □

Remarks 3.8.5 The change-of-base space isomorphisms suggest two interesting alternatives to considering the homology of spaces over a specified base space.

(1) If $p: X \to B$ is a G-space over B, we have $p = p_!(\text{id})$ where $\text{id}: X \to X$ is X considered as a space over itself. Thus, if γ is a representation of $\Pi_G B$ and $\underline{S}$ is a covariant $\widehat{\Pi}_{G,\delta}B$-module, we have

$$H_\gamma^{G,\delta}(X;\underline{S}) \cong H_{p^*\gamma}^{G,\delta}(X;p^*\underline{S}),$$

where the group on the right considers X as a space over itself, and similarly in cohomology. In other words, the homology and cohomology of X as a space over B are intrinsic to X and not dependent on B at all. B is simply a convenient carrier for γ and $\underline{S}$. If X is an ex-G-space over B, we can say similarly that the reduced homology and cohomology of X can be described in terms of the pair $(X, \sigma(B))$ over X, whose dependence on B may be reduced further using excision.

This is the way we originally described $RO(\Pi X)$-graded homology and cohomology in, for example, [14]. Continuing in that vein, we can define homology and cohomology as being defined on a category consisting of triples $(X, \gamma, \underline{S})$ and compatible maps, where γ is a representation of ΠX and $\underline{S}$ is a $\widehat{\Pi}_{G,\delta}X$-module. We simply take $H_{\gamma+*}^{G,\delta}(X;\underline{S})$ as the homology of X considered as parametrized over itself, and similarly for cohomology. That we can extend this to a functor follows from the results of this section. We've chosen to recast the theories in terms of parametrized spaces for several reasons, one being that parametrized spectra seem to be the most natural representing objects.

(2) In the other direction, we could ask whether there is some universal base space we could choose to use rather than define theories for each choice of base space. Ignoring coefficient systems, we have a positive answer from Theorem 2.1.7: There is a classifying space $Bv\mathscr{V}_G(n)$ such that, if B has the G-homotopy type of a G-CW complex, then $[B, Bv\mathscr{V}_G(n)]_G$ is in one-to-one correspondence with the set of n-dimensional virtual representations of $\Pi_G B$. Put another way, the space $Bv\mathscr{V}_G(n)$ carries a universal n-dimensional virtual representation. To consider homology graded on n-dimensional representations, it suffices then to consider spaces over $Bv\mathscr{V}_G(n)$.

To do this properly we should consider the disjoint union of these spaces over all n and the structures on this union implied by suspension and, more generally, addition of representations. We should also look at the spectrum over this space representing homology or cohomology, and investigate what additional structure it has.

This does not address the issue of coefficient systems—the classifying space for representations will certainly not carry all the coefficient systems we may be interested in. So, we could ask whether there is a space carrying a universal coefficient system. However, the collection of possible coefficient systems on a space is not a set but a proper class, so without some restriction on the possible coefficient systems this obviously can't work. We have not investigated whether there is a classifying space under such a restriction or what other approach might work.

3.9 Change of Groups

3.9.1 Subgroups

Let $\iota: K \to G$ be the inclusion of a (closed) subgroup. In [51, §14.3], May and Sigurdsson define the forgetful functor

$$\iota^*: G\mathscr{P}_B \to K\mathscr{P}_B$$

and, for A a K-space, an induction functor

$$\iota_!: K\mathscr{P}_A \to G\mathscr{P}_{G\times_K A},$$

an analogue of the space level functor that takes a K-space Z to the G-space $G \times_K Z$. In fact, $\iota_!$ is an equivalence of categories with inverse $\nu^* \circ \iota^*$, where $\nu: A \to G \times_K A$ is the K-map $\nu(a) = [e, a]$. If $\epsilon: G \times_K B \to B$ is the map $\epsilon[g, b] = gb$, we let

$$i_K^G = \epsilon_! \iota_!: K\mathscr{P}_B \to G\mathscr{P}_B.$$

(Note that this makes sense only when B is a G-space.) Then i_K^G is left adjoint to ι^*: If X is a K-spectrum over B and E is a G-spectrum over B, then

$$[i_K^G X, E]_{G,B} \cong [X, (\iota_!)^{-1}\epsilon^* E]_{K,B} \cong [X, \iota^* E]_{K,B}.$$

The second isomorphism follows from the natural isomorphism $(\iota_!)^{-1}\epsilon^* \cong \nu^*\iota^*\epsilon^* \cong \iota^*$, the last isomorphism being a special case of [51, 14.3.3]. We shall also write $G_+ \wedge_K X$ for $i_K^G X$ and $E|K$ for $\iota^* E$. With this notation we have

$$[G_+ \wedge_K X, E]_{G,B} \cong [X, E|K]_{K,B}.$$

Definition 3.9.1 Let B be a G-space, let δ be a dimension function for G, and let K be a subgroup of G. Let

$$i_K^G = G_+ \wedge_K \Sigma^{-\delta(G/K)}(-): \widehat{\Pi}_{K,\delta} B \to \widehat{\Pi}_{G,\delta} B.$$

Its effect on an object $b\colon K/L \to B$ is

$$i_K^G(b) = G \times_K b\colon G/L = G \times_K (K/L) \to G \times_K B \to B$$

and it takes a stable map $f\colon K_+ \wedge_L S^{-\delta(K/L),b} \to K_+ \wedge_M S^{-\delta(K/M),c}$ to the stable map

$$\begin{aligned} G_+ \wedge_L S^{-\delta(G/L),b} &= G_+ \wedge_L S^{-\delta(K/L)-\delta(G/K),b} \\ &= G_+ \wedge_K \Sigma^{-\delta(G/K)}(K_+ \wedge_L S^{-\delta(K/L),b}) \\ &\to G_+ \wedge_K \Sigma^{-\delta(G/K)}(K_+ \wedge_M S^{-\delta(K/M),c}) \\ &= G_+ \wedge_M S^{-\delta(G/M),c}. \end{aligned}$$

If $\overline{T}$ is a contravariant $\widehat{\Pi}_{G,\delta}B$-module, let

$$\overline{T}|K = (i_K^G)^*\overline{T} = \overline{T} \circ i_K^G,$$

a $\widehat{\Pi}_{K,\delta}B$-module, as in Definition 1.6.5. If $\overline{C}$ is a contravariant $\widehat{\Pi}_{K,\delta}B$-module, let

$$G \times_K \overline{C} = (i_K^G)_!\overline{C},$$

a $\widehat{\Pi}_{G,\delta}B$-module.

It follows from Proposition 1.6.6 that

$$\mathrm{Hom}_{\widehat{\Pi}_{G,\delta}B}(G \times_K \overline{C}, \overline{T}) \cong \mathrm{Hom}_{\widehat{\Pi}_{K,\delta}B}(\overline{C}, \overline{T}|K)$$

for any contravariant $\widehat{\Pi}_{K,\delta}B$-module $\overline{C}$ and contravariant $\widehat{\Pi}_{G,\delta}B$-module $\overline{T}$. The same adjunction holds for covariant modules. We also have the isomorphisms

$$(G \times_K \overline{C}) \otimes_{\widehat{\Pi}_{G,\delta}B} \underline{S} \cong \overline{C} \otimes_{\widehat{\Pi}_{K,\delta}B} (\underline{S}|K)$$

if $\underline{S}$ is a covariant $\widehat{\Pi}_{G,\delta}B$-module, and

$$(G \times_K \underline{D}) \otimes_{\widehat{\Pi}_{G,\delta}B} \overline{T} \cong \underline{D} \otimes_{\widehat{\Pi}_{K,\delta}B} (\overline{T}|K)$$

if $\underline{D}$ is a covariant $\widehat{\Pi}_{K,\delta}B$-module. Finally, we have the calculations

$$G \times_K \overline{A}_b \cong \overline{A}_{G \times_K b}$$

and

$$G \times_K \underline{A}^b \cong \underline{A}^{G\times_K b}.$$

Here, $b\colon K/L \to B$, $\overline{A}_b = \widehat{\Pi}_{K,\delta}B(-,b)$, and $\underline{A}^b = \widehat{\Pi}_{K,\delta}B(b,-)$.

Definition 3.9.2 Let γ be a representation of $\Pi_G B$ and let K be a subgroup of G. We define $\gamma|K$ to be the representation of $\Pi_K B$ whose value on $b\colon K/L \to B$ is the restriction to K/L of $\gamma(G \times_K b)$ over G/L. Its values on morphisms is defined by restriction similarly.

We can now state the Wirthmüller isomorphisms on the chain level and for homology and cohomology. For simplicity, we write γ again for $\gamma|K$.

We first note the following analogue of Proposition 1.4.6, shown in the same way.

Proposition 3.9.3 *Let δ be a dimension function for G and let $K \in \mathscr{F}(\delta)$. If X is a δ-K-$(\gamma - \delta(G/K))$-cell complex, then $G \times_K X$ is a δ-G-γ-cell complex with corresponding cells. If X is a CW complex, then so is $G \times_K X$ with this structure.* □

Proposition 3.9.4 *Let δ be a dimension function for G, let $K \in \mathscr{F}(\delta)$, and let γ be a virtual representation of $\Pi_G B$. Let X be a based δ-K-CW$(\gamma - \delta(G/K))$ complex and give $G_+ \wedge_K X$ the δ-G-CW(γ) structure from the preceding proposition. Then*

$$G \times_K \overline{C}^{K,\delta}_{\gamma-\delta(G/K)+*}(X,B) \cong \overline{C}^{G,\delta}_{\gamma+*}(G_+ \wedge_K X, B).$$

This isomorphism respects suspension in the sense that, if W is a representation of G, then the following diagram commutes:

$$\begin{array}{ccc}
G \times_K \overline{C}^{K,\delta}_{\gamma-\delta(G/K)+*}(X,B) & \xrightarrow{\ \cong\ } & \overline{C}^{G,\delta}_{\gamma+*}(G_+ \wedge_K X, B) \\
\sigma^W \downarrow & & \downarrow \sigma^W \\
G \times_K \overline{C}^{K,\delta}_{\gamma-\delta(G/K)+W+*}(\Sigma^W X,B) & \xrightarrow{\ \cong\ } & \overline{C}^{G,\delta}_{\gamma+W+*}(G_+ \wedge_K \Sigma^W X, B)
\end{array}$$

The proof is the same as that of Proposition 1.13.2.

Theorem 3.9.5 (Wirthmüller Isomorphisms) *Let δ be a dimension function for G, let $K \in \mathscr{F}(\delta)$, and let γ be a virtual representation of $\Pi_G B$. Then, for X a δ-K-CW$(\gamma - \delta(G/K))$ complex, there are natural isomorphisms*

$$\tilde{H}^{G,\delta}_{\gamma}(G_+ \wedge_K X; \underline{S}) \cong \tilde{H}^{K,\delta}_{\gamma-\delta(G/K)}(X; \underline{S}|K)$$

and

$$\tilde{H}^{\gamma}_{G,\delta}(G_+ \wedge_K X; \overline{T}) \cong \tilde{H}^{\gamma-\delta(G/K)}_{K,\delta}(X; \overline{T}|K).$$

These isomorphisms respect suspension in the sense that, if W is a representation of G, then the following diagram commutes:

$$\begin{array}{ccc} \tilde{H}^{G,\delta}_{\gamma}(G_+ \wedge_K X; \underline{S}) & \xrightarrow{\cong} & \tilde{H}^{K,\delta}_{\gamma-\delta(G/K)}(X; \underline{S}|K) \\ \sigma^W \downarrow & & \downarrow \sigma^W \\ \tilde{H}^{G,\delta}_{\gamma+W}(G_+ \wedge_K \Sigma^W X; \underline{S}) & \xrightarrow{\cong} & \tilde{H}^{K,\delta}_{\gamma+W-\delta(G/K)}(\Sigma^W X; \underline{S}|K) \end{array}$$

and similarly for cohomology. If δ is familial, for X an ex-K-space there are natural isomorphisms

$$\tilde{H}^{G,\delta}_{\gamma}(G_+ \wedge_K X; \underline{S}) \cong \tilde{H}^{K,\delta}_{\gamma-\delta(G/K)}(X; \underline{S}|K)$$

and

$$\tilde{H}_{G,\delta}^{\gamma}(G_+ \wedge_K X; \overline{T}) \cong \tilde{H}_{K,\delta}^{\gamma-\delta(G/K)}(X; \overline{T}|K).$$

These isomorphisms respect suspension in the sense above.

The proof is essentially the same as that of Theorem 1.13.3.

Definition 2.7.3 allows us to consider cellular homology and cohomology to be $\widehat{\Pi}_\delta B$-module-valued, with

$$\underline{H}^{G,\delta}_{\gamma}(X; \underline{S})(b: G/K \to B) = \tilde{H}^{G,\delta}_{\gamma}(X \wedge G_+ \wedge_K S^{-\delta(G/K),b}; \overline{S})$$

and

$$\overline{H}_{G,\delta}^{\gamma}(X; \overline{T})(b: G/K \to B) = \tilde{H}_{G,\delta}^{\gamma}(X \wedge G_+ \wedge_K S^{-\delta(G/K),b}; \overline{T}).$$

We then get the following analogue of Corollary 1.13.4, whose proof needs only the nonparametrized version of the Wirthmüller isomorphism. This result shows that the module-valued theories capture the local behavior of the theories on the fibers, not the global values.

Corollary 3.9.6 *If X is an ex-G-space, δ is a familial dimension function for G, $\underline{S}$ is a covariant $\widehat{\Pi}_\delta B$-module, and $\overline{T}$ is a contravariant $\widehat{\Pi}_\delta B$-module, then, for $K \in \mathscr{F}(\delta)$, we have*

$$\underline{H}^{G,\delta}_{\gamma}(X; \underline{S})(b: G/K \to B) \cong \tilde{H}^{K,\delta}_{\gamma_0(b)}(b^* X; b^* \underline{S})$$

and

$$\overline{H}^{\gamma}_{G,\delta}(X;\overline{T})(b\colon G/K \to B) \cong \tilde{H}^{\gamma_0(b)}_{K,\delta}(b^*X; b^*\overline{T}|K),$$

where we implicitly identify $G\mathscr{P}_{G/K}$ with $K\mathscr{P}$. Recall that b^*X must be understood as the derived functor, the homotopy pullback.

Proof We have

$$\begin{aligned}
\underline{H}^{G,\delta}_{\gamma}(X;\underline{S})(b\colon G/K \to B) &= \tilde{H}^{G,\delta}_{\gamma}(X \wedge G_+ \wedge_K S^{-\delta(G/K),b}; \overline{S}) \\
&= \tilde{H}^{G,\delta}_{\gamma}(X \wedge_B b_!(G \times_K S^{-\delta(G/K)}); \overline{S}) \\
&\cong \tilde{H}^{G,\delta}_{\gamma}(b_!(b^*X \wedge_{G/K} (G \times_K S^{-\delta(G/K)}); \overline{S}) \\
&\cong \tilde{H}^{G,\delta}_{\gamma(b)}(b^*X \wedge_{G/K} (G \times_K S^{-\delta(G/K)}); b^*\overline{S}) \\
&\cong \tilde{H}^{K,\delta}_{\gamma_0(b)}(b^*X; b^*\overline{S}).
\end{aligned}$$

The proof for cohomology is similar. □

Corollary 3.9.7 *Let δ be a familial dimension function for G and let $K \in \mathscr{F}(\delta)$. If X is an ex-G-space, $\underline{S}$ is a covariant $\widehat{\Pi}_\delta B$-module, and $\overline{T}$ is a contravariant $\widehat{\Pi}_\delta B$-module, then we have*

$$\underline{H}^{G,\delta}_{\gamma}(X;\underline{S})|K \cong \underline{H}^{K,\delta}_{\gamma}(X;\underline{S}|K)$$

and

$$\overline{H}^{\gamma}_{G,\delta}(X;\overline{T})|K \cong \overline{H}^{\gamma}_{K,\delta}(X;\overline{T}|K).$$

□

On the spectrum level, what underlies the Wirthmüller isomorphisms is the following fact.

Proposition 3.9.8 *Let $H_\delta\overline{T}^{\gamma}$ be a parametrized Eilenberg-MacLane spectrum and let K be a subgroup of G. Then*

$$(H_\delta\overline{T}^{\gamma})|K \simeq \Sigma_K^{-\delta(G/K)} H_\delta(\overline{T}|K)^{\gamma}.$$

Similarly, for a covariant $\underline{S}$, we have

$$(H^\delta\underline{S}_{\gamma})|K \simeq \Sigma_K^{-(\mathscr{L}(G/K)-\delta(G/K))} H^\delta(\underline{S}|K)_{\gamma}.$$

Proof We've already mentioned that parametrized Eilenberg-Mac Lane spectra are characterized by their fibers being non-parametrized Eilenberg-Mac Lane spectra. This proposition then follows from Proposition 1.13.6. □

On the represented level, the cohomology Wirthmüller isomorphism is then the adjunction

$$\begin{aligned}[G_+ \wedge_K \Sigma_B^\infty X, \Sigma_B^V H_\delta \overline{T}^\gamma]_{G,B} &\cong [\Sigma_B^\infty X, \Sigma_B^V (H_\delta \overline{T}^\gamma)|K]_{K,B} \\ &\cong [\Sigma_B^\infty X, \Sigma_B^{V-\delta(G/K)} H_\delta (\overline{T}|K)^\gamma]_{K,B}.\end{aligned}$$

For homology, we have

$$\begin{aligned}&[S^V, \rho_!(H^\delta \underline{S}_\gamma \wedge_B (G_+ \wedge_K X))]_G \\ &\quad\cong [S^V, G_+ \wedge_K \rho_!((H^\delta \underline{S}_\gamma)|K \wedge_B X)]_G \\ &\quad\cong [S^V, \Sigma^{\mathscr{L}(G/K)} \rho_!((H^\delta \underline{S}_\gamma)|K \wedge_B X)]_K \\ &\quad\cong [S^V, \rho_!(\Sigma^{\mathscr{L}(G/K)} \Sigma^{-(\mathscr{L}(G/K)-\delta(G/K))} H^\delta (\underline{S}|K)_\gamma \wedge_B X)]_K \\ &\quad= [S^V, \rho_!(\Sigma^{\delta(G/K)} H^\delta (\underline{S}|K)_\gamma \wedge_B X)]_K \\ &\quad\cong [S^{V-\delta(G/K)}, \rho_!(H^\delta (\underline{S}|K)_\gamma \wedge_B X)]_K.\end{aligned}$$

As in the non-parametrized case, we can use the Wirthmüller isomorphism to define restriction to subgroups. In the case of cohomology, we use the composite

$$\tilde{H}^\gamma_{G,\delta}(X; \overline{T}) \to \tilde{H}^\gamma_{G,\delta}(G/K_+ \wedge X; \overline{T}) \cong \tilde{H}^{\gamma-\delta(G/K)}_{K,\delta}(X; \overline{T}|K).$$

Here, the first map is induced by the projection $G/K_+ \wedge X \to X$ and the second is the Wirthmüller isomorphism.

In homology, we take an embedding of G/K in a representation V, with $c\colon S^V \to G_+ \wedge_K S^{V-\mathscr{L}(G/K)}$ being the induced collapse map, and then use the composite

$$\begin{aligned}\tilde{H}^{G,\delta}_\gamma(X; \underline{S}) &\cong \tilde{H}^{G,\delta}_{\gamma+V}(\Sigma^V X; \underline{S}) \\ &\to \tilde{H}^{G,\delta}_{\gamma+V}(G_+ \wedge_K \Sigma^{V-\mathscr{L}(G/K)} X; \underline{S}) \\ &\cong \tilde{H}^{K,\delta}_{\gamma+V-\delta(G/K)}(\Sigma^{V-\mathscr{L}(G/K)} X; \underline{S}|K) \\ &\cong \tilde{H}^{K,\delta}_{\gamma+\mathscr{L}(G/K)-\delta(G/K)}(X; \underline{S}|K).\end{aligned}$$

As before, we write $a \mapsto a|K$ for this restriction in either cohomology or homology.

On the represented level, these maps are given simply by restriction of stable G-maps to stable K-maps. Explicitly, the cohomology restriction is the following:

$$\begin{aligned}\tilde{H}^{\gamma}_{G,\delta}(X;\overline{T}) &\cong [\Sigma^{\infty}_{B}X, H_{\delta}\overline{T}^{\gamma}]_{G,B}\\ &\to [\Sigma^{\infty}_{B}X, (H_{\delta}\overline{T}^{\gamma})|K]_{K,B}\\ &\cong [\Sigma^{\infty}_{B}X, \Sigma^{-\delta(G/K)}H_{\delta}(\overline{T}|K)^{\gamma}]_{K,B}\\ &\cong \tilde{H}^{\gamma-\delta(G/K)}_{K,\delta}(X;\overline{T}|K).\end{aligned}$$

The homology restriction is the following:

$$\begin{aligned}\tilde{H}^{G,\delta}_{\gamma}(X;\underline{S}) &\cong [S, \rho_!(H^{\delta}\underline{S}_{\gamma} \wedge_B X)]_G\\ &\to [S, \rho_!((H^{\delta}\underline{S}_{\gamma})|K \wedge_B X)]_K\\ &\cong [S, \rho_!(\Sigma^{-(\mathscr{L}(G/K)-\delta(G/K)}H^{\delta}(\underline{S}|K)_{\gamma} \wedge_B X)]_K\\ &\cong [S^{\mathscr{L}(G/K)-\delta(G/K)}, \rho_!(H^{\delta}(\underline{S}|K)_{\gamma} \wedge_B X)]_K\\ &\cong \tilde{H}^{K,\delta}_{\gamma+\mathscr{L}(G/K)-\delta(G/K)}(X;\underline{S}|K).\end{aligned}$$

Remark 3.9.9 The specializations of these maps to the ordinary and dual theories give us the following restriction maps:

$$\begin{aligned}&\tilde{H}^{G}_{\gamma}(X;\underline{S}) \to \tilde{H}^{K}_{\gamma+\mathscr{L}(G/K)}(X;\underline{S}|K) \qquad &&\tilde{\mathscr{H}}^{\gamma}_{G}(X;\underline{S}) \to \tilde{\mathscr{H}}^{\gamma-\mathscr{L}(G/K)}_{K}(X;\underline{S}|K)\\ &\tilde{H}^{\gamma}_{G}(X;\overline{T}) \to \tilde{H}^{\gamma}_{K}(X;\overline{T}|K) &&\tilde{\mathscr{H}}^{G}_{\gamma}(X;\overline{T}) \to \tilde{\mathscr{H}}^{K}_{\gamma}(X;\overline{T}|K)\end{aligned}$$

3.9.2 *Quotient Groups*

Let N be a normal subgroup of G and let $\epsilon\colon G \to G/N$ denote the quotient map. See [51, §14.4] for some of the functors on parametrized spectra we use here.

Similar to Proposition 1.13.9, if δ is a dimension function for G, A is a G/N-space, γ is a virtual representation of $\Pi_{G/N}A$, and Y is a based δ-G/N-CW(γ) complex over A, then ϵ^*Y is a based δ-G-CW(γ) complex over ϵ^*A. We will usually not write ϵ^* when the meaning is clear without it.

Similar to Proposition 1.13.10, if B is a G-space, γ is a virtual representation of $\Pi_G B$, and X is a based δ-G-CW(γ) complex over B, then X^N is a based δ-G/N-CW(γ^N) complex over B^N.

To describe the associated algebra on chains, let

$$\theta\colon \widehat{\Pi}_{G/N,\delta}A \to \widehat{\Pi}_{G,\delta}A$$

be the restriction of $\epsilon^\sharp$, the functor that assigns G/N-spectra over A the induced G action and expands from an N-trivial universe to a complete G-universe.

If E is a G-spectrum over a G-space B, then E^N is a G/N-spectrum over B^N. We define the geometric fixed-point construction

$$\Phi^N(E) = (\tilde{E}\mathscr{F}[N] \wedge E)^N$$

as in Definition 1.13.12. In this context, Φ^N continues to enjoy the properties listed after that definition. As in Remark 1.6.7, consider $\widehat{\Pi}_{G,\delta}B$ as augmented with a zero object $*$ given by the trivial spectrum. We write

$$\Phi^N\colon \widehat{\Pi}_{G,\delta}B \to \widehat{\Pi}_{G/N,\delta}B^N$$

for the restriction of Φ^N to the augmented stable fundamental groupoid.

Definition 3.9.10 If $\underline{S}$ is a covariant and $\overline{T}$ a contravariant $\widehat{\Pi}_{G,\delta}B$-module, let

$$\underline{S}^N = \Phi^N_!\underline{S} \quad \text{and}$$
$$\overline{T}^N = \Phi^N_!\overline{T}.$$

We call this the *N-fixed point functor.* If $\underline{U}$ is a covariant and $\overline{V}$ is a contravariant $\widehat{\Pi}_{G/N,\delta}B^N$-module, let

$$\mathrm{Inf}^G_{G/N}\underline{U} = (\Phi^N)^*\underline{U} \quad \text{and}$$
$$\mathrm{Inf}^G_{G/N}\overline{V} = (\Phi^N)^*\overline{V}.$$

We call this the *inflation functor.*

For $b\colon G/L \to B$, write

$$\underline{A}^b_G = \widehat{\Pi}_{G,\delta}B(b,-)$$

and

$$\overline{A}^G_b = \widehat{\Pi}_{G,\delta}B(-,b).$$

We have the following analogue of Proposition 1.13.15.

Proposition 3.9.11 *If N is normal in G and b: G/L → B, then*

$$(\underline{A}_G^b)^N \cong \begin{cases} \underline{A}_{G/N}^b & \text{if } N \le L \\ 0 & \text{if } N \not\le L \end{cases}$$

and

$$(\overline{A}_b^G)^N \cong \begin{cases} \overline{A}_b^{G/N} & \text{if } N \le L \\ 0 & \text{if } N \not\le L. \end{cases}$$

If B is a G/N-space, $N \le L \le G$, and $b: (G/N)/(G/L) \to B$, then

$$\theta_! \underline{A}_{G/N}^b \cong \underline{A}_G^b$$

and

$$\theta_! \overline{A}_b^{G/N} \cong \overline{A}_b^G,$$

where we write b again for $\epsilon^* b: G/L \to B$.

Proof These are special cases of Proposition 1.6.6(5). □

We now have the following chain-level calculations, proved in the same way as Propositions 1.13.16 and 1.13.17.

Proposition 3.9.12 *Let N be a normal subgroup of G, let δ be a dimension function for G, let B be a G/N-space, and let γ be a virtual representation of $\Pi_{G/N}B$. Let Y be a based δ-G/N-CW(γ) complex over B and give $\epsilon^* Y$ the corresponding δ-G-CW(γ) structure. Then we have a natural chain isomorphism*

$$\theta_! \overline{C}_{\gamma+*}^{G/N,\delta}(Y,*) \cong \overline{C}_{\gamma+*}^{G,\delta}(Y,*),$$

writing Y for $\epsilon^* Y$ on the right as usual. This isomorphism respects suspension in the sense that, if W is a representation of G/N, then the following diagram commutes:

$$\begin{array}{ccc} \theta_! \overline{C}_{\gamma+*}^{G/N,\delta}(Y,*) & \xrightarrow{\cong} & \overline{C}_{\gamma+*}^{G,\delta}(Y,*) \\ {\scriptstyle \sigma^W}\big\downarrow & & \big\downarrow{\scriptstyle \sigma^W} \\ \theta_! \overline{C}_{\gamma+W+*}^{G/N,\delta}(\Sigma^W Y,*) & \xrightarrow[\cong]{} & \overline{C}_{\gamma+W+*}^{G,\delta}(\Sigma^W Y,*). \end{array}$$

□

Proposition 3.9.13 *Let N be a normal subgroup of G, let δ be a dimension function for G, let B be a G-space, and let γ be a virtual representation of $\Pi_G B$. Let X be a based δ-G-CW(γ) complex over B and give X^N the corresponding δ-G/N-CW(γ^N) structure. With this structure, we have a natural chain isomorphism*

$$\overline{C}^{G,\delta}_{\gamma+*}(X,*)^N \cong \overline{C}^{G/N,\delta}_{\gamma^N+*}(X^N,*).$$

This isomorphism respects suspension in the sense that, if W is a representation of G, then the following diagram commutes:

$$\begin{array}{ccc} \overline{C}^{G,\delta}_{\gamma+*}(X,*)^N & \xrightarrow{\cong} & \overline{C}^{G/N,\delta}_{\gamma^N+*}(X^N,*) \\ {\scriptstyle (\sigma^W)^N}\downarrow & & \downarrow{\scriptstyle \sigma^{W^N}} \\ \overline{C}^{G,\delta}_{\gamma+W+*}(\Sigma^W X,*)^N & \xrightarrow[\cong]{} & \overline{C}^{G/N,\delta}_{\gamma^N+W^N+*}(\Sigma^{W^N} X^N,*) \end{array}$$

□

The chain isomorphisms lead to the following analogues of Theorems 1.13.20 and 1.13.22, with similar proofs.

Theorem 3.9.14 *Let N be a normal subgroup of G, let δ be a dimension function for G, let B be a G/N-space, let γ be a virtual representation of $\Pi_{G/N}B$, let $\underline{S}$ be a covariant $\widehat{\Pi}_{G,\delta}B$-module, and let $\overline{T}$ be a contravariant $\widehat{\Pi}_{G,\delta}B$-module. Then, for Y a δ-G/N-CW(γ) complex over B, we have natural isomorphisms*

$$\tilde{H}^{G,\delta}_\gamma(Y;\underline{S}) \cong \tilde{H}^{G/N,\delta}_\gamma(Y;\theta^*\underline{S}) \qquad \text{and}$$
$$\tilde{H}^\gamma_{G,\delta}(Y;\overline{T}) \cong \tilde{H}^\gamma_{G/N,\delta}(Y;\theta^*\overline{T}).$$

These isomorphisms respect suspension in the sense that, if W is a representation of G/N, then the following diagram commutes:

$$\begin{array}{ccc} \tilde{H}^{G,\delta}_\gamma(Y;\underline{S}) & \xrightarrow{\cong} & \tilde{H}^{G/N,\delta}_\gamma(Y;\theta^*\underline{S}) \\ {\scriptstyle \sigma^W}\downarrow & & \downarrow{\scriptstyle \sigma^W} \\ \tilde{H}^{G,\delta}_{\gamma+W}(\Sigma^W Y;\underline{S}) & \xrightarrow[\cong]{} & \tilde{H}^{G/N,\delta}_{\gamma+W}(\Sigma^W Y;\theta^*\underline{S}) \end{array}$$

and similarly for cohomology. If δ is N-closed (as in Definition 1.13.18) and familial, then, for Y an ex-G/N-space over B, we have natural isomorphisms

$$\tilde{H}_\gamma^{G,\delta}(Y;\underline{S}) \cong \tilde{H}_\gamma^{G/N,\delta}(Y;\theta^*\underline{S}) \quad \text{and}$$
$$\tilde{H}^\gamma_{G,\delta}(Y;\overline{T}) \cong \tilde{H}^\gamma_{G/N,\delta}(Y;\theta^*\overline{T}).$$

These isomorphisms respect suspension in the sense that, if Y is well-based and W is a representation of G/N, then the following diagram commutes:

$$\begin{array}{ccc} \tilde{H}_\gamma^{G,\delta}(Y;\underline{S}) & \xrightarrow{\cong} & \tilde{H}_\gamma^{G/N,\delta}(Y;\theta^*\underline{S}) \\ \sigma^W \downarrow & & \downarrow \sigma^W \\ \tilde{H}_{\gamma+W}^{G,\delta}(\Sigma^W Y;\underline{S}) & \xrightarrow[\cong]{} & \tilde{H}_{\gamma+W}^{G/N,\delta}(\Sigma^W Y;\theta^*\underline{S}) \end{array}$$

and similarly for cohomology. □

Theorem 3.9.15 *Let δ be a dimension function for G, let N be a normal subgroup of G, let B be a G-space, let $\underline{S}$ be a covariant $\widehat{\Pi}_{G/N,\delta}B^N$-module, and let $\overline{T}$ be a contravariant $\widehat{\Pi}_{G/N,\delta}B^N$-module. Then, for X a based δ-G-CW(γ) complex over B, we have natural isomorphisms*

$$\tilde{H}_\gamma^{G,\delta}(X;\mathrm{Inf}_{G/N}^G\,\underline{S}) \cong \tilde{H}_{\gamma^N}^{G/N,\delta}(X^N;\underline{S}) \quad \text{and}$$
$$\tilde{H}^\gamma_{G,\delta}(X;\mathrm{Inf}_{G/N}^G\,\overline{T}) \cong \tilde{H}^{\gamma^N}_{G/N,\delta}(X^N;\overline{T}).$$

These isomorphisms respect suspension in the sense that, if W is a representation of G, then the following diagram commutes:

$$\begin{array}{ccc} \tilde{H}_\gamma^{G,\delta}(X;\mathrm{Inf}_{G/N}^G\,\underline{S}) & \xrightarrow{\cong} & \tilde{H}_{\gamma^N}^{G/N,\delta}(X^N;\underline{S}) \\ \sigma^W \downarrow & & \downarrow \sigma^{W^N} \\ \tilde{H}_{\gamma+W}^{G,\delta}(\Sigma^W X;\mathrm{Inf}_{G/N}^G\,\underline{S}) & \xrightarrow[\cong]{} & \tilde{H}_{\gamma^N+W^N}^{G/N,\delta}(\Sigma^{W^N} X^N;\underline{S}) \end{array}$$

and similarly for cohomology. If δ is familial, then, for X an ex-G-space over B, we have natural isomorphisms

$$\tilde{H}_\gamma^{G,\delta}(X;\mathrm{Inf}_{G/N}^G\,\underline{S}) \cong \tilde{H}_{\gamma^N}^{G/N,\delta}(X^N;\underline{S}) \quad \text{and}$$
$$\tilde{H}^\gamma_{G,\delta}(X;\mathrm{Inf}_{G/N}^G\,\overline{T}) \cong \tilde{H}^{\gamma^N}_{G/N,\delta}(X^N;\overline{T}).$$

These isomorphisms respect suspension in the sense that, if X is well-based and W is a representation of G, then the following diagram commutes:

$$\begin{array}{ccc}
\tilde{H}^{G,\delta}_{\gamma}(X;\mathrm{Inf}^G_{G/N}\underline{S}) & \xrightarrow{\cong} & \tilde{H}^{G/N,\delta}_{\gamma^N}(X^N;\underline{S}) \\
\downarrow{\scriptstyle \sigma^W} & & \downarrow{\scriptstyle \sigma^{W^N}} \\
\tilde{H}^{G,\delta}_{\gamma+W}(\Sigma^W X;\mathrm{Inf}^G_{G/N}\underline{S}) & \xrightarrow[\cong]{} & \tilde{H}^{G/N,\delta}_{\gamma^N+W^N}(\Sigma^{W^N} X^N;\underline{S})
\end{array}$$

and similarly for cohomology. □

Remarks 3.9.16

(1) As in the nonparametrized case, these isomorphisms are compatible, in the sense that, if B is a G/N-space, γ is a virtual representation of $\Pi_{G/N}B$, X is an ex-G/N-space over B, and $\overline{T}$ is a $\widehat{\Pi}_{G/N,\delta}B$-module, then the following composite is the identity:

$$\begin{aligned}
\tilde{H}_{G/N,\delta}^{\gamma}(X;\overline{T}) &= \tilde{H}_{G/N,\delta}^{\gamma^N}(X^N;\overline{T}) \\
&\cong \tilde{H}_{G,\delta}^{\gamma}(X;\mathrm{Inf}^G_{G/N}\overline{T}) \\
&\cong \tilde{H}_{G/N,\delta}^{\gamma}(X;\theta^*\mathrm{Inf}^G_{G/N}\overline{T}) \\
&\cong \tilde{H}_{G/N,\delta}^{\gamma}(X;\overline{T}).
\end{aligned}$$

(Here, we use the isomorphism $\theta^*\mathrm{Inf}^G_{G/N}\overline{T} \cong \overline{T}$ that we noted first in the nonparametrized case.) The similar statement for homology is also true.

(2) The two isomorphisms combine to give a third isomorphism, if B is a G-space, γ is a virtual representation of $\Pi_G B$, X is an ex-G-space over B, and $\overline{T}$ is a $\widehat{\Pi}_{G/N,\delta}B^N$-module:

$$\begin{aligned}
\tilde{H}_{G,\delta}^{\gamma}(X;\mathrm{Inf}^G_{G/N}\overline{T}) &\cong \tilde{H}_{G/N,\delta}^{\gamma^N}(X^N;\overline{T}) \\
&\cong \tilde{H}_{G/N,\delta}^{\gamma^N}(X^N;\theta^*\mathrm{Inf}^G_{G/N}\overline{T}) \\
&\cong \tilde{H}_{G,\delta}^{\gamma^N}(X^N;\mathrm{Inf}^G_{G/N}\overline{T}).
\end{aligned}$$

In the last group, we consider X^N as an ex-G-space over B^N. We get a similar isomorphism in homology.

We can now define induction and restriction to fixed sets.

Definition 3.9.17 Let δ be an N-closed familial dimension function for G, let B be a G/N-space, let γ be a virtual representation of $\Pi_{G/N}B$, let $\underline{S}$ be a covariant $\widehat{\Pi}_{G/N,\delta}B$-module, and let $\overline{T}$ be a contravariant $\widehat{\Pi}_{G/N,\delta}B$-module. If Y is an ex-G/N-space over B, we define *induction from G/N to G* to be the composites

$$\epsilon^*: \tilde{H}_\gamma^{G/N,\delta}(Y;\underline{S}) \to \tilde{H}_\gamma^{G/N,\delta}(Y;\theta^*\theta_!\underline{S}) \cong \tilde{H}_\gamma^{G,\delta}(Y;\theta_!\underline{S})$$

and

$$\epsilon^*: \tilde{H}^\gamma_{G/N,\delta}(Y;\overline{T}) \to \tilde{H}^\gamma_{G/N,\delta}(Y;\theta^*\theta_!\overline{T}) \cong \tilde{H}^\gamma_{G,\delta}(Y;\theta_!\overline{T}).$$

The first map in each case is induced by the unit of the $\theta_!$-θ^* adjunction.

Definition 3.9.18 Let δ be a familial dimension function for G, let B be a G-space, let γ be a virtual representation of $\Pi_G B$, let $\underline{S}$ be a covariant $\widehat{\Pi}_{G,\delta}B$-module, and let $\overline{T}$ be a contravariant $\widehat{\Pi}_{G,\delta}B$-module If X is an ex-G-space over B and N is a normal subgroup of G, define *restriction to fixed sets* to be the composites

$$(-)^N: \tilde{H}_\gamma^{G,\delta}(X;\underline{S}) \to \tilde{H}_\gamma^{G,\delta}(X;\mathrm{Inf}_{G/N}^G\,\underline{S}^N) \xrightarrow{\cong} \tilde{H}_{\gamma^N}^{G/N,\delta}(X^N;\underline{S}^N)$$

and

$$(-)^N: \tilde{H}^\gamma_{G,\delta}(X;\overline{T}) \to \tilde{H}^\gamma_{G,\delta}(X;\mathrm{Inf}_{G/N}^G\,\overline{T}^N) \xrightarrow{\cong} \tilde{H}^{\gamma^N}_{G/N,\delta}(X^N;\overline{T}^N).$$

The first map in each case is induced by the unit of the $(-)^N$-$\mathrm{Inf}_{G/N}^G$ adjunction.

As in the nonparametrized case, Theorems 3.9.14 and 3.9.15 imply that both induction and restriction to fixed sets respect suspension. We also have that induction followed by restriction to fixed sets is the identity.

On the represented level, we get the following results.

Proposition 3.9.19 *Let N be a normal subgroup of G, let δ be an N-closed familial dimension function for G, let B be a G/N-space, let γ be a virtual representation of $\Pi_{G/N}B$, let $\underline{S}$ be a covariant $\widehat{\Pi}_{G,\delta}B$-module and let $\overline{T}$ be a contravariant $\widehat{\Pi}_{G,\delta}B$-module. Then*

$$(H^\delta \underline{S}_\gamma)^N \simeq H^\delta(\theta^*\underline{S})_\gamma$$

and

$$(H_\delta \overline{T}^\gamma)^N \simeq H_\delta(\theta^*\overline{T})^\gamma.$$

Proof This follows from Proposition 1.13.28 applied fiberwise. For example, if $N \leq L \leq G$ and $b\colon G/L \to B$, we have

$$\begin{aligned} b^*((H_\delta \overline{T}^\gamma)^N) &\simeq (b^* H_\delta \overline{T}^\gamma)^N \\ &\simeq (\Sigma^{\gamma_0(b)} H_\delta(b^* \overline{T}))^N \\ &\simeq \Sigma^{\gamma_0(b)} H_\delta(\theta^* b^* \overline{T}) \\ &\simeq \Sigma^{\gamma_0(b)} H_\delta(b^* \theta^* \overline{T}) \\ &\simeq b^*(H_\delta(\theta^* \overline{T})^\gamma). \end{aligned}$$

The rest of the characterization of Eilenberg-MacLane spectra goes as in the proof of Proposition 1.13.28. □

The isomorphisms of Theorem 3.9.14 are then represented as follows, for Y an ex-G/N-space over the G/N-space B:

$$\begin{aligned} [S, \rho_!(H^\delta \underline{S}_\gamma \wedge_B Y)]_G &\cong [S, (\rho_!(H^\delta \underline{S}_\gamma \wedge_B Y))^N]_{G/N} \\ &\cong [S, \rho_!((H^\delta \underline{S}_\gamma \wedge_B Y)^N)]_{G/N} \\ &\cong [S, \rho_!((H^\delta \underline{S}_\gamma)^N \wedge_B Y)]_{G/N} \\ &\cong [S, \rho_!(H^\delta(\theta^* \underline{S})_\gamma \wedge_B Y)]_{G/N} \end{aligned}$$

for homology, where the second isomorphism comes from [51, 14.4.4], and

$$\begin{aligned} [\Sigma_B^\infty Y, H_\delta \overline{T}^\gamma]_{G,B} &\cong [\Sigma_B^\infty Y, (H_\delta \overline{T}^\gamma)^N]_{G/N,B} \\ &\cong [\Sigma_B^\infty Y, H_\delta(\theta^* \overline{T})^\gamma]_{G/N,B} \end{aligned}$$

for cohomology.

Proposition 3.9.20 *Let δ be a familial dimension function for G, let N be a normal subgroup of G, let B be a G-space, let γ be a virtual representation of $\Pi_G B$, let $\overline{T}$ be a contravariant $\widehat{\Pi}_{G/N,\delta} B^N$-module, and let $\underline{S}$ be a covariant $\widehat{\Pi}_{G/N,\delta} B^N$-module. If $i\colon B^N \to B$ is the inclusion, then*

$$i_!(\epsilon^\sharp H^\delta \underline{S}_{\gamma^N}) \wedge \tilde{E}\mathscr{F}[N] \simeq H^\delta(\mathrm{Inf}_{G/N}^G \underline{S})_\gamma$$

and

$$i_!(\epsilon^\sharp H_\delta \overline{T}^{\gamma^N}) \wedge \tilde{E}\mathscr{F}[N] \simeq H_\delta(\mathrm{Inf}_{G/N}^G \overline{T})^\gamma.$$

Therefore,

$$(H^\delta(\mathrm{Inf}^G_{G/N}\underline{S})_\gamma)^N \simeq \Phi^N H^\delta(\mathrm{Inf}^G_{G/N}\underline{S})_\gamma \simeq H^\delta \underline{S}_{\gamma^N}$$

and

$$(H_\delta(\mathrm{Inf}^G_{G/N}\overline{T})^\gamma)^N \simeq \Phi^N H_\delta(\mathrm{Inf}^G_{G/N}\overline{T})^\gamma \simeq H_\delta \overline{T}^{\gamma^N}.$$

Proof If $b\colon G/K \to B$ with $N \not\leq K$, then $(\mathrm{Inf}^G_{G/N}\underline{S})(b) = 0$. Combined with the equivalence

$$S^{\gamma(b)} \wedge \tilde{E}\mathscr{F}[N] \simeq S^{\gamma(b)^N} \wedge \tilde{E}\mathscr{F}[N],$$

the first equivalence follows from Proposition 1.13.29 applied fiberwise. A similar argument applies to the second. These imply the last two equivalences as in the proof of Proposition 1.13.29. □

The isomorphisms of Theorem 3.9.15 are then represented as

$$\begin{aligned}
[S, \rho_!(H^\delta(\mathrm{Inf}^G_{G/N}\underline{S})_\gamma \wedge_B X)]_G &\cong [S, \rho_!(H^\delta(\mathrm{Inf}^G_{G/N}\underline{S})_\gamma \wedge_B X)^N]_{G/N} \\
&\cong [S, \rho_!\Phi^N((H^\delta \,\mathrm{Inf}^G_{G/N}\underline{S})_\gamma \wedge_B X)]_{G/N} \\
&\cong [S, \rho_!(\Phi^N H^\delta(\mathrm{Inf}^G_{G/N}\underline{S})_\gamma \wedge_{B^N} X^N)]_{G/N} \\
&\cong [S, \rho_!(H^\delta \underline{S}_{\gamma^N} \wedge_{B^N} X^N)]_{G/N}
\end{aligned}$$

and

$$\begin{aligned}
[\Sigma^\infty_G X, (H_\delta \,\mathrm{Inf}^G_{G/N}\overline{T})^\gamma]_{G,B} &\cong [\Phi^N(\Sigma^\infty_G X), \Phi^N H_\delta(\mathrm{Inf}^G_{G/N}\overline{T})^\gamma]_{G/N,B^N} \\
&\cong [\Sigma^\infty_{G/N} X^N, H_\delta \overline{T}^{\gamma^N}]_{G/N,B^N}.
\end{aligned}$$

Finally, we can define restriction to K-fixed sets when K is not normal by first restricting to the normalizer NK and then taking K-fixed sets. As in the nonparametrized case, the possible dimension shift goes away and we get the following fixed set maps:

$$(-)^K\colon \tilde{H}^{G,\delta}_\gamma(X;\underline{S}) \to \tilde{H}^{WK,\delta}_{\gamma^K}(X^K;\underline{S}^K) \qquad \text{and}$$

$$(-)^K\colon \tilde{H}^\gamma_{G,\delta}(X;\overline{T}) \to \tilde{H}^{\gamma^K}_{WK,\delta}(X^K;\overline{T}^K)$$

where $WK = NK/K$.

3.9.3 Subgroups of Quotient Groups

As in the nonparametrized case, we now look at how induction and restriction to fixed sets interact with restriction to subgroups. Once again, the main result needed is that the coefficient systems involved agree.

Suppose that N is a normal subgroup of G, $N \le L \le G$, and δ is familial with $L \in \mathscr{F}(\delta)$. We then have the following commutative diagrams:

$$\begin{array}{ccc} \widehat{\Pi}_{L/N,\delta}B^N & \xrightarrow{\theta} & \widehat{\Pi}_{L,\delta}B \\ {\scriptstyle i^{G/N}_{L/N}}\big\downarrow & & \big\downarrow{\scriptstyle i^G_L} \\ \widehat{\Pi}_{G/N,\delta}B^N & \xrightarrow[\theta]{} & \widehat{\Pi}_{G,\delta}B \end{array}$$

and

$$\begin{array}{ccc} \widehat{\Pi}_{L,\delta}B & \xrightarrow{\Phi^N} & \widehat{\Pi}_{L/N,\delta}B^N \\ {\scriptstyle i^G_L}\big\downarrow & & \big\downarrow{\scriptstyle i^{G/N}_{L/N}} \\ \widehat{\Pi}_{G,\delta}B & \xrightarrow[\Phi^N]{} & \widehat{\Pi}_{G/N,\delta}B^N. \end{array}$$

By Proposition 1.6.8, we have a natural map $\xi\colon \theta_! i^* \to i^*\theta_!$.

Lemma 3.9.21 *The natural transformation $\xi\colon \theta_! i^* \to i^*\theta_!$ is an isomorphism.*

The impatient reader may skip the proof, which is simply an elaboration of the proof of Lemma 1.13.31.

Proof We give the argument for contravariant Mackey functors; the proof for covariant functors is similar or we can appeal to duality (replacing δ with $\mathscr{L} - \delta$).

By Proposition 1.6.8, the result will follow if we show that

$$\xi\colon \int^{z\in\widehat{\Pi}_{L/N,\delta}B^N} \widehat{\Pi}_{G/N,\delta}B^N(G\times_L z, b)\otimes \widehat{\Pi}_{L,\delta}B(x,\theta z) \to \widehat{\Pi}_{G,\delta}B(G\times_L x, \theta b),$$

given by $\xi(f\otimes g) = \theta f\circ (G_+ \wedge_L g)$, is an isomorphism for all $x \in \widehat{\Pi}_{L,\delta}B$ and $b \in \widehat{\Pi}_{G/N,\delta}B^N$. Fix $x\colon L/H \to B$ and $b\colon G/K \to B^N$, where $N \le K \le G$.

Define a map ζ inverse to ξ as follows: Using Theorem 2.6.4, let

$$[G\times_L x \xleftarrow{\;p\;} w \xRightarrow{\;q\;} \theta b]$$

be a generator of $\widehat{\Pi}_{G,\delta}B(G\times_L x, \theta b)$, with $w: G/M \to B$, so $M \leq H \leq L$ and p is the projection. Then $w = G\times_L w'$ with $w: L/M \to B$ and $p = G\times_L p'$ where $p': w' \to x$. By assumption, $N \leq K$, so q factors as $w \Rightarrow \theta v \to \theta b$ with $v: G/MN \to B^N$; because $MN \leq L$ we can write $v = G\times_L v'$ with $v': L/MN \to B^N$. We then let

$$\zeta[G\times_L x \leftarrow w \Rightarrow \theta b] = [G\times_L v' \to b] \otimes [x \xleftarrow{p'} w' \Rightarrow \theta v'].$$

Clearly, $\xi \circ \zeta$ is the identity. On the other hand, a typical element in the coend is a sum of elements of the form

$$[G\times_L z \xleftarrow{p} w \Rightarrow b] \otimes g,$$

where $z: L/J \to B$ and $w: G/M \to B$ with $N \leq M \leq J \leq L$ and p the projection. Write $w = G\times_L w'$ so $p = G\times_L p'$ with $p': w' \to z$. Finally, factor $w \Rightarrow b$ as $G\times_L w' \Rightarrow G\times_L v' \to b$ as above and let $v = G\times_L v'$, so

$$[G\times_L z \xleftarrow{p} w \Rightarrow b] = [G\times_L z \xleftarrow{G\times_L p'} G\times_L w' \Rightarrow G\times_L v' \to b].$$

We then have

$$\begin{aligned}
[G\times_L z \xleftarrow{p} w \Rightarrow b] \otimes g &\sim [v \to b] \otimes [z \xleftarrow{p'} w' \Rightarrow v'] \circ g \\
&= \sum_i [v \to b] \otimes [x \leftarrow u_i \Rightarrow v'] \\
&= \sum_i [v \to b] \otimes [x \leftarrow u_i \Rightarrow t_i \to v'] \\
&\sim \sum_i [G\times_L t_i \to v \to b] \otimes [x \leftarrow u_i \Rightarrow t_i]
\end{aligned}$$

which is in the image of ζ. (Here, $u_i: L/P_i \to B$ for some P_i and $t_i: L/P_iN \to B$.) So, ζ is an epimorphism, hence an isomorphism and the inverse of ξ. □

Proposition 3.9.22 *Let N be a normal subgroup of G, let δ be an N-closed familial dimension function for G, let γ be a virtual representation of $\Pi_{G/N}B^N$, let $\underline{S}$ be a covariant $\widehat{\Pi}_{G/N,\delta}B^N$-module, and let $\overline{T}$ be a contravariant $\widehat{\Pi}_{G/N,\delta}B^N$-module. Let $N \leq L \leq G$ with $L \in \mathscr{F}(\delta)$. If Y is an ex-G/N-space over B^N and $y \in \tilde{H}_\gamma^{G/N,\delta}(Y;\underline{S})$, then*

$$\epsilon^*(y|L/N) = (\epsilon^* y)|L \in \tilde{H}_\gamma^{L,\delta}(Y; \theta_!(\underline{S}|L/N)) \cong \tilde{H}_\gamma^{L,\delta}(Y; (\theta_!\underline{S})|L).$$

Similarly, if $y \in \tilde{H}^{\gamma}_{G/N,\delta}(Y;\overline{T})$*, then*

$$\epsilon^*(y|L/N) = (\epsilon^* y)|L \in \tilde{H}^{\gamma}_{L,\delta}(Y;\theta_!(\overline{T}|L/N)) \cong \tilde{H}^{\gamma}_{L,\delta}(Y;(\theta_!\overline{T})|L).$$

Proof The proof is formally the same as the proof of Proposition 1.13.32. □

We now turn to the relationship between restriction to fixed sets and restriction to subgroups. The development is very similar to the nonparametrized case. Referring back to a diagram at the start of this section, by Proposition 1.6.8, we have a natural map

$$\xi \colon \Phi^N_! i^* \to i^* \Phi^N_! .$$

As in the nonparametrized case, we first use Theorem 2.6.4 to get an explicit description of Φ^N.

Proposition 3.9.23 *Let* $x \colon G/H \to B$ *and* $y \colon G/K \to B$ *be objects of* $\widehat{\Pi}_{G,\delta}B$*. Consider the map*

$$\Phi^N \colon \widehat{\Pi}_{G,\delta}B(x,y) \to \widehat{\Pi}_{G/N,\delta}B^N(x^N,y^N).$$

If either $N \not\leq H$ *or* $N \not\leq K$*, the target is the trivial group. If* $N \leq H$ *and* $N \leq K$*, then, on generators,* Φ^N *is given by*

$$\begin{aligned}\Phi^N[x \leftarrow z \Rightarrow y] &= [x^N \leftarrow z^N \Rightarrow y^N] \\ &= \begin{cases} [x \leftarrow z \Rightarrow y] & \text{if } N \leq J \\ 0 & \text{otherwise,} \end{cases}\end{aligned}$$

where $z \colon G/J \to B$*. Therefore, it is a split epimorphism with kernel generated by those diagrams* $[x \leftarrow z \Rightarrow y]$ *with* $z \colon G/J \to B$ *and* $N \not\leq J$*.*

Proof The proof is essentially the same as the proof of Proposition 1.13.33. □

Now we can prove that ξ is an isomorphism.

Lemma 3.9.24 *If* N *is a normal subgroup of* G*,* $N \leq L \leq G$*, and* $\overline{T}$ *and* $\underline{S}$ *are, respectively, contravariant and covariant* $\widehat{\Pi}_\delta B$*-modules, then* $\xi \colon (\overline{T}|L)^N \to \overline{T}^N|(L/N)$ *and* $\xi \colon (\underline{S}|L)^N \to \underline{S}^N|(L/N)$ *are isomorphisms.*

Again, the proof is an elaboration of the proof of Lemma 1.13.34.

Proof By Proposition 1.6.8, the result will follow if we show that

$$\xi \colon \int^{z \in \widehat{\Pi}_{L,\delta}B} \widehat{\Pi}_{G,\delta}B(G \times_L z, b) \otimes \widehat{\Pi}_{L/N,\delta}B^N(x, z^N) \to \widehat{\Pi}_{G/N,\delta}B^N(G \times_L x, b^N),$$

given by $\xi(f \otimes g) = f^N \circ (G_+ \wedge_L g)$, is an isomorphism for all $x \in \widehat{\Pi}_{L/N,\delta}B^N$ and $b \in \widehat{\Pi}_{G,\delta}B$. Fix $x: L/H \to B^N$ and $b: G/K \to B$.

If $N \not\leq K$, then $b^N = *$ and the target of ξ is 0. For the source, consider a typical generator

$$f = [G \times_L z \xleftarrow{p} w \xrightarrow{q} b]$$

of $\widehat{\Pi}_{G,\delta}B(G \times_L z, b)$. Because p is a projection, we can write $p = G \times_L p'$ where $p': L/M \to L/J$ is a projection. Because M is subconjugate to K, $N \not\leq M$, hence $(L/M)^N = 0$. Therefore, for any g,

$$f \otimes g = t(p')^* q \otimes g = q \otimes t((p')^N)g = 0$$

in the coend.

So, assume that $N \leq K$ so $(G/K)^N = G/K$. Define a map ζ inverse to ξ as follows: If

$$[G \times_L x \xleftarrow{p} z \xrightarrow{q} b]$$

is a generator of $\widehat{\Pi}_{G/N,\delta}B^N(G \times_L x, b^N)$ with $z: G/J \to B^N$ and $N \leq J \leq H$, let

$$\zeta[G \times_L x \xleftarrow{p} z \xrightarrow{q} b] = q \otimes t(p')$$

where $p = G \times_L p'$ and $p': L/J \to L/H$ is the projection. Clearly, $\xi \circ \zeta$ is the identity. On the other hand, a typical element in the coend is a sum of elements of the form

$$[z \xleftarrow{p} w \Rightarrow b] \otimes g,$$

where we may assume $z: G/J \to B$ with $N \leq J$, because otherwise the element would live in a 0 group. If $w: G/M \to B$, this is equivalent to $[w \Rightarrow b] \otimes t(p')g$ where $p': L/M \to L/J$ (and we may assume $N \leq M$, otherwise this element is 0). In turn, such an element can be written as a sum of elements of the form $[w \Rightarrow b] \otimes [x \leftarrow v \Rightarrow w']$ (where $w = G \times_L w'$), which is equivalent to

$$[G \times_L v \Rightarrow w \Rightarrow b] \otimes [x \leftarrow v],$$

which is in the image of ζ. Thus, ζ is an isomorphism inverse to ξ.

The argument for covariant modules is similar, or we can appeal to duality (replacing δ with $\mathscr{L} - \delta$). □

The following now follows as it did in the nonparametrized case.

Proposition 3.9.25 *Let N be a normal subgroup of G, let δ be a familial dimension function for G, let γ be a virtual representation of $\Pi_G B$, let $\underline{S}$ be a covariant $\widehat{\Pi}_{G,\delta}B$-*

module and let $\overline{T}$ be a contravariant $\widehat{\Pi}_{G,\delta}B$-module. Let $N \leq L \leq G$ with $L \in \mathscr{F}(\delta)$. If X is an ex-G-space over B and $x \in \tilde{H}^{G,\delta}_{\gamma}(X; \underline{S})$, then

$$(x|L)^N = x^N|L/N \in \tilde{H}^{L/N,\delta}_{\gamma^N-\delta(G/L)}(X^N; (\underline{S}|L)^N) \cong \tilde{H}^{L/N,\delta}_{\gamma^N-\delta(G/L)}(X^N; \underline{S}^N|L/N).$$

Similarly, if $x \in \tilde{H}^{\gamma}_{G,\delta}(X; \overline{T})$, then

$$(x|L)^N = x^N|L/N \in \tilde{H}^{\gamma^N-\delta(G/L)}_{L/N,\delta}(X^N; (\overline{T}|L)^N) \cong \tilde{H}^{\gamma^N-\delta(G/L)}_{L/N,\delta}(X^N; \overline{T}^N|L/N).$$

□

3.10 Products

We now turn to various pairings that we saw in the nonparametrized case.

3.10.1 Cup Products

As in the nonparametrized case, we start with the external cup product. The appropriate tensor product of modules is defined as follows.

Definition 3.10.1 Let G and K be two compact Lie groups. Let δ be a dimension function for G, let ϵ be a dimension function for K, and let $\delta \times \epsilon$ denote their product as in Example 1.2.2(3). Let ζ be a dimension function for $G \times K$ with $\zeta \succcurlyeq \delta \times \epsilon$, as in Definition 1.2.8. Let A be a G-space and let B be a K-space.

(1) Let $\widehat{\Pi}_{G,\delta}A \otimes \widehat{\Pi}_{K,\epsilon}B$ denote the preadditive category whose objects are pairs of objects (a, b), as in the product category, with

$$(\widehat{\Pi}_{G,\delta}A \otimes \widehat{\Pi}_{K,\epsilon}B)((a, b), (c, d)) = \widehat{\Pi}_{G,\delta}A(a, c) \otimes \widehat{\Pi}_{K,\epsilon}B(b, d).$$

Let

$$p: \widehat{\Pi}_{G,\delta}A \otimes \widehat{\Pi}_{K,\epsilon}B \to \text{Ho}(G \times K)\mathscr{P}_{A\times B}$$

be the restriction of the external smash product.

Let $i: \widehat{\Pi}_{G\times K,\zeta}(A \times B) \to \text{Ho}(G \times K)\mathscr{P}_{A\times B}$ denote the inclusion of the subcategory.

(2) If $\overline{T}$ is a contravariant $\widehat{\Pi}_{G,\delta}A$-module and $\overline{U}$ is a contravariant $\widehat{\Pi}_{K,\epsilon}B$-module, let $\overline{T} \otimes \overline{U}$ denote the $(\widehat{\Pi}_{G,\delta}A \otimes \widehat{\Pi}_{K,\epsilon}B)$-module defined by

$$(\overline{T} \otimes \overline{U})(a, b) = \overline{T}(a) \otimes \overline{U}(b).$$

Define $\underline{S} \otimes \underline{V}$ similarly for covariant modules.

(3) Let $\overline{T} \boxtimes \overline{U} = \overline{T} \boxtimes_\zeta \overline{U}$ (the *external box product*) be the contravariant $\widehat{\Pi}_{G\times K,\zeta}(A \times B)$-module defined by

$$\overline{T} \boxtimes \overline{U} = \overline{T} \boxtimes_\zeta \overline{U} = i^* p_!(\overline{T} \otimes \overline{U}),$$

where i^* and $p_!$ are defined in Definition 1.6.5. Define the external box product of covariant modules similarly.

(4) If $G = K$ and $A = B$, and $\Delta \in \mathscr{F}(\zeta)$, where $\Delta \leq G \times G$ is the diagonal subgroup, define the *(internal) box product* to be

$$\overline{T} \,\square\, \overline{U} = \overline{T} \,\square_\zeta\, \overline{U} = (\overline{T} \boxtimes \overline{U})|\Delta(B)$$

where $\Delta(B) \subset B \times B$ is the diagonal subspace and we also restrict to $\Delta \leq G\times G$, so that $\overline{T} \,\square\, \overline{U}$ is a $\widehat{\Pi}_{G,\zeta}B$-module. Define the box product of covariant modules similarly.

Proposition 3.10.2 *Let δ be a dimension function for G, let ϵ be a dimension function for K, and let $\zeta \succcurlyeq \delta \times \epsilon$. Let $a: G/H \to A$ be an object of $\widehat{\Pi}_{G,\delta}A$ and let $b: K/L \to B$ be an object of $\widehat{\Pi}_{K,\epsilon}B$. Then*

$$\overline{A}_a \boxtimes_\zeta \overline{A}_b \cong \mathrm{Ho}(G \times K)\mathscr{P}_{A\times B}(-, a \times b)$$

and

$$\underline{A}^a \boxtimes_\zeta \underline{A}^b \cong \mathrm{Ho}(G \times K)\mathscr{P}_{A\times B}(a \times b, -).$$

Proof It's clear from the definitions that

$$\overline{A}_a \otimes \overline{A}_b \cong (\widehat{\Pi}_{G,\delta}A \otimes \widehat{\Pi}_{K,\epsilon}B)(-, (a, b)).$$

Applying $p_!$ and using Proposition 1.6.6 gives the result. The proof for covariant functors is similar. □

Let $\overline{A}_{G/G}$ denote the free G-Mackey functor over a point as in Proposition 1.14.7 and also denote its pullback to a $\widehat{\Pi}_{G,\delta}B$-module along the projection $B \to *$. The following proposition is proved in the same way as Proposition 1.14.7.

Proposition 3.10.3 *Let δ be a complete dimension function for G and let $\overline{T}$ be a contravariant $\widehat{\Pi}_{G,\delta}B$-module. Then*

$$\overline{A}_{G/G} \square_{\delta_\Delta} \overline{T} \cong \overline{T}. \qquad \square$$

The following result identifies the chain complex of a product of CW complexes. Its proof is the same as that of Proposition 1.14.8.

Proposition 3.10.4 *Let δ be a dimension function for G and let ϵ be a dimension function for K. Let A be a G-space and let B be a K-space. If X is a based δ-G-CW(β) complex over A and Y is a based ϵ-K-CW(γ) complex over B, then $X \wedge_{A\times B} Y$ is a based $(\delta \times \epsilon)$-$(G \times K)$-CW$(\beta + \gamma)$ complex over $A \times B$ with*

$$\overline{C}^{G,\delta}_{\beta+*}(X, *) \boxtimes_{\delta\times\epsilon} \overline{C}^{K,\epsilon}_{\gamma+*}(Y, *) \cong \overline{C}^{G\times K,\delta\times\epsilon}_{\beta+\gamma+*}(X \wedge_{A\times B} Y, *).$$

Moreover, this isomorphism respects suspension in each of X and Y. $\square$

As we noted in the nonparametrized case, because of the limitations of $\delta \times \epsilon$, this result by itself is not as useful as we would like. To get more useful results, we first note that level-wise smash product induces a map

$$\wedge : G\mathscr{P}\mathscr{V}_A \otimes K\mathscr{P}\mathscr{W}_B \to (G \times K)\mathscr{P}(\mathscr{V} \oplus \mathscr{W})_{A\times B}$$

where $\mathscr{V} \oplus \mathscr{W} = \{V_i \oplus W_i\}$. This pairing extends to semistable maps as well.

We also need the following.

Definition 3.10.5 Suppose that X is a G-space over B. We say that X is *compactly supported* if there exists a compact subspace C of B and a G-space X' over C such that $X = i_! X'$, where $i: C \to B$ is the inclusion. Similarly, suppose that $E \in G\mathscr{P}\mathscr{V}_B$ is a spectrum over B. We say that E is *compactly supported* if there exists a compact subspace $i: C \to B$ and a spectrum E' over C such that $E = i_! E'$.

Proposition 3.10.6 *Let δ be a familial dimension function for G and let ϵ be a familial dimension function for K. Let X be a based G-space over A and let Y be a based K-space over B. Take a δ-G-CW(α) approximation $\Gamma^\delta \Sigma^\infty_{G,A} X \to \Sigma^\infty_{G,A} X$ and an ϵ-K-CW(β) approximation $\Gamma^\epsilon \Sigma^\infty_{K,B} Y \to \Sigma^\infty_{K,B} Y$. If $j: C \to A \times B$ is the inclusion of a compact subspace, then there exists an integer N (depending on C) such that*

$$j^*(\Gamma^\delta \Sigma^\infty_{G,A} X \wedge \Gamma^\epsilon \Sigma^\infty_{K,B} Y)(V_i \oplus W_i) \to j^*(\Sigma^{V_i}_{G,A} X \wedge \Sigma^{W_i}_{K,B} Y)$$

is an $\overline{\mathscr{F}(\delta) \times \mathscr{F}(\epsilon)}$-equivalence for all $i \geq N$ (where we first make the spectra fibrant before taking j^*; equivalently, we take the homotopy pullback).

Proof Weak equivalence is determined (homotopy) fiberwise. Because C is compact, its fundamental groupoid has a skeleton with only finitely many objects. Therefore, there are only finitely many fibers to consider, and a sufficiently large suspension allows us to apply the analogue of Proposition 1.14.1. $\square$

Lemma 3.10.7 *Let E be a δ-G-CW(α) spectrum over a σ-compact basespace B. Then there is a sequence of compactly supported subcomplexes $\{E_i\}$ such that $E = \operatorname{colim}_i E_i$ and $\overline{C}^{G,\delta}_{\alpha+*}(E) \cong \operatorname{colim}_i \overline{C}^{G,\delta}_{\alpha+*}(E_i)$.*

Proof Write $B = \cup_i B_i$ where $B_1 \subset B_2 \subset \cdots$ and each B_i is compact. Construct E'_k over B_k as follows: Take as its vertices all 0-dimensional cells in E that lie over B_k. Inductively, take as its n-cells all n-cells of E that lie over B_k and attach to lower-dimensional cells already chosen. Let $E_k = i_! E'_k$ where $i: B_1 \to B$ is the inclusion.

Because $B_{k-1} \subset B_k$, we get that $E_{k-1} \subset E_k$. Because every cell in E is contained in a finite subcomplex, it will appear in some E_k. Therefore, E is the sequential colimit of the E_k and the statement about chain complexes is clear. □

Proposition 3.10.8 *Let δ be a familial dimension function for G, let ϵ be a familial dimension function for K, and let ζ be a familial dimension function for $G \times K$ with $\zeta \succcurlyeq \delta \times \epsilon$. Let A be a σ-compact G-space and let B be a σ-compact K-space. Let X be a based G-space over A and let Y be a based K-space over B. Let $\Gamma^\delta \Sigma^\infty_{G,A} X \to \Sigma^\infty_{G,A} X$ and $\Gamma^\epsilon \Sigma^\infty_{K,B} Y \to \Sigma^\infty_{K,B} Y$ be, respectively, δ-G-CW(α) and ϵ-K-CW(β) approximations and let $\Gamma^\zeta \Sigma^\infty_{G\times K, A\times B}(X \wedge Y) \to \Sigma^\infty_{G\times K, A\times B}(X \wedge Y)$ be a ζ-($G \times K$)-CW($\alpha + \beta$) approximation. Write*

$$\Gamma^\zeta \Sigma^\infty_{G\times K, A\times B}(X \wedge Y) = \operatorname*{colim}_i \Gamma_i$$

where each Γ_i is compactly supported, by the preceding lemma. Then there exist compatible semistable cellular lax maps

$$\mu_i : \Gamma_i \Rightarrow \Gamma^\delta \Sigma^\infty_{G,A} X \wedge \Gamma^\epsilon \Sigma^\infty_{K,B} Y$$

over $\Sigma^\infty_{G\times K, A\times B}(X \wedge Y)$, the collection of maps being unique up to semistable cellular homotopy.

Proof Write $A \times B = \cup C_i$, where each C_i is compact, and $\Gamma_i = j_! \Gamma'_i$ where $j: C_i \to A \times B$ is the inclusion. Consider the following diagram, in which j^* indicates homotopy pullback.

$$\begin{array}{ccc}
\Gamma'_{i-1} & \longrightarrow & j^*(\Gamma^\delta \Sigma^\infty_{G,A} X \wedge \Gamma^\epsilon \Sigma^\infty_{K,B} Y) \\
\downarrow & \nearrow & \downarrow \\
\Gamma'_i & \longrightarrow & j^*(\Sigma^\infty_{G,A} X \wedge \Sigma^\infty_{K,B} Y)
\end{array}$$

A semistable lift exists by the relative Whitehead theorem, using Proposition 3.10.6. The adjoint is the map

$$\mu_i : \Gamma_i = j_! \Gamma_i \Rightarrow \Gamma^\delta \Sigma^\infty_{G,A} X \wedge \Gamma^\epsilon \Sigma^\infty_{K,B} Y.$$

Uniqueness up to semistable cellular homotopy follows from cellular approximation. □

Definition 3.10.9 Let δ be a familial dimension function for G, let ϵ be a familial dimension function for K, and let ζ be a familial dimension function for $G \times K$ with $\zeta \succcurlyeq \delta \times \epsilon$. If X is an ex-G-space over A and Y is an ex-K-space over B, where A and B are σ-compact, let

$$\mu_*: \overline{C}^{G\times K,\zeta}_{\alpha+\beta+*}(X \wedge Y) \to \overline{C}^{G,\delta}_{\alpha+*}(X) \boxtimes_\zeta \overline{C}^{K,\epsilon}_{\beta+*}(Y)$$

be the chain map induced by the family $\{\mu_i\}$ from the preceding corollary (using Proposition 3.10.4 to identify the chain complex on the right). It is well-defined up to chain homotopy.

We can now define pairings in cohomology. Let δ, ϵ, and ζ be familial with $\zeta \succcurlyeq \delta \times \epsilon$, as above, let X be an ex-G-space over A, let Y be an ex-K-space over B, with A and B σ-compact, let $\overline{T}$ be a $\widehat{\Pi}_{G,\delta}A$-module, and let $\overline{U}$ be a $\widehat{\Pi}_{K,\epsilon}B$-module. The external box product $\boxtimes = \boxtimes_\zeta$ and the map μ_* induce a natural chain map

$$\begin{aligned}
\mathrm{Hom}_{\widehat{\Pi}_{G,\delta}A}(\overline{C}^{G,\delta}_{\alpha+*}(X), \overline{T}) \otimes \mathrm{Hom}_{\widehat{\Pi}_{K,\epsilon}B}(\overline{C}^{K,\epsilon}_{\beta+*}(Y), \overline{U}) \\
\to \mathrm{Hom}_{\widehat{\Pi}_{G\times K,\zeta}(A\times B)}(\overline{C}^{G,\delta}_{\alpha+*}(X) \boxtimes \overline{C}^{K,\epsilon}_{\beta+*}(Y), \overline{T} \boxtimes \overline{U}) \\
\to \mathrm{Hom}_{\widehat{\Pi}_{G\times K,\zeta}(A\times B)}(\overline{C}^{G\times K,\zeta}_{\alpha+\beta+*}(X \wedge Y), \overline{T} \boxtimes \overline{U}).
\end{aligned}$$

This induces the (external) cup product

$$-\cup-: \tilde{H}^\alpha_{G,\delta}(X;\overline{T}) \otimes \tilde{H}^\beta_{K,\epsilon}(Y;\overline{U}) \to \tilde{H}^{\alpha+\beta}_{G\times K,\zeta}(X \wedge_{A\times B} Y; \overline{T} \boxtimes \overline{U}).$$

When $G = K$, $A = B$, and $\mathscr{F}(\zeta)$ contains the diagonal subgroup $\Delta \leq G \times G$, we can follow the external cup product with the restriction to $\Delta(B) \subset B \times B$ and $\Delta \leq G \times G$. This gives the internal cup product

$$-\cup-: \tilde{H}^\alpha_{G,\delta}(X;\overline{T}) \otimes \tilde{H}^\beta_{G,\epsilon}(Y;\overline{U}) \to \tilde{H}^{\alpha+\beta-\zeta(G\times G/\Delta)}_{G,\zeta|\Delta}(X \wedge_B Y; \overline{T} \square \overline{U}).$$

Of course, when $X = Y$ we can apply restriction along the diagonal $X \to X \wedge_B X$ to get

$$-\cup-: \tilde{H}^\alpha_{G,\delta}(X;\overline{T}) \otimes \tilde{H}^\beta_{G,\epsilon}(X;\overline{U}) \to \tilde{H}^{\alpha+\beta-\zeta(G\times G/\Delta)}_{G,\zeta|\Delta}(X; \overline{T} \square \overline{U}).$$

A useful special case is $\delta_\Delta \succcurlyeq 0 \times \delta$ with δ complete, which gives us pairings

$$-\cup-: \tilde{H}^\alpha_{G}(X;\overline{T}) \otimes \tilde{H}^\beta_{G,\delta}(Y;\overline{U}) \to \tilde{H}^{\alpha+\beta}_{G,\delta}(X \wedge_B Y; \overline{T} \square \overline{U})$$

and

$$- \cup -\colon \tilde{H}^{\alpha}_{G}(X;\overline{T}) \otimes \tilde{H}^{\beta}_{G,\delta}(X;\overline{U}) \to \tilde{H}^{\alpha+\beta}_{G,\delta}(X;\overline{T} \,\Box\, \overline{U}).$$

Say that a contravariant $\widehat{\Pi}_G B$-module $\overline{T}$ is a *ring* if there is an associative multiplication $\overline{T} \,\Box\, \overline{T} \to \overline{T}$. This then makes $\tilde{H}^{*}_{G}(X;\overline{T})$ a ring. For example, recall that we let $\overline{A}_{G/G} = \overline{A}_{G/G,0}$, pulled back to B along the projection $B \to *$. $\overline{A}_{G/G}$ is a ring by Proposition 3.10.3. By that same proposition, $\overline{A}_{G/G} \,\Box_{\delta_\Delta}\, \overline{T} \cong \overline{T}$ for any contravariant δ-G-Mackey functor $\overline{T}$. This makes $\overline{T}$ a *module* over $\overline{A}_{G/G}$, in the sense that there is an associative pairing $\overline{A}_{G/G} \Box_{\delta_\Delta} \overline{T} \to \overline{T}$ (namely, the isomorphism). This makes every cellular cohomology theory a module over ordinary cohomology with coefficients in $\overline{A}_{G/G}$, using the cup product

$$- \cup -\colon \tilde{H}^{\alpha}_{G}(X;\overline{A}_{G/G}) \otimes \tilde{H}^{\beta}_{G,\delta}(X;\overline{T}) \to \tilde{H}^{\alpha+\beta}_{G,\delta}(X;\overline{T}).$$

The following theorem records the main properties of the external cup product, from which similar properties of the other products follow by naturality.

Theorem 3.10.10 *Let δ be a familial dimension function for G, let ϵ be a familial dimension function for K, and let ζ be a familial dimension function for $G \times K$ with $\zeta \succcurlyeq \delta \times \epsilon$. The external cup product*

$$- \cup -\colon \tilde{H}^{\alpha}_{G,\delta}(X;\overline{T}) \otimes \tilde{H}^{\beta}_{K,\epsilon}(Y;\overline{U}) \to \tilde{H}^{\alpha+\beta}_{G\times K,\zeta}(X \wedge_{A\times B} Y;\overline{T} \boxtimes \overline{U})$$

satisfies the following.

(1) *It is natural: $f^*(x) \cup g^*(y) = (f \wedge g)^*(x \cup y)$.*
(2) *It respects suspension: For any representation V of G, the following diagram commutes:*

$$\begin{array}{ccc}
\tilde{H}^{\alpha}_{G,\delta}(X;\overline{T}) \otimes \tilde{H}^{\beta}_{K,\epsilon}(Y;\overline{U}) & \xrightarrow{\ \cup\ } & \tilde{H}^{\alpha+\beta}_{G\times K,\zeta}(X \wedge_{A\times B} Y;\overline{T} \boxtimes \overline{U}) \\
\Big\downarrow{\scriptstyle \sigma^V \otimes \mathrm{id}} & & \\
\tilde{H}^{\alpha+V}_{G,\delta}(\Sigma^V X;\overline{T}) \otimes \tilde{H}^{\beta}_{K,\epsilon}(Y;\overline{U}) & & \Big\downarrow{\scriptstyle \sigma^V} \\
\Big\downarrow{\scriptstyle \cup} & & \\
\tilde{H}^{\alpha+V+\beta}_{G\times K,\zeta}(\Sigma^V X \wedge_{A\times B} Y;\overline{T} \boxtimes \overline{U}) & \xrightarrow[\cong]{} & \tilde{H}^{\alpha+\beta+V}_{G\times K,\zeta}(\Sigma^V (X \wedge_{A\times B} Y);\overline{T} \boxtimes \overline{U})
\end{array}$$

The horizontal isomorphism at the bottom of the diagram comes from the identification $\alpha + V + \beta \cong \alpha + \beta + V$. The similar diagram for suspension of Y also commutes.

(3) *It is associative:* $(x \cup y) \cup z = x \cup (y \cup z)$ *when we identify gradings using the obvious identification* $(\alpha + \beta) + \gamma \cong \alpha + (\beta + \gamma)$.
(4) *It is commutative: If* $x \in \tilde{H}^{\alpha}_{G,\delta}(X; \overline{T})$ *and* $y \in \tilde{H}^{\beta}_{K,\epsilon}(Y; \overline{U})$ *then* $x \cup y = \iota(y \cup x)$ *where* ι *is the evident isomorphism*

$$\iota\colon \tilde{H}^{\alpha+\beta}_{G\times K,\zeta}(X \wedge_{A\times B} Y; \overline{T} \boxtimes \overline{U}) \cong \tilde{H}^{\beta+\alpha}_{K\times G,\tilde{\zeta}}(Y \wedge_{B\times A} X; \overline{U} \boxtimes \overline{T});$$

$\tilde{\zeta}$ *is the dimension function on* $K \times G$ *induced by* ζ *and* ι *uses the isomorphism of* $\alpha + \beta$ *and* $\beta + \alpha$ *that switches the direct summands.*
(5) *It is unital: The map*

$$\tilde{H}^{0}_{G}(S^0; \overline{A}_{G/G}) \otimes \tilde{H}^{\alpha}_{G,\delta}(X; \overline{T}) \to \tilde{H}^{\alpha}_{G,\delta}(X; \overline{T})$$

takes $1 \otimes x \mapsto x$, *where* $1 \in \tilde{H}^{0}_{G}(S^0; \overline{A}_{G/G}) \cong A(G)$ *is the unit.*
(6) *It respects restriction to subgroups:* $(x|J) \cup (y|L) = (x \cup y)|(J \times L)$. *(But see Remark 3.10.11.)*
(7) *It respects restriction to fixed sets:* $x^J \cup y^L = (x \cup y)^{J\times L}$. *(But, again, see Remark 3.10.11.)*

The proofs are all standard except for the last two points, but the proofs of these are just the obvious generalizations of the proofs of the last two parts of Theorem 1.14.12.

Remark 3.10.11 Part (6) of Theorem 3.10.10 is stated a bit loosely. In fact,

$$(x|J) \cup (y|L) \in H^{*}_{J\times L}(X \wedge Y; (\overline{T}|J) \boxtimes (\overline{U}|L))$$

while

$$(x \cup y)|(J \times L) \in H^{*}_{J\times L}(X \wedge Y; (\overline{T} \boxtimes \overline{U})|(J \times L)).$$

The theorem should say that, when we apply the map induced by the homomorphism $(\overline{T}|J) \boxtimes (\overline{U}|L) \to (\overline{T} \boxtimes \overline{U})|(J \times L)$, the element $(x|J) \cup (y|L)$ maps to $(x \cup y)|(J \times L)$.

A similar comment applies to part (7) in general, when we have to first restrict to normalizers. However, if J is normal in G and L is normal in K, we can use the isomorphism $\overline{T}^J \boxtimes \overline{U}^L \cong (\overline{T} \boxtimes \overline{U})^{J\times L}$ to identify the two cohomology groups.

Now we look at how the cup product is represented. Let A be a G-space and let B be a K-space. Let α be a virtual representation of $\Pi_G A$ and let β be a virtual representation of $\Pi_K B$. Given the G-spectrum $H_\delta \overline{T}^{\alpha}$ over A and the K-spectrum $H_\epsilon \overline{U}^{\beta}$ over B, we can form the $(G \times K)$-spectrum $H_\delta \overline{T} \wedge H_\epsilon \overline{U}$ over $A \times B$. The external cup product should then be represented by a $(G \times K)$-map

$$H_\delta \overline{T}^{\alpha} \wedge H_\epsilon \overline{U}^{\beta} \to H_\zeta(\overline{T} \boxtimes \overline{U})^{\alpha+\beta}$$

that is an isomorphism in $\overline{\pi}^{G\times K,\zeta}_{\alpha+\beta}$. An explicit construction is based on the following calculation.

Proposition 3.10.12 *Let δ be a familial dimension function for G, let ϵ be a familial dimension function for K, and let $\zeta \succcurlyeq \delta \times \epsilon$. Let $\overline{T}$ be a contravariant $\widehat{\Pi}_{G,\delta}A$-module and let $\overline{U}$ be a contravariant $\widehat{\Pi}_{K,\epsilon}B$-module. Then we have*

$$\overline{\pi}^{G\times K,\zeta}_{\alpha+\beta+n}(H_\delta \overline{T}^\alpha \wedge H_\epsilon \overline{U}^\beta) = \begin{cases} 0 & \text{if } n < 0 \\ \overline{T} \boxtimes_\zeta \overline{U} & \text{if } n = 0. \end{cases}$$

Proof Let $c\colon (G\times K)/J \to A\times B$ be a $(G\times K)$-map with $J \in \mathscr{F}(\zeta)$. Let $H = p_1(J) \leq G$ and $L = p_2(J) \leq K$; because $\zeta \succcurlyeq \delta \times \epsilon$, we must have $H \in \mathscr{F}(\delta)$ and $L \in \mathscr{F}(\epsilon)$. If $c(eJ) = (x, y)$, then we have maps $a\colon G/H \to A$ with $a(eH) = x$ and $b\colon K/L \to B$ with $b(eL) = y$, and c factors as $(a \times b) \circ p$ where $p\colon (G \times K)/J \to G/H \times K/L$ is the projection. Further, $a \times b$ is initial with the properties that it is a product and c factors through it. We need to compute

$$\begin{aligned} \overline{\pi}^{G\times K,\zeta}_{\alpha+\beta+n}(H_\delta \overline{T}^\alpha \wedge H_\epsilon \overline{U}^\beta)(c) &\cong \pi^{J,\zeta}_{\alpha_0(a)+\beta_0(b)+n}(c^*(H_\delta \overline{T}^\alpha \wedge H_\epsilon \overline{U}^\beta)) \\ &\cong \pi^{H\times L,\zeta}_{\alpha_0(a)+\beta_0(b)+n}((a\times b)^*(H_\delta \overline{T}^\alpha \wedge H_\epsilon \overline{U}^\beta)), \end{aligned}$$

where we implicitly consider a $(G\times K)$-spectrum over $(G\times K)/J$ to be a J-spectrum, and similarly for spectra over $G/H \times K/L$. Using the calculation of the fibers given at the beginning of Sect. 3.7 we get

$$(a\times b)^*(H_\delta \overline{T}^\alpha \wedge H_\epsilon \overline{U}^\beta) \simeq \Sigma^{\alpha_0(a)} H_\delta(a^* \overline{T}) \wedge \Sigma^{\beta_0(b)} H_\epsilon(b^* \overline{U}).$$

The calculation then follows from Proposition 1.14.16. □

As usual, for example, by killing higher homotopy groups, it follows that there is a map of $(G \times K)$-spectra over $A \times B$

$$H_\delta \overline{T}^\alpha \wedge H_\epsilon \overline{U}^\beta \to H_\zeta(\overline{T} \boxtimes_\zeta \overline{U})^{\alpha+\beta}$$

that is an isomorphism in $\overline{\pi}^{G\times K,\zeta}_{\alpha+\beta}$. That this represents the cup product in cohomology that we constructed on the chain level follows by considering its effect on products of cells, which reduces to the statement that the maps of fibers represent nonparametrized cup products.

When $G = K$ and $A = B$ we can restrict to diagonals to get a map of G-spectra

$$H_\delta \overline{T}^\alpha \wedge H_\epsilon \overline{U}^\beta \to \Sigma^{-\zeta(G\times G/\Delta)} H_\zeta(\overline{T} \,\Box_\zeta\, \overline{U})^{\alpha+\beta}$$

that is an isomorphism in $\overline{\pi}^{G,\zeta}_{\alpha+\beta}$.

Finally, let's point out the specializations to several interesting choices of δ and ϵ, using the internal cup products.

Remark 3.10.13 The general cup product gives us the following special cases.

(1) Taking $\delta = \epsilon = 0$ on G and $\zeta = 0$ on $G \times G$, we have the cup product

$$\tilde{H}_G^\alpha(X; \overline{T}) \otimes \tilde{H}_G^\beta(Y; \overline{U}) \to \tilde{H}_G^{\alpha+\beta}(X \wedge Y; \overline{T} \square \overline{U}).$$

This product is represented by a G-map

$$H\overline{T}^\alpha \wedge H\overline{U}^\beta \to H(\overline{T} \square \overline{U})^{\alpha+\beta}.$$

(2) Taking $\delta = \epsilon = \mathscr{L}$ on G and $\zeta = \mathscr{L}$ on $G \times G$, we have the cup product

$$\tilde{\mathscr{H}}_G^\alpha(X; \underline{R}) \otimes \tilde{\mathscr{H}}_G^\beta(Y; \underline{S}) \to \tilde{\mathscr{H}}_G^{\alpha+\beta-\mathscr{L}(G)}(X \wedge Y; \underline{R} \square_{\mathscr{L}} \underline{S}).$$

This product is represented by a G-map

$$H_{\mathscr{L}}\underline{R}^\alpha \wedge H_{\mathscr{L}}\underline{S}^\beta \to \Sigma^{-\mathscr{L}(G)} H_{\mathscr{L}}(\underline{R} \square_{\mathscr{L}} \underline{S})^{\alpha+\beta}.$$

(3) Taking $\delta = 0$, $\epsilon = \mathscr{L}$, and $\zeta = \mathscr{L}_\Delta$, we have the cup product

$$\tilde{H}_G^\alpha(X; \overline{T}) \otimes \tilde{\mathscr{H}}_G^\beta(Y; \underline{S}) \to \tilde{\mathscr{H}}_G^{\alpha+\beta}(X \wedge Y; \overline{T} \square_{\mathscr{L}_\Delta} \underline{S}).$$

This product is represented by a G-map

$$H\overline{T}^\alpha \wedge H_{\mathscr{L}}\underline{S}^\beta \to H_{\mathscr{L}}(\overline{T} \square_{\mathscr{L}_\Delta} \underline{S})^{\alpha+\beta}.$$

Note that $\mathscr{L}_\Delta$ restricts to $\mathscr{L}$ on the diagonal, so $\overline{T} \square_{\mathscr{L}_\Delta} \underline{S}$ is a contravariant $\widehat{\Pi}_{G,\mathscr{L}}B$-module, which we may also consider as a covariant $\widehat{\Pi}_{G,0}B$-module.

(4) Taking $\delta = 0$, $\epsilon = \mathscr{L}$, and $\zeta = \mathscr{L} - \mathscr{L}_\Delta$, we have the cup product

$$\tilde{H}_G^\alpha(X; \overline{T}) \otimes \tilde{\mathscr{H}}_G^\beta(Y; \underline{S}) \to \tilde{H}_G^{\alpha+\beta-\mathscr{L}(G)}(X \wedge Y; \overline{T} \square_{\mathscr{L}-\mathscr{L}_\Delta} \underline{S}).$$

This product is represented by a G-map

$$H\overline{T}^\alpha \wedge H_{\mathscr{L}}\underline{S}^\beta \to \Sigma^{-\mathscr{L}(G)} H(\overline{T} \square_{\mathscr{L}-\mathscr{L}_\Delta} \underline{S})^{\alpha+\beta}.$$

Note that $\overline{T} \square_{\mathscr{L}-\mathscr{L}_\Delta} \underline{S}$ is a contravariant $\widehat{\Pi}_{G,0}B$-module because $\mathscr{L} - \mathscr{L}_\Delta$ restricts to 0 on the diagonal.

3.10.2 Slant Products, Evaluations, and Cap Products

We now consider evaluation maps and cap products. See the comments at the beginning of Sect. 1.14.3.

Definition 3.10.14 Let δ be a dimension function for G, let ϵ be a dimension function for K, and let ζ be a dimension function for $G \times K$ with $\zeta \succcurlyeq \delta \times \epsilon$. Let A be a G-space and let B be a K-space. Suppose that $\overline{T}$ is a contravariant $\widehat{\Pi}_{K,\epsilon}B$-module and $\underline{U}$ is a covariant $\widehat{\Pi}_{G\times K,\zeta}(A \times B)$-module. Then we define a covariant $\widehat{\Pi}_{G,\delta}A$-module $\overline{T} \triangleleft \underline{U}$ by

$$(\overline{T} \triangleleft \underline{U})(a) = \overline{T} \otimes_{\widehat{\Pi}_{K,\epsilon}B} (p^* i_! \underline{U})(a \otimes -)$$

where

$$p\colon \widehat{\Pi}_{G,\delta}A \otimes \widehat{\Pi}_{K,\epsilon}B \to \mathrm{Ho}(G \times K)\mathscr{P}_{A\times B}$$

and

$$i\colon \widehat{\Pi}_{G\times K,\zeta}(A \times B) \to \mathrm{Ho}(G \times K)\mathscr{P}_{A\times B}$$

are the functors given in Definition 3.10.1.

Example 3.10.15 As an example and as a calculation we'll need later, we show that

$$\overline{A}_b \triangleleft \underline{A}^c \cong \mathrm{Ho}(G \times K)\mathscr{P}_{A\times B}(c, - \wedge b)$$

when $b \in \widehat{\Pi}_{K,\epsilon}B$ and $c \in \widehat{\Pi}_{G\times K,\zeta}(A \times B)$. For, if a is an object in $\widehat{\Pi}_{G,\delta}A$, we have

$$\begin{aligned}(\overline{A}_b \triangleleft \underline{A}^c)(a) &= \overline{A}_b \otimes_{\widehat{\Pi}_{K,\epsilon}B} (p^* i_! \underline{A}^c)(a \otimes -)\\ &= \overline{A}_b \otimes_{\widehat{\Pi}_{K,\epsilon}B} \mathrm{Ho}(G \times K)\mathscr{P}_{A\times B}(c, a \wedge -)\\ &\cong \mathrm{Ho}(G \times K)\mathscr{P}_{A\times B}(c, a \wedge b).\end{aligned}$$

For X an ex-G-space over A and Y an ex-K-space over B, the external slant product will be a map

$$- \backslash -\colon \tilde{H}^{\beta}_{K,\epsilon}(Y; \overline{T}) \otimes \tilde{H}^{G\times K,\zeta}_{\alpha+\beta}(X \wedge Y; \underline{U}) \to \tilde{H}^{G,\delta}_{\alpha}(X; \overline{T} \triangleleft \underline{U}).$$

On the chain level, we take the following map:

$$\mathrm{Hom}_{\widehat{\Pi}_{K,\epsilon}B}(\overline{C}^{K,\epsilon}_{\beta+*}(Y), \overline{T}) \otimes (\overline{C}^{G\times K,\zeta}_{\alpha+\beta+*}(X \wedge Y) \otimes_{\widehat{\Pi}_{G\times K,\zeta}(A\times B)} \underline{U})$$

$$\to \mathrm{Hom}_{\widehat{\Pi}_{K,\epsilon}B}(\overline{C}^{K,\epsilon}_{\beta+*}(Y), \overline{T}) \otimes ((\overline{C}^{G,\delta}_{\alpha+*}(X) \boxtimes_\zeta \overline{C}^{K,\epsilon}_{\beta+*}(Y)) \otimes_{\widehat{\Pi}_{G\times K,\zeta}(A\times B)} \underline{U})$$
$$\xrightarrow{\nu} (\overline{C}^{G,\delta}_{\alpha+*}(X) \boxtimes_\zeta \overline{T}) \otimes_{\widehat{\Pi}_{G\times K,\zeta}(A\times B)} \underline{U}$$
$$\cong (\overline{C}^{G,\delta}_{\alpha+*}(X) \otimes \overline{T}) \otimes_{\widehat{\Pi}_{G,\delta}A\otimes\widehat{\Pi}_{K,\epsilon}B} p^* i_! \underline{U}$$
$$\cong \overline{C}^{G,\delta}_{\alpha+*}(X) \otimes_{\widehat{\Pi}_{G,\delta}A} (\overline{T} \otimes_{\widehat{\Pi}_{K,\epsilon}B} p^* i_! \underline{U})$$
$$= \overline{C}^{G,\delta}_{\alpha+*}(X) \otimes_{\widehat{\Pi}_{G,\delta}A} (\overline{T} \lhd \underline{U}).$$

The first arrow is induced by the map μ_* from Definition 3.10.9. The map ν is evaluation, with the sign

$$\nu(y \otimes a \otimes b \otimes u) = (-1)^{pq} a \otimes y(b) \otimes u$$

if $y \in \mathrm{Hom}_{\widehat{\Pi}_{K,\epsilon}B}(\overline{C}^{K,\epsilon}_{\beta+p}(Y), \overline{T})$ and $a \in \overline{C}^{G,\delta}_{\alpha+q}(X)$. Our various sign conventions imply that the composite above is a chain map. Taking homology defines our slant product. The following properties follow easily from the definition.

Theorem 3.10.16 *Let δ be a familial dimension function for G, let ϵ be a familial dimension function for K, and let ζ be a familial dimension function for $G \times K$ with $\zeta \succcurlyeq \delta \times \epsilon$. Let α be a virtual representation of $\Pi_G A$ and let β be a virtual representation of $\Pi_K B$. The slant product*

$$- \setminus -\colon \tilde{H}^\beta_{K,\epsilon}(Y; \overline{T}) \otimes \tilde{H}^{G\times K,\zeta}_{\alpha+\beta}(X \wedge Y; \underline{U}) \to \tilde{H}^{G,\delta}_\alpha(X; \overline{T} \lhd \underline{U}).$$

has the following properties.

(1) *It is natural in the following sense: Given a G-map $f\colon X \Rightarrow X'$ over A, a K-map $g\colon Y \Rightarrow Y'$ over B, and elements $y' \in \tilde{H}^\beta_{K,\epsilon}(Y'; \overline{T})$ and $z \in \tilde{H}^{G\times K,\zeta}_{\alpha+\beta}(X \wedge Y; \underline{U})$, we have*

$$y' \setminus (f \wedge g)_*(z) = f_*(g^*(y') \setminus z).$$

Put another way, the slant product is a natural transformation in its adjoint form

$$\tilde{H}^{G\times K,\zeta}_{\alpha+\beta}(X \wedge Y; \underline{U}) \to \mathrm{Hom}(\tilde{H}^\beta_{K,\epsilon}(Y; \overline{T}), \tilde{H}^{G,\delta}_\alpha(X; \overline{T} \lhd \underline{U})).$$

(2) *It respects suspension in the sense that*

$$(\sigma^W y) \setminus (\sigma^{V+W} z) = \sigma^V (y \setminus z).$$

(3) *It is associative in the following sense. Given* $y \in \tilde{H}^{\beta}_{K,\epsilon}(Y;\overline{R})$, $z \in \tilde{H}^{\gamma}_{L,\zeta}(Z;\overline{S})$, *and* $w \in \tilde{H}^{G\times K\times L,\eta}_{\alpha+\beta+\gamma}(X\wedge Y\wedge Z;\underline{U})$, *where* $\eta \succcurlyeq \delta\times\epsilon\times\zeta$, *we have*

$$(y\cup z)\setminus w = y\setminus(z\setminus w).$$

□

We're most interested in the internalization of the slant product to the diagonal $\delta \le G\times G$. So, we also call the following a slant product.

Definition 3.10.17 Let δ and ϵ be familial dimension functions for G and let $\zeta \succcurlyeq \delta\times\epsilon$ be a familial dimension function for $G\times G$; assume that $\Delta \in \mathscr{F}(\zeta)$ and write ζ again for $\zeta|\Delta$. Let A and B be G-spaces, let X be an ex-G-space over A and let Y be an ex-G-space over B. If $\overline{T}$ is a contravariant $\widehat{\Pi}_{G,\epsilon}B$-module and $\underline{U}$ is a covariant $\widehat{\Pi}_{G,\zeta}(A\times B)$-module, write

$$\overline{T}\lhd\underline{U} = \overline{T}\lhd[(G\times G)\times_{\Delta}\underline{U}],$$

a covariant $\widehat{\Pi}_{G,\delta}A$-module defined using the external version of $\lhd$ on the right. The *internal slant product*

$$-\setminus -\colon \tilde{H}^{\beta}_{G,\epsilon}(Y;\overline{T})\otimes\tilde{H}^{G,\zeta}_{\alpha+\beta-\zeta(G\times G/\Delta)}(X\wedge Y;\underline{U}) \to \tilde{H}^{G,\delta}_{\alpha}(X;\overline{T}\lhd\underline{U})$$

is then the composite

$$\begin{aligned}
&\tilde{H}^{\beta}_{G,\epsilon}(Y;\overline{T})\otimes\tilde{H}^{G,\zeta}_{\alpha+\beta-\zeta(G\times G/\Delta)}(X\wedge Y;\underline{U})\\
&\quad\to \tilde{H}^{\beta}_{G,\epsilon}(Y;\overline{T})\otimes\tilde{H}^{G,\zeta}_{\alpha+\beta-\zeta(G\times G/\Delta)}(X\wedge Y;((G\times G)\times_{\Delta}\underline{U})|\Delta)\\
&\quad\cong \tilde{H}^{\beta}_{G,\epsilon}(Y;\overline{T})\otimes\tilde{H}^{G\times G,\zeta}_{\alpha+\beta}((G\times G)_+\wedge_{\Delta}(X\wedge Y);(G\times G)\times_{\Delta}\underline{U})\\
&\quad\to \tilde{H}^{\beta}_{G,\epsilon}(Y;\overline{T})\otimes\tilde{H}^{G\times G,\zeta}_{\alpha+\beta}(X\wedge Y;(G\times G)\times_{\Delta}\underline{U})\\
&\quad\to \tilde{H}^{G,\delta}_{\alpha}(X;\overline{T}\lhd\underline{U}).
\end{aligned}$$

The first map is the unit $\underline{U}\to((G\times G)\times_{\Delta}\underline{U})|\Delta$, the second is the Wirthmüller isomorphism, the third is induced by the $(G\times G)$-map $(G\times G)_+\wedge_{\Delta}(X\wedge Y)\to X\wedge Y$ over $A\times B$, and the last map is the external slant product.

We can now use the internal slant product to define evaluation and the cap product.

Definition 3.10.18 Let δ be a familial dimension function for G, let α be a virtual representation of G, and let β be a virtual representation of $\Pi_G B$. Let $\overline{T}$ be a contravariant $\widehat{\Pi}_{G,\delta}B$-module and let $\overline{U}$ be a covariant $\widehat{\Pi}_{G,\delta}B$-module. The *evaluation map*

$$\langle -,-\rangle\colon \tilde{H}^{\beta}_{G,\delta}(X;\overline{T})\otimes\tilde{H}^{G,\delta}_{\alpha+\beta}(X;\underline{U}) \to \tilde{H}^{G,0}_{\alpha}(S^0;\overline{T}\lhd\underline{U})$$

is the slant product

$$-\backslash -: \tilde{H}^{\beta}_{G,\delta}(X;\overline{T}) \otimes \tilde{H}^{G,\delta_\Delta}_{\alpha+\beta}(S^0 \wedge X;\underline{U}) \to \tilde{H}^{G,0}_{\alpha}(S^0;\overline{T} \triangleleft \underline{U})$$

where X is an ex-G-space over B and S^0 is considered as an ex-G-space over $*$. Here we are using $\delta_\Delta \succcurlyeq 0 \times \delta$ and the fact that $\delta_\Delta(G \times G/\Delta) = 0$ and $\delta_\Delta|\Delta = \delta$.

Note that we can express the naturality of evaluation by saying that the adjoint map

$$\tilde{H}^{\beta}_{G,\delta}(X;\overline{T}) \to \mathrm{Hom}(\tilde{H}^{G,\delta}_{\alpha+\beta}(X;\underline{U}), \tilde{H}^{G,0}_{\alpha}(S^0;\overline{T} \triangleleft \underline{U}))$$

is contravariant in X. When $\alpha = n \in \mathbb{Z}$, $\tilde{H}^{G,0}_{n}(S^0;\overline{T} \triangleleft \underline{U})$ is nonzero only when $n = 0$ and is then $(\overline{T} \triangleleft \underline{U})(G/G) \cong \overline{T} \otimes_{\widehat{\Pi}_{G,\delta}B} \underline{U}$, giving the evaluation

$$\tilde{H}^{\beta}_{G,\delta}(X;\overline{T}) \to \mathrm{Hom}(\tilde{H}^{G,\delta}_{\beta}(X;\underline{U}), \overline{T} \otimes_{\widehat{\Pi}_{G,\delta}B} \underline{U}).$$

There are many other interesting variations available, which we leave to the imagination of the reader.

Definition 3.10.19 Let δ and ϵ be familial dimension functions for G and let $\zeta \succcurlyeq \delta \times \epsilon$ be a familial dimension function for $G \times G$; assume that $\Delta \in \mathscr{F}(\zeta)$ and write ζ again for $\zeta|\Delta$. Let B be a G-space and let X be an ex-G-space over B. Let $\overline{T}$ be a contravariant $\widehat{\Pi}_{G,\epsilon}B$-module and let $\underline{U}$ be a covariant $\widehat{\Pi}_{G,\zeta}B$-module. The *cap product*

$$-\cap -: \tilde{H}^{\beta}_{G,\epsilon}(X;\overline{T}) \otimes \tilde{H}^{G,\zeta}_{\alpha+\beta-\zeta(G\times G/\Delta)}(X;\underline{U}) \to \tilde{H}^{G,\delta}_{\alpha}(X;\overline{T} \triangleleft \Delta^B_!\underline{U})$$

is the composite

$$\begin{aligned}
&\tilde{H}^{\beta}_{G,\epsilon}(X;\overline{T}) \otimes \tilde{H}^{G,\zeta}_{\alpha+\beta-\zeta(G\times G/\Delta)}(X;\underline{U}) \\
&\qquad \to \tilde{H}^{\beta}_{G,\epsilon}(X;\overline{T}) \otimes \tilde{H}^{G,\zeta}_{\alpha+\beta-\zeta(G\times G/\Delta)}(X \wedge X;\Delta^B_!\underline{U}) \\
&\qquad \to \tilde{H}^{G,\delta}_{\alpha}(X;\overline{T} \triangleleft \Delta^B_!\underline{U})
\end{aligned}$$

where the first map is induced by the diagonal $X \to X \wedge X$ and the unit $\underline{U} \to (\Delta^B)^*\Delta^B_!\underline{U}$, $\Delta^B: B \to B \times B$ being the diagonal.

We get interesting special cases by considering particular choices of δ, ϵ, and ζ. One we'll use later is the case $\delta = \mathscr{L} - \epsilon$ and $\zeta = \mathscr{L}_\Delta$. Using the facts that $\mathscr{L}_\Delta(G \times G/\Delta) = 0$ and $\mathscr{L}_\Delta|\Delta = \mathscr{L}$, the cap product in this case takes the form

$$-\cap -: \tilde{H}^{\beta}_{G,\epsilon}(X;\overline{T}) \otimes \tilde{\mathscr{H}}^{G}_{\alpha+\beta}(X;\underline{U}) \to \tilde{H}^{G,\mathscr{L}-\epsilon}_{\alpha}(X;\overline{T} \triangleleft \Delta^B_!\underline{U}).$$

Specializing further, we can use $\underline{U} = \overline{A}_{G/G} = \rho^*\overline{A}_{G/G}$ (considered as a covariant $\widehat{\Pi}_{G,\mathscr{L}}B$-module) and the following map.

Proposition 3.10.20 *If $\overline{T}$ is a contravariant $\widehat{\Pi}_{G,\epsilon}B$-module, then there is a natural map $\overline{T} \lhd \Delta_!^B\overline{A}_{G/G} \to \overline{T}$ of covariant $\widehat{\Pi}_{G,\mathscr{L}-\epsilon}B$-modules, which agrees with the isomorphism of Proposition 1.14.24 when $B = *$.*

Proof Write $\delta = \mathscr{L} - \epsilon$. Let

$$\Delta\colon \widehat{\Pi}_{G,\mathscr{L}}B \to \widehat{\Pi}_{G\times G,\mathscr{L}_\Delta}(B \times B)$$

be the combination of the inclusion of the diagonal of $B \times B$ and extension from the diagonal copy of G to $G \times G$. For $a \in \widehat{\Pi}_{G,\delta}B$, let $D(a)$ denote the homological dual of a, considered as an element of $\widehat{\Pi}_{G,\epsilon}B$. We first define a map

$$(p^* i_! \Delta_! \overline{A}_{G/G})(a \otimes -) \to \widehat{\Pi}_{G,\epsilon}B(D(a), -)$$

as follows, for $b \in \widehat{\Pi}_{G,\epsilon}B$:

$$\begin{aligned}
&(p^* i_! \Delta_! \overline{A}_{G/G})(a \otimes b) \\
&\quad\cong \int^{c\in\widehat{\Pi}_{G,\mathscr{L}}B} \operatorname{Ho}(G\times G)\mathscr{P}_{B\times B}(\Delta_!(c), a \wedge_{B\times B} b) \otimes \operatorname{Ho} G\mathscr{P}(S, \rho_!(c)) \\
&\quad\cong \int^{c\in\widehat{\Pi}_{G,\mathscr{L}}B} \operatorname{Ho} G\mathscr{P}_B(c, \Delta^*(a \wedge_{B\times B} b)) \otimes \operatorname{Ho} G\mathscr{P}(S, \rho_!(c)) \\
&\quad\cong \int^{c\in\widehat{\Pi}_{G,\mathscr{L}}B} \operatorname{Ho} G\mathscr{P}_B(c, a \wedge_B b) \otimes \operatorname{Ho} G\mathscr{P}(S, \rho_!(c)) \\
&\quad\to \operatorname{Ho} G\mathscr{P}(S, \rho_!(a \wedge_B b)) \\
&\quad\cong \operatorname{Ho} G\mathscr{P}_B(D(a), b) \\
&\quad= \widehat{\Pi}_{G,\epsilon}B(D(a), b),
\end{aligned}$$

where the map is given by composition. The map of the proposition is then

$$\begin{aligned}
(\overline{T} \lhd \Delta_!^B\overline{A}_{G/G})(a) &= \overline{T} \otimes_{\widehat{\Pi}_{G,\epsilon}B} (p^* i_! \Delta_! \overline{A}_{G/G})(a \otimes -) \\
&\to \overline{T} \otimes_{\widehat{\Pi}_{G,\epsilon}B} \widehat{\Pi}_{G,\epsilon}B(D(a), -) \\
&\cong \overline{T}(D(a)).
\end{aligned}$$

∎

Using this map, we have a cap product

$$-\cap -: \tilde{H}^{\beta}_{G,\epsilon}(X;\overline{T}) \otimes \tilde{\mathscr{H}}^{G}_{\alpha+\beta}(X;\overline{A}_{G/G}) \to \tilde{H}^{G,\mathscr{L}-\epsilon}_{\alpha}(X;\overline{T}).$$

Both the evaluation map and the cap product inherit properties from Theorem 3.10.16. In particular, the associativity property gives us the following:

$$(x \cup y) \cap z = x \cap (y \cap z)$$

and

$$\langle x \cup y, z\rangle = \langle x, y \cap z\rangle$$

when x, y, and z lie in appropriate groups.

We now look at these pairings on the spectrum level. The main calculation is the following.

Proposition 3.10.21 *Let δ be a familial dimension function for G, let ϵ be a familial dimension function for K, and let ζ be a familial dimension function for $G \times K$ with $\zeta \succcurlyeq \delta \times \epsilon$. Let A be a G-space, let α be a virtual representation of $\Pi_G A$, let B be a K-space, and let β be a virtual representation of $\Pi_K B$. Let $\overline{T}$ be a contravariant $\widehat{\Pi}_{K,\epsilon}B$-module and let $\underline{U}$ be a covariant $\widehat{\Pi}_{G\times K,\zeta}(A\times B)$-module. Then, if $q: A\times B \to A$ is the projection, we have*

$$\overline{\pi}^{G,\mathscr{L}-\delta}_{-\alpha+n}(q_!(H_\epsilon \overline{T}^{\beta} \wedge_B H^{\zeta}\underline{U}_{\alpha+\beta})^K) \cong \begin{cases} 0 & \text{if } n < 0 \\ \overline{T} \triangleleft \underline{U} & \text{if } n = 0. \end{cases}$$

Proof The proof is similar to that of Proposition 1.14.25. If $a: G/H \to A$, write as shorthand

$$S^{\alpha,\delta,a} = G_+ \wedge_H S^{\alpha_0(a)-\delta(G/H),a}.$$

The necessary calculational input is that, for $a: G/H \to A$, $b: K/L \to B$, and $c: G \times K/M \to A \times B$,

$$\begin{aligned} \mathrm{Ho}\, G\mathscr{P}_A(&S^{-\alpha,\mathscr{L}-\delta,a}, q_!(S^{\beta,\epsilon,b} \wedge_B S^{-\alpha-\beta,\mathscr{L}-\zeta,c})^K) \\ &\cong \mathrm{Ho}(G\times K)\mathscr{P}_A(S^{-\alpha,\mathscr{L}-\delta,a}, q_!(S^{\beta,\epsilon,b} \wedge_B S^{-\alpha-\beta,\mathscr{L}-\zeta,c})) \\ &\cong \mathrm{Ho}(G\times K)\mathscr{P}_{A\times B}(S^{\alpha+\beta,\zeta,c}, S^{\alpha,\delta,a} \wedge S^{\beta,\epsilon,b}) \end{aligned}$$

using homological duality, so that

$$\mathrm{Ho}\, G\mathscr{P}_A(S^{-\alpha,\mathscr{L}-\delta,a}, q_!(S^{\beta,\epsilon,b} \wedge_B S^{-\alpha-\beta,\mathscr{L}-\zeta,c})^K) \cong (\overline{A}_b \triangleleft \underline{A}^c)(a)$$

as functors in $a \in \widehat{\Pi}_{G,\delta}A$, using Example 3.10.15 and Theorem 2.6.4. The result now follows by analyzing the structure provided by Construction 3.7.1. □

It follows that there is a map

$$q_!(H_\epsilon \overline{T}^\beta \wedge_B H^\zeta \underline{U}_{\alpha+\beta})^K \to P^\delta(\overline{T} \lhd \underline{U})_\alpha$$

of G-spectra over A that is an isomorphism on $\overline{\pi}_{-\alpha}^{G,\mathscr{L}-\delta}$, where $P^\delta(\overline{T} \lhd \underline{U})_\alpha$ is a spectrum with $\overline{\pi}_{-\alpha+n}^{G,\mathscr{L}-\delta}$ concentrated at $n = 0$, as in Construction 3.7.1. We then use that $E\mathscr{F}(\zeta) \times p_1^* E\mathscr{F}(\delta) \simeq E\mathscr{F}(\zeta)$ because $\mathscr{F}(\zeta) \subset \mathscr{F}(\delta) \times \mathscr{A}(K)$, where $\mathscr{A}(K)$ is the collection of all subgroups of K, so that

$$H^\zeta \underline{U}_{\alpha+\beta} \wedge p_1^* E\mathscr{F}(\delta)_+ \simeq H^\zeta \underline{U}_{\alpha+\beta},$$

to get a map

$$\begin{aligned} q_!(H_\epsilon \overline{T}^\beta \wedge_B H^\zeta \underline{U}_{\alpha+\beta})^K &\simeq q_!(H_\epsilon \overline{T}^\beta \wedge_B H^\zeta \underline{U}_{\alpha+\beta} \wedge p_1^* E\mathscr{F}(\delta)_+)^K \\ &\simeq q_!(H_\epsilon \overline{T}^\beta \wedge_B H^\zeta \underline{U}_{\alpha+\beta})^K \wedge E\mathscr{F}(\delta)_+ \\ &\to P^\delta(\overline{T} \lhd \underline{U})_\alpha \wedge E\mathscr{F}(\delta)_+ \\ &\simeq H^\delta(\overline{T} \lhd \underline{U})_\alpha. \end{aligned}$$

The slant product

$$- \backslash -: \tilde{H}^\beta_{K,\epsilon}(Y; \overline{T}) \otimes \tilde{H}^{G\times K,\zeta}_{\alpha+\beta}(X \wedge Y; \underline{U}) \to \tilde{H}^{G,\delta}_\alpha(X; \overline{T} \lhd \underline{U}).$$

is then represented as follows:

$$\begin{aligned} &[Y, H_\epsilon \overline{T}^\beta]_{K,B} \otimes [S, \rho_!(H^\zeta \underline{U}_{\alpha+\beta} \wedge_{A\times B} (X \wedge Y))]_{G\times K} \\ &\quad \to [S, \rho_!(H^\zeta \underline{U}_{\alpha+\beta} \wedge_{A\times B} (X \wedge H_\epsilon \overline{T}^\beta))]_{G\times K} \\ &\quad \cong [S, \rho_!((H_\epsilon \overline{T}^\beta \wedge_B H^\zeta \underline{U}_{\alpha+\beta}) \wedge_A X)]_{G\times K} \\ &\quad \cong [S, \rho_!((H_\epsilon \overline{T}^\beta \wedge_B H^\zeta \underline{U}_{\alpha+\beta}) \wedge_A X)^K]_G \\ &\quad \cong [S, \rho_!(q_!(H_\epsilon \overline{T}^\beta \wedge_B H^\zeta \underline{U}_{\alpha+\beta})^K \wedge_A X)]_G \\ &\quad \to [S, \rho_!(H^\delta(\overline{T} \lhd \underline{U})_\alpha \wedge_A X)]_G. \end{aligned}$$

3.11 The Thom Isomorphism and Poincaré Duality

3.11.1 The Thom Isomorphism

Definition 3.11.1 Let $p: E \to B$ be a G-vector bundle and let γ be the associated representation of $\Pi_G B$. Let $T(p) = D(p)/_B S(p)$ be the *fiberwise Thom space* of p. A *Thom class* for p is an element $t \in \tilde{H}_G^{\gamma}(T(p); \overline{A}_{G/G})$ such that, for each G-map $b: G/K \to B$,

$$\begin{aligned} b^*(t) &\in \tilde{H}_G^{\gamma(b)}(b^*T(p); \overline{A}_{G/G}) \\ &\cong \tilde{H}_G^{\gamma(b)}(G \times_K S^V; \overline{A}_{G/G}) \\ &\cong \tilde{H}_K^V(S^V; \overline{A}_{K/K}) \\ &\cong A(K) \end{aligned}$$

is a generator. Here, V is a representation of K such that $\gamma(b) \simeq G \times_K S^V$.

As in Remark 1.15.2, the Thom class must live in ordinary cohomology $\tilde{H}^*_{G,\delta}$ with $\delta = 0$.

Since a global Thom class is characterized by its local behavior, and locally a G-vector bundle is a V-bundle, the next proposition follows from the $RO(G)$-graded analogue, Proposition 1.15.3.

Proposition 3.11.2 *The following are equivalent for a cohomology class* $t \in \tilde{H}_G^{\gamma}(T(p); \overline{A}_{G/G})$.

(1) *t is a Thom class for p.*
(2) *For every subgroup $K \subset G$, $t|K \in \tilde{H}_K^{\gamma|K}(T(p); \overline{A}_{K/K})$ is a Thom class for p as a K-bundle.*
(3) *For every subgroup $K \subset G$, $t^K \in \tilde{H}_{WK}^{\gamma^K}(T(p)^K; \overline{A}_{WK/WK})$ is a Thom class for $p^K: E^K \to B^K$ as a WK-bundle.*
(4) *For every subgroup $K \subset G$, $t^K|e \in \tilde{H}^{|\gamma^K|}(T(p)^K; \mathbb{Z})$ is a Thom class for p^K as a nonequivariant bundle.*

□

Note that, in the last part of the proposition, $|\gamma^K|$ must be interpreted as a representation of the nonequivariant fundamental groupoid of B^K, so that the cohomology may be twisted.

The following generalizes Theorem C of [14].

Theorem 3.11.3 (Thom Isomorphism) *If $p: E \to B$ is any G-vector bundle and γ is the associated representation of $\Pi_G B$, then there exists a Thom class $t \in \tilde{H}^{\gamma}_{G}(T(p); \overline{A}_{G/G})$. For any Thom class t, the map*

$$t \cup -: H^{\alpha}_{G,\delta}(B; \overline{T}) \to \tilde{H}^{\alpha+\gamma}_{G,\delta}(T(p); \overline{T})$$

is an isomorphism for any familial δ.

Proof As in the proof of Theorem 1.15.5, the theorem is clear in the special case that p is a bundle over an orbit. The general case follows, as it does in the nonequivariant case for twisted coefficients, by a Mayer-Vietoris patching argument. The key point is that we can choose a compatible collection of local classes because the action of $\Pi_G B$ on γ is the same as the action on the fibers of p. □

3.11.2 Poincaré Duality

We are now in a position to describe Poincaré duality for arbitrary compact smooth G-manifolds. (Again, the noncompact case can be handled using cohomology with compact supports.)

Definition 3.11.4 Let M be a closed smooth G-manifold and let τ be the tangent representation of $\Pi_G M$, i.e., the representation associated with the tangent bundle. Think of M as a G-space over itself in the following. A *fundamental class* of M is a class $[M] \in \mathscr{H}^G_\tau(M; \overline{A}_{G/G})$ such that, for each point $m \in M$, thought of as the map $m: G/G_m \to M$ with image Gm, and tangent plane $\tau(m) = G \times_{G_m} V$, the image of $[M]$ in

$$\begin{aligned}\mathscr{H}^G_\tau(M, M - Gm; \overline{A}_{G/G}) &\cong \tilde{\mathscr{H}}^G_V(G_+ \wedge_{G_m} S^{V - \mathscr{L}(G/G_m)}; \overline{A}_{G/G}) \\ &\cong \tilde{\mathscr{H}}^{G_m}_V(S^V; A_{G/G}) \\ &\cong A(G_m)\end{aligned}$$

is a generator.

The fundamental class $[M]$ is related to fundamental classes of the fixed submanifolds M^K as in Proposition 1.15.7.

Proposition 3.11.5 *The following are equivalent for a dual homology class $\mu \in \mathscr{H}^G_\tau(M; \overline{A}_{G/G})$.*

(1) *μ is a fundamental class for M as a G-manifold.*
(2) *For every subgroup $K \subset G$, $\mu|K \in \mathscr{H}^K_{\tau|K}(M; \overline{A}_{K/K})$ is a fundamental class for M as a K-manifold.*
(3) *For every subgroup $K \subset G$, $\mu^K \in \mathscr{H}^{WK}_{\tau^K}(M^K; \overline{A}_{WK/WK})$ is a fundamental class for M^K as a WK-manifold.*

(4) *For every subgroup $K \subset G$, $\mu^K|e \in H_{|\tau^K|}(M^K;\mathbb{Z})$ is a fundamental class for M^K as a nonequivariant manifold.* □

Note that, in the last part of the proposition, $|\tau^K|$ must be interpreted as a representation of the nonequivariant fundamental groupoid of M^K, so that the homology may be twisted.

We now have sufficient machinery in place to prove Poincaré duality for arbitrary G-manifolds either geometrically, along the lines of [52] or [14], or homotopically as in Sect. 1.15. The homotopical approach uses, of course, the duality of Theorem 2.9.11.

Theorem 3.11.6 (Poincaré Duality) *Every closed smooth G-manifold M has a fundamental class $[M] \in \mathscr{H}_\tau^G(M;\overline{A}_{G/G})$, and*

$$- \cap [M]: H^\gamma_{G,\delta}(M;\overline{T}) \to H^{G,\mathscr{L}-\delta}_{\tau-\gamma}(M;\overline{T})$$

is an isomorphism for every familial δ such that M is an $\mathscr{F}(\delta)$-manifold (i.e., $\mathscr{F}(\delta)$ contains every isotropy subgroup of M). □

In particular, when $\delta = 0$ we get the isomorphism

$$- \cap [M]: H^\gamma_G(M;\overline{T}) \to \mathscr{H}^G_{\tau-\gamma}(M;\overline{T})$$

and when $\delta = \mathscr{L}$ we get the isomorphism

$$- \cap [M]: \mathscr{H}^\gamma_G(M;\overline{T}) \to H^G_{\tau-\gamma}(M;\overline{T}).$$

If M is a compact G-manifold with boundary, then we get relative, or Lefschetz, duality.

Definition 3.11.7 Let M be a compact G-manifold with boundary, with tangent representation τ. A *fundamental class* of M is a dual homology class $[M, \partial M] \in \mathscr{H}_\tau^G(M, \partial M;\overline{A}_{G/G})$ such that, for each point $m \in M - \partial M$, thought of as the map $m: G/G_m \to M$ with image Gm, and tangent plane $\tau(m) = G \times_{G_m} V$, the image of $[M, \partial M]$ in

$$\begin{aligned}\mathscr{H}_\tau^G(M, M - Gm;\overline{A}_{G/G}) &\cong \tilde{\mathscr{H}}_V^G(G_+ \wedge_{G_m} S^{V-\mathscr{L}(G/G_m)};\overline{A}_{G/G})\\ &\cong \tilde{\mathscr{H}}_V^{G_m}(S^V;\overline{A}_{G/G})\\ &\cong A(G_m)\end{aligned}$$

is a generator.

There is an obvious relative version of Proposition 3.11.5.

Theorem 3.11.8 (Lefschetz Duality) *Every compact smooth G-manifold M has a fundamental class* $[M, \partial M] \in \mathscr{H}_\tau^G(M, \partial M; \overline{A}_{G/G})$, *and the following are isomorphisms for every familial dimension function* δ *such that M is an* $\mathscr{F}(\delta)$*-manifold:*

$$- \cap [M, \partial M]: H^\gamma_{G,\delta}(M; \overline{T}) \to H^{G,\mathscr{L}-\delta}_{\tau-\gamma}(M, \partial M; \overline{T})$$

and

$$- \cap [M, \partial M]: H^\alpha_{G,\delta}(M, \partial M; \overline{T}) \to H^{G,\mathscr{L}-\delta}_{\tau-\alpha}(M; \overline{T}).$$

□

As in the nonparametrized version, we get the following special cases when $\delta = 0$ or $\delta = \mathscr{L}$:

$$\begin{aligned}
&- \cap [M, \partial M]: H^\gamma_G(M; \overline{T}) \to \mathscr{H}^G_{\tau-\gamma}(M, \partial M; \overline{T}),\\
&- \cap [M, \partial M]: H^\gamma_G(M, \partial M; \overline{T}) \to \mathscr{H}^G_{\tau-\gamma}(M; \overline{T}),\\
&- \cap [M, \partial M]: \mathscr{H}^\gamma_G(M; \overline{T}) \to H^G_{\tau-\gamma}(M, \partial M; \overline{T}), \qquad \text{and}\\
&- \cap [M, \partial M]: \mathscr{H}^\gamma_G(M, \partial M; \overline{T}) \to H^G_{\tau-\gamma}(M; \overline{T}).
\end{aligned}$$

3.12 A Calculation

In this section, all cohomology is assumed to have coefficients in $\overline{A}_{G/G}$, which we will drop from the notation for simplicity.

In [66], the second author defined equivariant Chern classes in $RO(G)$-graded cohomology, for complex vector bundles modeled on a single representation. We can generalize that definition now as follows. Let ω be a complex vector bundle over the G-space B, and also write ω for the associated (real) representation of ΠB. By the Thom isomorphism (3.11.3), there exists a Thom class $t(\omega) \in \tilde{H}^\omega_G(T(\omega))$, and we let $c_\omega(\omega) = e(\omega) \in H^\omega_G(B)$ be its Euler class, the restriction of the Thom class to the zero section. Now consider the Gysin sequence of ω, the long exact sequence induced by the cofibration $S(\omega)_+ \to D(\omega)_+ \to T(\omega)$ over B. Part of that sequence is the following:

$$0 = H^{-2}_G(B) \to H^{\omega-2}_G(B) \to H^{\omega-2}_G(S(\omega)) \to H^{-1}_G(B) = 0.$$

When we pull back the bundle ω along the projection $p: S(\omega) \to B$, we get $p^*\omega \cong \omega' \oplus \mathbb{C}$, so we get the Euler class $e(\omega') \in H^{\omega-2}_G(B)$, which we write as $c_{\omega-2}(\omega)$.

Definition 3.12.1 Let ω be a complex vector bundle over B. We define the *equivariant Chern classes* of ω by letting $c_\omega(\omega)$ be its Euler class, and then defining classes

$$c_{\omega-2i}(\omega) \in H_G^{\omega-2i}(B) \qquad 0 \le i \le \dim_{\mathbb{C}} \omega$$

inductively as above.

Recall that Lewis calculated the $RO(G)$-graded cohomology of $B_G U(1)$ for $G = \mathbb{Z}/p$; the result for $G = \mathbb{Z}/2$ was stated in Theorem 1.17.2. By expanding the grading to $RO(\Pi_G B_G U(1))$, we can get a calculation that includes the Chern classes of the canonical line bundle. The following calculation for $G = \mathbb{Z}/2$ was made by the first author (using a different notation for the elements) in [12]; an extension to the case $G = \mathbb{Z}/p$ is work in progress.

For the remainder of this section, let $G = \mathbb{Z}/2$. Recall that we write Λ for the nontrivial irreducible representation of G. As a model of $B_G U(1)$ we take the infinite projective space $\mathbb{C}P(\mathbb{C}^\infty \oplus \Lambda_c^\infty)$, where $\Lambda_c = \Lambda \otimes_{\mathbb{R}} \mathbb{C}$. Its fixed set has two components, $\mathbb{C}P(\mathbb{C}^\infty)$ and $\mathbb{C}P(\Lambda_c^\infty)$, each a copy of the nonequivariant $BU(1)$.

There is a G-involution χ on $\mathbb{C}P(\mathbb{C}^\infty \oplus \Lambda_c^\infty)$ that swaps $\mathbb{C}^\infty$ and Λ_c^∞; its effect on fixed sets is to swap the two components. (This involution represents the operation of tensoring a line bundle with Λ_c.)

Proposition 3.12.2 *Let* $G = \mathbb{Z}/2$. $RO(\Pi_G B_G U(1))$ *is a free abelian group with generators* 1, Λ, *and* Ω, *where* 1 *and* Λ *are the classes of the constant representations at* $\mathbb{R}$ *and* Λ, *respectively, and* Ω *is a representation that restricts to* $\mathbb{R} - \Lambda$ *on* $\mathbb{C}P(\mathbb{C}^\infty)$ *and* $\Lambda - \mathbb{R}$ *on* $\mathbb{C}P(\Lambda_c^\infty)$. *The involution* χ *induces an involution* χ^* *on* $RO(\Pi_G B_G U(1))$ *with* $\chi^*(a + b\Lambda + c\Omega) = a + b\Lambda - c\Omega$. □

Let ω be the canonical line bundle over $B_G U(1)$. The associated element of $RO(\Pi_G B_G U(1))$ is then

$$\omega = 1 + \Lambda + \Omega.$$

The element associated to $\chi^*\omega = \omega \otimes \Lambda_c$ is

$$\chi\omega = 1 + \Lambda - \Omega.$$

The calculation is then the following. It uses the calculation of the cohomology of a point, as given in Theorem 1.17.1.

Theorem 3.12.3 *Let* $G = \mathbb{Z}/2$. *Let* c_ω *and* $c_{\omega-2}$ *be the Chern classes of the canonical line bundle over* $B_G U(1)$ *and let* $c_{\chi\omega}$ *and* $c_{\chi\omega-2}$ *be the Chern classes of* $\chi^*\omega$. *Then, as an* $RO(\Pi_G B_G U(1))$*-graded ring,* $H_G^*(B_G U(1))$ *is a free* $H_G^*(*)$*-module generated multiplicatively by the elements* c_ω, $c_{\omega-2}$, $c_{\chi\omega}$, *and* $c_{\chi\omega-2}$, *subject* *to the following two relations:*

$$c_{\omega-2}c_{\chi\omega-2} = \xi \quad \textit{and}$$

$$c_{\omega-2}c_{\chi\omega} = \epsilon^2 - (1-g)c_{\chi\omega-2}c_{\omega},$$

where $\epsilon \in H_G^{\Lambda}(*)$, $\xi \in H_G^{-2+2\Lambda}(*)$, *and* $g \in H_G^0(*)$ *are elements in the cohomology of a point; in particular,* $g = [G/e] \in A(G)$. *The involution* χ^* *acts as a ring map with*

$$\chi^*(c_\omega) = c_{\chi\omega} \quad \textit{and}$$

$$\chi^*(c_{\omega-2}) = c_{\chi\omega-2}.$$

□

The connection with the $RO(G)$-graded calculation, Theorem 1.17.2, is that the elements γ and Γ used there are given by

$$\gamma = c_{\chi\omega-2}c_{\omega} \quad \text{and} \quad \Gamma = c_{\omega}c_{\chi\omega}.$$

It is our hope that more calculations like these can be done and will contribute to a theory of equivariant characteristic classes.

Bibliography

1. J.F. Adams, *Stable Homotopy and Generalised Homology*. Chicago Lectures in Mathematics (University of Chicago Press, Chicago, IL, 1995). Reprint of the 1974 original. MR 1324104 (96a:55002)
2. J.M. Boardman, Conditionally convergent spectral sequences, in *Homotopy Invariant Algebraic structures (Baltimore, MD, 1998)*. Contemporary Mathematics, vol. 239 (American Mathematical Society, Providence, RI, 1999), pp. 49–84. MR MR1718076 (2000m:55024)
3. A. Borel, in *Seminar on Transformation Groups*, With Contributions ed. By G. Bredon, E.E. Floyd, D. Montgomery, R. Palais. Annals of Mathematics Studies, vol. 46 (Princeton University Press, Princeton, NJ, 1960). MR 0116341 (22 #7129)
4. G.E. Bredon, *Equivariant Cohomology Theories*. Lecture Notes in Mathematics, vol. 34 (Springer, Berlin, 1967). MR MR0214062 (35 #4914)
5. R. Brown, *Elements of Modern Topology* (McGraw-Hill Book Co., New York, 1968). MR MR0227979 (37 #3563)
6. J.L. Caruso, Operations in equivariant $\mathbf{Z}/p$-cohomology. Math. Proc. Camb. Philos. Soc. **126**(3), 521–541 (1999). MR 1684248 (2000d:55028)
7. M. Clapp, Duality and transfer for parametrized spectra. Arch. Math. (Basel) **37**(5), 462–472 (1981). MR MR643290 (83i:55010)
8. M. Clapp, D. Puppe, The homotopy category of parametrized spectra. Manuscr. Math. **45**(3), 219–247 (1984). MR MR734840 (85c:55007)
9. S.R. Costenoble, The equivariant Conner-Floyd isomorphism. Trans. Am. Math. Soc. **304**(2), 801–818 (1987). MR 911096 (88h:57033)
10. S.R. Costenoble, The structure of some equivariant Thom spectra. Trans. Am. Math. Soc. **315**(1), 231–254 (1989). MR MR958887 (89m:57038)
11. S.R. Costenoble, Classifying spaces of bundles of monoids (preprint, 2002)
12. S.R. Costenoble, The $\mathbb{Z}/2$ cohomology of $\mathbb{C}P^\infty$, preprint, arXiv:1312.0926 [math.AT] (2013)
13. S.R. Costenoble, S. Waner, Fixed set systems of equivariant infinite loop spaces. Trans. Am. Math. Soc. **326**(2), 485–505 (1991). MR MR1012523 (91k:55015)
14. S.R. Costenoble, S. Waner, Equivariant Poincaré duality. Mich. Math. J. **39**(2), 325–351 (1992). MR MR1162040 (93d:55007)
15. S.R. Costenoble, S. Waner, The equivariant Spivak normal bundle and equivariant surgery. Mich. Math. J. **39**(3), 415–424 (1992). MR MR1182497 (94d:57058)
16. S.R. Costenoble, S. Waner, The equivariant Thom isomorphism theorem. Pac. J. Math. **152**(1), 21–39 (1992). MR MR1139971 (93c:55005)
17. S.R. Costenoble, S. Waner, Equivariant simple Poincaré duality. Mich. Math. J. **40**(3), 577–604 (1993). MR MR1236180 (94m:55006)

S.R. Costenoble, S. Waner, *Equivariant Ordinary Homology and Cohomology*,
Lecture Notes in Mathematics 2178, DOI 10.1007/978-3-319-50448-3

18. S.R. Costenoble, S. Waner, Equivariant ordinary homology and cohomology, preprint, arXiv:math/0310237v1 [math.AT] (2003)
19. S.R. Costenoble, S. Waner, Equivariant vector fields and self-maps of spheres. J. Pure Appl. Algebra **187**(1–3), 87–97 (2004). MR MR2027897 (2005a:57026)
20. S.R. Costenoble, J.P. May, S. Waner, Equivariant orientation theory. Homology Homotopy Appl. **3**(2), 265–339 (2001). (electronic), Equivariant stable homotopy theory and related areas (Stanford, CA, 2000). MR MR1856029 (2002j:55016)
21. A. Dold, Chern classes in general cohomology, in *Symposia Mathematica (INDAM, Rome, 1969/1970)*, vol. V (Academic, London, 1971), pp. 385–410. MR MR0276968 (43 #2707)
22. A.W.M. Dress, Contributions to the theory of induced representations, in *Algebraic K-Theory, II: "Classical" Algebraic K-Theory and Connections with Arithmetic (Proceeding Conference, Battelle Memorial Institute, Seattle, Washington, 1972)*. Lecture Notes in Mathematics, vol. 342 (Springer, Berlin, 1973), pp. 183–240. MR MR0384917 (52 #5787)
23. D. Dugger, An Atiyah-Hirzebruch spectral sequence for *KR*-theory. *K*-Theory **35**(3–4), 213–256 (2005/2006). MR 2240234 (2007g:19004)
24. D. Dugger, Bigraded cohomology of $\mathbb{Z}/2$-equivariant Grassmannians. Geom. Topol. **19**(1), 113–170 (2015). MR 3318749
25. A.D. Elmendorf, Systems of fixed point sets. Trans. Am. Math. Soc. **277**(1), 275–284 (1983). MR 690052
26. A.D. Elmendorf, I. Kříž, M.A. Mandell, J.P. May, Modern foundations for stable homotopy theory, in *Handbook of Algebraic Topology* (North-Holland, Amsterdam, 1995), pp. 213–253. MR 1361891 (97d:55016)
27. J.P.C. Greenlees, J.P. May, Equivariant stable homotopy theory, in *Handbook of Algebraic Topology* (North-Holland, Amsterdam, 1995), pp. 277–323. MR MR1361893 (96j:55013)
28. A. Grothendieck, *Revêtements Étales et Groupe Fondamental* (Springer, Berlin, 1971). Séminaire de Géométrie Algébrique du Bois Marie 1960–1961 (SGA 1), Dirigé par Alexandre Grothendieck. Augmenté de deux exposés de M. Raynaud, Lecture Notes in Mathematics, vol. 224. MR MR0354651 (50 #7129)
29. H. Hauschild, Äquivariante Transversalität und äquivariante Bordismentheorien. Arch. Math. (Basel) **26**(5), 536–546 (1975). MR 0402787 (53 #6601)
30. H. Hauschild, Zerspaltung äquivarianter Homotopiemengen. Math. Ann. **230**(3), 279–292 (1977). MR 0500966 (58 #18451)
31. S. Illman, Equivariant singular homology and cohomology. I. Mem. Am. Math. Soc. **1**(2), 156, ii+74 (1975). MR MR0375286 (51 #11482)
32. S. Illman, The equivariant triangulation theorem for actions of compact Lie groups. Math. Ann. **262**(4), 487–501 (1983). MR MR696520 (85c:57042)
33. H. Kleisli, Every standard construction is induced by a pair of adjoint functors. Proc. Am. Math. Soc. **16**, 544–546 (1965). MR MR0177024 (31 #1289)
34. C. Kosniowski, A note on *RO*(*G*) graded *G* bordism theory. Q. J. Math. Oxford Ser. (2) **26**(104), 411–419 (1975). MR MR0388418 (52 #9254)
35. W.C. Kronholm, The RO(*G*)-graded Serre spectral sequence. Homology Homotopy Appl. **12**(1), 75–92 (2010). MR 2607411 (2011e:55023)
36. L.G. Lewis Jr., The theory of Green functors, unpublished manuscript (1981)
37. L.G. Lewis Jr., The *RO*(*G*)-graded equivariant ordinary cohomology of complex projective spaces with linear $\mathbf{Z}/p$ actions, in *Algebraic topology and transformation groups (Göttingen, 1987)*. Lecture Notes in Mathematics, vol. 1361 (Springer, Berlin, 1988), pp. 53–122. MR 979507 (91h:55003)
38. L.G. Lewis Jr., Equivariant Eilenberg-Mac Lane spaces and the equivariant Seifert-van Kampen and suspension theorems. Topol. Appl. **48**(1), 25–61 (1992). MR MR1195124 (93i:55016)
39. L.G. Lewis Jr., The equivariant Hurewicz map. Trans. Am. Math. Soc. **329**(2), 433–472 (1992). MR 92j:55024

40. L.G. Lewis Jr., M.A. Mandell, Equivariant universal coefficient and Künneth spectral sequences. Proc. London Math. Soc. (3) **92**(2), 505–544 (2006). MR 2205726 (2008b:55011)
41. L.G. Lewis Jr., J.P. May, J.E. McClure, Ordinary $RO(G)$-graded cohomology. Bull. Am. Math. Soc. (N.S.) **4**(2), 208–212 (1981). MR MR598689 (82e:55008)
42. L.G. Lewis Jr., J.P. May, M. Steinberger, *Equivariant Stable Homotopy Theory*, With contributions by J.E. McClure. Lecture Notes in Mathematics, vol. 1213 (Springer, Berlin, 1986). MR MR866482 (88e:55002)
43. S. Mac Lane, *Categories for the Working Mathematician*. Graduate Texts in Mathematics, vol. 5 (Springer, New York, 1971). MR MR0354798 (50 #7275)
44. M.A. Mandell, J.P. May, Equivariant orthogonal spectra and S-modules. Mem. Am. Math. Soc. **159**(755), x+108 (2002). MR 2003i:55012
45. J.P. May, Homotopical foundations of algebraic topology, unpublished manuscript (1974)
46. J.P. May, Characteristic classes in Borel cohomology, in *Proceedings of the Northwestern Conference on Cohomology of Groups (Evanston, 1985)*, vol. 44 (1987), pp. 287–289. MR 885112 (88i:55004)
47. J.P. May, *Equivariant Homotopy and Cohomology Theory*, With contributions by M. Cole, G. Comezaña, S. Costenoble, A.D. Elmendorf, J.P.C. Greenlees, L.G. Lewis Jr., R.J. Piacenza, G. Triantafillou, S. Waner. CBMS Regional Conference Series in Mathematics, vol. 91 (Published for the Conference Board of the Mathematical Sciences, Washington, DC, 1996). MR MR1413302 (97k:55016)
48. J.P. May, *A Concise Course in Algebraic Topology*. Chicago Lectures in Mathematics (University of Chicago Press, Chicago, 1999). MR MR1702278 (2000h:55002)
49. J.P. May, Equivariant orientations and Thom isomorphisms, in *Tel Aviv Topology Conference: Rothenberg Festschrift (1998)*. Contemporary Mathematics, vol. 231 (American Mathematical Society, Providence, RI, 1999), pp. 227–243. MR MR1707345 (2000i:55042)
50. J.P. May, The additivity of traces in triangulated categories. Adv. Math. **163**(1), 34–73 (2001). MR 1867203 (2002k:18019)
51. J.P. May, J. Sigurdsson, *Parametrized Homotopy Theory*. Mathematical Surveys and Monographs, vol. 132 (American Mathematical Society, Providence, RI, 2006). MR MR2271789
52. J.W. Milnor, J.D. Stasheff, *Characteristic Classes*. Annals of Mathematics Studies, vol. 76 (Princeton University Press, Princeton, NJ, 1974). MR MR0440554 (55 #13428)
53. B. Mitchell, Rings with several objects. Adv. Math. **8**, 1–161 (1972). MR 0294454 (45 #3524)
54. B. Mitchell, Some applications of module theory to functor categories. Bull. Am. Math. Soc. **84**(5), 867–885 (1978). MR 499732 (81i:18010)
55. T. Petrie, Pseudoequivalences of G-manifolds, in *Algebraic and geometric topology (Proceedings of the Symposium in Pure Mathematics, Stanford University, Stanford, 1976), Part 1, Proceedings of the Symposium in Pure Mathematics, XXXII* (American Mathematical Society, Providence, RI, 1978), pp. 169–210. MR MR520505 (80e:57039)
56. W. Pulikowski, $RO(G)$-graded G-bordism theory. Bull. Acad. Pol. Sci. Sér. Sci. Math. Astron. Phys. **21**, 991–995 (1973). MR MR0339234 (49 #3996a)
57. D. Puppe, On the stable homotopy category, in *Proceedings of the International Symposium on Topology and its Applications (Budva, 1972)* (Savez Društava Matematičara, Fizičara i Astronoma Jugoslavije, Belgrade, 1973), pp. 200–212. MR 0383401 (52 #4282)
58. R.E. Stong, *Notes on Cobordism Theory*. Mathematical Notes (Princeton University Press, Princeton, NJ; University of Tokyo Press, Tokyo, 1968). MR 0248858 (40 #2108)
59. R. Street, The formal theory of monads. J. Pure Appl. Algebra **2**(2), 149–168 (1972). MR MR0299653 (45 #8701)
60. J. Thévenaz, P.J. Webb, Simple Mackey functors, in *Proceedings of the Second International Group Theory Conference (Bressanone, 1989)*, vol. 23 (1990), pp. 299–319. MR MR1068370 (91g:20011)
61. T. tom Dieck, *Transformation Groups*. de Gruyter Studies in Mathematics, vol. 8 (Walter de Gruyter and Co., Berlin, 1987). MR MR889050 (89c:57048)
62. R. Vogt, *Boardman's Stable Homotopy Category*. Lecture Notes Series, vol. 21 (Matematisk Institut, Aarhus Universitet, Aarhus, 1970). MR 0275431 (43 #1187)

63. C.T.C. Wall, *Surgery on Compact Manifolds*. London Mathematical Society Monographs, vol. 1 (Academic, London, 1970). MR 55 #4217
64. S. Waner, G-CW(V) complexes, unpublished manuscript, 1979
65. S. Waner, Equivariant homotopy theory and Milnor's theorem. Trans. Am. Math. Soc. **258**(2), 351–368 (1980). MR 558178
66. S. Waner, Equivariant Chern classes, unpublished manuscript, 1983
67. S. Waner, Equivariant $RO(G)$-graded bordism theories. Topol. Appl. **17**(1), 1–26 (1984). MR 85e:57042
68. S. Waner, Periodicity in the cohomology of universal G-spaces. Ill. J. Math. **30**(3), 468–478 (1986). MR 850344 (87g:55007)
69. S. Waner, G-CW(V) complexes and $RO(G)$-graded cohomology, in *Equivariant Homotopy and Cohomology Theory*. CBMS Regional Conference Series in Mathematics, vol. 91 (American Mathematical Society, Providence, RI, 1996), pp. 89–96
70. A.G. Wasserman, Equivariant differential topology. Topology **8**, 127–150 (1969). MR MR0250324 (40 #3563)
71. K. Wirthmüller, Equivariant homology and duality. Manuscr. Math. **11**, 373–390 (1974). MR MR0343260 (49 #8004)

Index of Notations

Some symbols are used generically in this index, including: G for the ambient group; H and K for subgroups of G; N for a normal subgroup of G; $\mathscr{F}$ for a collection of subgroups of G; δ for a dimension function for G; V and W for representations of G; α for a virtual representation of G; X and Y for spaces; B for a parametrizing space; M for a manifold; γ for a virtual representation of ΠB; D, E, and F for spectra; $\underline{S}$ and $\overline{T}$ for Mackey functors or ΠB – modules.

S.R. Costenoble, S. Waner, *Equivariant Ordinary Homology and Cohomology*, Lecture Notes in Mathematics 2178, DOI 10.1007/978-3-319-50448-3

Index

S.R. Costenoble, S. Waner, *Equivariant Ordinary Homology and Cohomology*,
Lecture Notes in Mathematics 2178, DOI 10.1007/978-3-319-50448-3

LECTURE NOTES IN MATHEMATICS

Springer

Editors in Chief: J.-M. Morel, B. Teissier;

Editorial Policy

1. Lecture Notes aim to report new developments in all areas of mathematics and their applications – quickly, informally and at a high level. Mathematical texts analysing new developments in modelling and numerical simulation are welcome.

   Manuscripts should be reasonably self-contained and rounded off. Thus they may, and often will, present not only results of the author but also related work by other people. They may be based on specialised lecture courses. Furthermore, the manuscripts should provide sufficient motivation, examples and applications. This clearly distinguishes Lecture Notes from journal articles or technical reports which normally are very concise. Articles intended for a journal but too long to be accepted by most journals, usually do not have this "lecture notes" character. For similar reasons it is unusual for doctoral theses to be accepted for the Lecture Notes series, though habilitation theses may be appropriate.

2. Besides monographs, multi-author manuscripts resulting from SUMMER SCHOOLS or similar INTENSIVE COURSES are welcome, provided their objective was held to present an active mathematical topic to an audience at the beginning or intermediate graduate level (a list of participants should be provided).

   The resulting manuscript should not be just a collection of course notes, but should require advance planning and coordination among the main lecturers. The subject matter should dictate the structure of the book. This structure should be motivated and explained in a scientific introduction, and the notation, references, index and formulation of results should be, if possible, unified by the editors. Each contribution should have an abstract and an introduction referring to the other contributions. In other words, more preparatory work must go into a multi-authored volume than simply assembling a disparate collection of papers, communicated at the event.

3. Manuscripts should be submitted either online at www.editorialmanager.com/lnm to Springer's mathematics editorial in Heidelberg, or electronically to one of the series editors. Authors should be aware that incomplete or insufficiently close-to-final manuscripts almost always result in longer refereeing times and nevertheless unclear referees' recommendations, making further refereeing of a final draft necessary. The strict minimum amount of material that will be considered should include a detailed outline describing the planned contents of each chapter, a bibliography and several sample chapters. Parallel submission of a manuscript to another publisher while under consideration for LNM is not acceptable and can lead to rejection.

4. In general, **monographs** will be sent out to at least 2 external referees for evaluation.

   A final decision to publish can be made only on the basis of the complete manuscript, however a refereeing process leading to a preliminary decision can be based on a pre-final or incomplete manuscript.

   Volume Editors of **multi-author works** are expected to arrange for the refereeing, to the usual scientific standards, of the individual contributions. If the resulting reports can be forwarded to the LNM Editorial Board, this is very helpful. If no reports are forwarded or if other questions remain unclear in respect of homogeneity etc, the series editors may wish to consult external referees for an overall evaluation of the volume.

5. Manuscripts should in general be submitted in English. Final manuscripts should contain at least 100 pages of mathematical text and should always include

   - a table of contents;
   - an informative introduction, with adequate motivation and perhaps some historical remarks: it should be accessible to a reader not intimately familiar with the topic treated;
   - a subject index: as a rule this is genuinely helpful for the reader.
   - For evaluation purposes, manuscripts should be submitted as pdf files.

6. Careful preparation of the manuscripts will help keep production time short besides ensuring satisfactory appearance of the finished book in print and online. After acceptance of the manuscript authors will be asked to prepare the final LaTeX source files (see LaTeX templates online: https://www.springer.com/gb/authors-editors/book-authors-editors/manuscriptpreparation/5636) plus the corresponding pdf- or zipped ps-file. The LaTeX source files are essential for producing the full-text online version of the book, see http://link.springer.com/bookseries/304 for the existing online volumes of LNM). The technical production of a Lecture Notes volume takes approximately 12 weeks. Additional instructions, if necessary, are available on request from lnm@springer.com.

7. Authors receive a total of 30 free copies of their volume and free access to their book on SpringerLink, but no royalties. They are entitled to a discount of 33.3 % on the price of Springer books purchased for their personal use, if ordering directly from Springer.

8. Commitment to publish is made by a *Publishing Agreement*; contributing authors of multiauthor books are requested to sign a *Consent to Publish form*. Springer-Verlag registers the copyright for each volume. Authors are free to reuse material contained in their LNM volumes in later publications: a brief written (or e-mail) request for formal permission is sufficient.

Addresses:
Professor Jean-Michel Morel, CMLA, École Normale Supérieure de Cachan, France
E-mail: moreljeanmichel@gmail.com

Professor Bernard Teissier, Equipe Géométrie et Dynamique,
Institut de Mathématiques de Jussieu – Paris Rive Gauche, Paris, France
E-mail: bernard.teissier@imj-prg.fr

Springer: Ute McCrory, Mathematics, Heidelberg, Germany,
E-mail: lnm@springer.com

Printed in the United States
By Bookmasters

Printed in the United States
By Bookmasters